WILLIAM F. MAAG LIBRARY
YOUNGSTOWN STATE UNIVERSITY

SEMICONDUCTORS AND SEMIMETALS

VOLUME 19

Deep Levels, GaAs, Alloys, Photochemistry

SEMICONDUCTORS AND SEMIMETALS

Edited by   *R. K. WILLARDSON*
WILLARDSON CONSULTING
SPOKANE, WASHINGTON

ALBERT C. BEER
BATTELLE COLUMBUS LABORATORIES
COLUMBUS, OHIO

VOLUME 19
Deep Levels, GaAs, Alloys, Photochemistry

1983

ACADEMIC PRESS
A Subsidiary of Harcourt Brace Jovanovich, Publishers

New York   London
Paris   San Diego   San Francisco   São Paulo   Sydney   Tokyo   Toronto

COPYRIGHT © 1983, BY ACADEMIC PRESS, INC.
ALL RIGHTS RESERVED.
NO PART OF THIS PUBLICATION MAY BE REPRODUCED OR
TRANSMITTED IN ANY FORM OR BY ANY MEANS, ELECTRONIC
OR MECHANICAL, INCLUDING PHOTOCOPY, RECORDING, OR ANY
INFORMATION STORAGE AND RETRIEVAL SYSTEM, WITHOUT
PERMISSION IN WRITING FROM THE PUBLISHER.

ACADEMIC PRESS, INC.
111 Fifth Avenue, New York, New York 10003

United Kingdom Edition published by
ACADEMIC PRESS, INC. (LONDON) LTD.
24/28 Oval Road, London NW1 7DX

Library of Congress Cataloging in Publication Data
(Revised for volume 19)
Main entry under title:

Semiconductors and semimetals.

   Includes bibliographical references and indexes.
   Contents: v. 1–2. Physics of III–V compounds--
v. 3. Optical properties of III–V compounds--
[etc.] -- v. 19. Deep levels, GaAs, alloys,
photochemistry.
   1. Semiconductors--Collected works. 2. Semimetals--
Collected works. I. Willardson, Robert K.
II. Beer, Albert C.
QC610.9.S47   537.6'22   65–26048
ISBN 0–12–752119–4

PRINTED IN THE UNITED STATES OF AMERICA

83 84 85 86   9 8 7 6 5 4 3 2 1

Contents

LIST OF CONTRIBUTORS vii
PREFACE ix

Chapter 1   Deep Levels in Wide Band-Gap III–V Semiconductors
G. F. Neumark and K. Kosai

I. Introduction. 1
II. Experimental Characterization of Deep Levels 7
III. Observation and Intercomparison of Deep Levels. Results for GaAs and GaP 20
IV. Theory of Recombination 30
V. Analysis of Optical Cross Sections 52
References 65

Chapter 2   The Electrical and Photoelectronic Properties of Semi-Insulating GaAs
David C. Look

List of Symbols 76
I. Introduction. 77
II. Hall Effect and Magnetoresistance 77
III. Thermal Equilibrium Processes 86
IV. Nonequilibrium Processes 96
V. Comparison of Techniques 121
Appendix A. Charge Transport Theory 127
Appendix B. Semiconductor Statistics 148
Appendix C. Derivation of TSC and PITS Equations 162
References 167

Chapter 3   Associated Solution Model for Ga–In–Sb and Hg–Cd–Te
R. F. Brebrick, Ching-Hua Su, and Pok-Kai Liao

List of Symbols 172
I. Introduction. 173
II. Thermodynamic Equations for the Liquidus Surface of $(A_{1-u}B_u)C(s)$ . . 178
III. Solution Thermodynamics 181
IV. Associated Solution Model for the Liquid Phase 186

V. Ga–In–Sb Ternary	197
VI. Hg–Cd–Te Ternary	214
VII. Summary	230
Appendix A	231
Appendix B	234
Appendix C	242
References	251

Chapter 4 Photoelectrochemistry of Semiconductors
Yu. Ya. Gurevich and Yu. V. Pleskov

List of Symbols	256
I. Introduction	257
II. General Concepts of the Electrochemistry of Semiconductors	259
III. The Theory of Processes Caused by Photoexcitation of Semiconductors	273
IV. Photocorrosion and Its Prevention	282
V. Light-Sensitive Etching of Semiconductors	294
VI. Processes Caused by Photoexcitation of Reactants in the Solution	303
VII. Selected Problems in the Photoelectrochemistry of Semiconductors	310
VIII. Conclusion: Problems and Prospects	323
References	324

INDEX	329
CONTENTS OF PREVIOUS VOLUMES	339

List of Contributors

Numbers in parentheses indicate the pages on which the authors' contributions begin.

R. F. BREBRICK, *Materials Science and Metallurgy Program, College of Engineering, Marquette University, Milwaukee, Wisconsin 53233* (171)

YU. YA. GUREVICH, *Institute of Electrochemistry, Academy of Sciences of the USSR, 117071 Moscow, USSR* (255)

K. KOSAI,[1] *Philips Laboratories, Briarcliff Manor, New York 10510* (1)

POK-KAI LIAO, *Materials Science and Metallurgy Program, College of Engineering, Marquette University, Milwaukee, Wisconsin 53233* (171)

DAVID C. LOOK, *University Research Center, Wright State University, Dayton, Ohio 45435* (75)

G. F. NEUMARK, *Philips Laboratories, Briarcliff Manor, New York 10510* (1)

YU. V. PLESKOV, *Institute of Electrochemistry, Academy of Sciences of the USSR, 117071 Moscow, USSR* (255)

CHING-HUA SU, *Materials Science and Metallurgy Program, College of Engineering, Marquette University, Milwaukee, Wisconsin 53233* (171)

---

[1] Present address: Santa Barbara Research Center, Goleta, California 93117.

Preface

This treatise reflects the rapid growth in the field of semiconductors and semimetals. Chapters 1 and 2 of the present volume are concerned with deep levels in wide band-gap III–V compounds and the measurement techniques used to characterize such levels. These chapters provide a background for the next volume, which will address the use of semi-insulating GaAs in digital and microwave integrated circuits (ICs) as well as providing valuable insight into methods of improving solar cells, lasers, and light-emitting diodes (LEDs). Chapter 3 in the present volume is concerned with thermodynamic modeling to improve Hg–Cd–Te phase diagrams applicable to the infrared detector arrays discussed in Volume 18 of this treatise. Chapter 4 treats photoelectrochemistry (PEC), an emerging technology.

Chapter 1 focuses on the characteristics of deep states in wide band-gap III–V compound semiconductors, particularly the recombination properties which control minority-carrier lifetime and luminescence efficiency. These properties are significant for many optoelectronic devices, including lasers, LEDs, and solar cells. While this review emphasizes areas of extensive recent development, it also provides references to previous comprehensive reviews. The compilation of levels reported in GaAs and GaP since 1974 is an important contribution, as is the discussion of the methods used to characterize these levels.

Recombination at and excitation from deep levels are emphasized. Nonradiative transitions at defect levels—Auger, cascade capture, and multiphonon emission processes—are discussed in detail. Factors to be considered in the analysis of optical cross sections which can give information about the parity of the impurity wave function and thus about the symmetry of a particular center are reviewed.

In Chapter 2, not only are the electrical and optical properties of semi-insulating GaAs discussed, but also the measurement techniques are critically reviewed. GaAs digital IC and monolithic microwave IC technologies are dependent on the development of reliable substrate materials and an understanding of the deep levels of GaAs. Thus these first two chapters provide an important background to the discussion of the crystal growth and device fabrication techniques for GaAs, which are the subjects of Volume 20 of this treatise "Semiconductors and Semimetals."

Automated computer-controlled measurement techniques for Hall effect, photoconductivity, and photo-Hall measurements in semi-insulating materials as well as methods for the calculation of electron mobilities for this case of mixed conduction are presented. Particularly significant is the comparison of the results of the various techniques to determine carrier lifetimes, impurity concentrations, and activation energies in addition to the identification of impurity and defect levels. The appendixes contain the derivation of pertinent charge transport theory for mixed conduction as well as the derivation of occupation factors and thermally stimulated (TSC) and photoinduced transient current spectroscopy (PITS) equations.

Mercury–cadmium–telluride is the principal semiconductor now being used in advanced infrared systems, both for military and other surveillance applications. Its preparation and use in infrared detectors and arrays was the subject of Volume 18 of this treatise. New generations of detectors and arrays require sophisticated epitaxial growth, which in turn requires precise phase diagram data.

It is therefore essential that thermodynamic modeling be applied to obtain a simultaneous quantitative fit to the phase diagram and thermodynamic data in order to evaluate the internal consistency of the various published data. Then a reliable framework can be established for smoothing, interpolating, and extrapolating experimental data that are costly and laborious to obtain. In Chapter 3 an associated solution model is presented. This model provides a good fit to the data for the Hg–Cd–Te system as well as for the Ga–In–Sb system, which is closer to the simpler picture of an ideal solution.

Photoelectrochemistry (PEC) is emerging from the research laboratories with the promise of significant practical applications. One application of PEC systems is the conversion and storage of solar energy. Chapter 4 reviews the main principles of the theory of PEC processes at semiconductor electrodes and discusses the most important experimental results of interactions at an illuminated semiconductor–electrolyte interface. In addition to the fundamentals of electrochemistry and photoexcitation of semiconductors, the phenomena of photocorrosion and photoetching are discussed. Other PEC phenomena treated are photoelectron emission, electrogenerated luminescence, and electroreflection. Relationships among the various PEC effects are established.

The editors are indebted to the many contributors and their employers who make this treatise possible. They wish to express their appreciation to Willardson Consulting and Battelle Memorial Institute for providing the facilities and the environment necessary for such an endeavor. Special thanks are also due the editors' wives for their patience and understanding.

R. K. WILLARDSON
ALBERT C. BEER

CHAPTER 1

Deep Levels in Wide Band-Gap III–V Semiconductors

G. F. Neumark and K. Kosai*

PHILIPS LABORATORIES
BRIARCLIFF MANOR, NEW YORK

I.	INTRODUCTION	1
II.	EXPERIMENTAL CHARACTERIZATION OF DEEP LEVELS	7
	1. General Discussion.	7
	2. Experimental Techniques	15
	3. Identification of Deep Levels.	19
III.	OBSERVATION AND INTERCOMPARISON OF DEEP LEVELS. RESULTS FOR GaAs AND GaP	20
	4. Introduction	20
	5. GaP	26
	6. GaAs	26
IV.	THEORY OF RECOMBINATION	30
	7. Radiative Recombination	30
	8. Nonradiative Recombination: General Aspects	30
	9. Nonradiative Recombination: Auger and Cascade	32
	10. Nonradiative Recombination: Multiphonon Emission	35
V.	ANALYSIS OF OPTICAL CROSS SECTIONS	52
	11. General Discussion. Complications	52
	12. Electronic Cross Section	56
	REFERENCES	65

I. Introduction

This chapter focuses on the properties of deep states in wide band-gap III–V semiconductors. Deep states are important in semiconductors in general and very often determine the recombination properties. Wide band-gap III–V compounds are now commonly used for many important optoelectronic devices (for example, lasers, light-emitting diodes, and Gunn diodes), and are being suggested for additional interesting applications (for example, solar cells). In view of the importance of such work there are a number of earlier

---

* Present address: Santa Barbara Research Center, Goleta, California.

reviews on various aspects (see below for a partial listing), and these will be referred to when appropriate. However, those on deep states are generally not slanted specifically towards III–V's, and those on the III–V materials either do not emphasize deep states or are fairly brief. We thus feel that the present work fills a novel and appropriate slot.

In discussing deep levels in wide band-gap semiconductors, the first requirement is to define "deep" and "wide." The latter can be done relatively easily, although arbitrarily. We list in Table I the 4°K band gaps of the various III–V semiconductors, based on the tabulation by Strehlow and Cook (1973). We shall call those with $E_g > 1.5$ eV the wide band-gap ones. In practice, our review will present data only on GaAs and on GaP as prototypes of direct and indirect gap materials of this class. These are also the only two materials of this class that have been extensively studied and that are in common use. Discussion of deep levels in ternary and quarternary alloys of III–V semiconductors are omitted since treating these in detail might well have doubled the size of this chapter.

The next step is the definition of "deep." The choice of quantitative values will again involve some arbitrariness. However, a further complication is that there is at present no universally accepted *qualitative* criterion. For instance, from the point of view of energy level calculations it is often convenient to define deep states as noneffective-mass-like or as those with a localized potential (see, for example, Bassani and Pastori Parravicini, 1975; Jaros, 1980). However, the disadvantage of this definition of deep is that it includes many isoelectronic states that are very shallow on an energy scale. On the other hand, if one uses an energy criterion, should the states be deep with respect to some fraction of the band gap, with respect to kT, or with respect to some shallow levels? In this chapter we shall adopt an energy criterion for "deep," and we shall require that our states be deep enough to be important in recombination. The importance of deep levels in recombination under many conditions of practical interest was already realized in the early work of Hall

TABLE I

BAND GAPS (E_g) AT 0 OR 4°K[a]

	N	P	As	Sb
B	3.4			
Al		2.5	2.2	1.6
Ga		2.2	1.5	0.8
In		1.4	0.4	0.2

[a] From Strehlow and Cook (1973).

(1952) and of Shockley and Read (1952), based on a one-defect model. Qualitatively, if a carrier, say an electron, is captured by a center, the subsequent step can either be recombination or reemission to the band. The latter does not lead to recombination. Moreover, it is well known that the latter becomes increasingly more probable the closer a level is to the band. This aspect has, for instance, been used to separate levels into "traps," with reemission more probable, or "recombination centers" [see, for example, Rose (1955, 1957), Bube (1960), or Blakemore (1962)]. A convenient way of quantifying this importance of "deep" levels to recombination is via the "occupancy" factor f of a level, i.e., the probability that a level is filled. This approach has often been used to analyze the decay in GaP (see, for example, Jayson et al., 1970; Dishman and DiDomenico, 1970; Neumark, 1974). Under quasiequilibrium conditions this occupancy factor gives the fraction of levels available for recombination. It can be shown that it is given (Neumark, 1974), e.g., for n-type material, by

$$f = [1 + (N_c/gn)\exp(-E_A/kT)]^{-1}. \tag{1}$$

Here N_c is the density of states in the conduction band, g the level degeneracy factor, n the carrier concentration in the band, E_A the activation energy of the level, k Boltzmann's constant, and T the temperature. Now, in general, except at fairly low temperatures, the occupancy for shallow levels (with $f \equiv f^S$) will be small, i.e., $f^S \ll 1$, and consequently

$$f^S \approx (gn/N_c)\exp(E_A^S/kT), \tag{2}$$

where E_A^S is the energy of the shallow level. Now if we wish a deep level (with $f \equiv f^D$) to be appreciably more effective in recombination, the requirement is $f^D \gg f^S$. A sufficient condition is obtained by omitting the factor of unity in Eq. (1), i.e., for

$$\exp[(E_A^D - E_A^S)/kT] \gg 1. \tag{3}$$

For an arbitrary criterion that, at room temperature,

$$f^D \approx 10 f^S \tag{4}$$

this gives

$$E_A^D - E_A^S \approx 0.06 \quad \text{eV}. \tag{5}$$

In GaAs, the shallow acceptors (Zn, Cd) have an energy of ≈ 0.03 eV, and we thus regard acceptors in GaAs as deep for $E_A^D \geq 0.1$ eV. The donors are more shallow, but we shall nevertheless also use the same criterion for the donors. In GaP, one commonly used acceptor, Zn, is at $E_A \approx 0.07$ eV, but the common donors (S, Se, Te) as well as the Cd acceptor are at $E_A \approx 0.1$ eV. Thus for GaP we use the criterion $E_A^D \gtrsim 0.15$ eV.

The next factor we consider about deep levels is why they are important. Here, a particular reason why we defined "deep" in terms of the recombination properties is because of the profound influence of such recombination on material behavior. Properties specifically affected can include the minority-carrier lifetime and the luminescence efficiency. If deep levels dominate the recombination, these properties will be degraded from the value obtained from the "dopants." (The dopants are deliberately added impurities, used to provide either desired electrical and/or luminescent properties. Dopants usually introduce shallow levels, although there are some exceptions—for example, red GaP diodes utilize a deep level.) Moreover, degradation of a diode or laser during operation can be caused by formation of deep levels. In this connection it can be noted that really deep levels—which at midgap in GaP would be located at ~ 1.1 eV from the band edge—can dominate the behavior at very low concentrations, possibly even at $10^{12}-10^{14}$ cm^{-3} [see Eq. (1) and related discussion]. It was thus soon realized that a study and understanding of deep levels is required for improved recombination performance. A second important role of such levels is in giving high resistivity material. This aspect has been covered, for instance, in Rees (1980) and in a recent detailed quantitative analysis by Neumark (1982); the latter emphasizes that the condition for obtaining high resistivity, in, e.g., n-type material, is $N_{D1} < N_A < (N_{D1} + N_{D2})$, where N_{D1} is the concentration of shallow donors, N_{D2} is that of deep donors, and N_A is that of acceptors.

As regards this chapter, we shall emphasize areas of extensive recent development. However, this emphasis is subject to the constraint that we cover only briefly subjects that have been treated in adequately comprehensive recent reviews. Progress in experimental studies on deep levels has been both extensive and rapid. In Part II we cover the various methods available for characterization and in Part III give a compilation of levels reported in GaAs and GaP since 1974 (together with a few important levels that were already well characterized previously). This particular year was chosen since Milnes (1973) gives a good review of earlier values. On theoretical aspects, there have been corresponding numerous treatments. Such work has included calculations on both the energies of deep levels and on their properties. In this chapter we emphasize the properties, i.e., recombination at (Part IV) and excitation from (Part V) the deep levels. The reason is partly that it is these properties that determine the behavior of the material (the role of recombination has already been emphasized above). However, a second reason is due to several recent comprehensive reviews on level calculations. These are by Masterov and Samorukov (1978), Pantelides (1978), Jaros (1980, 1982), Lannoo and Bourgoin (1981), and Vogl (1981), and among them they cover the field quite adequately. Consequently, we do not cover this topic in the main body of our chapter, but just give a comment here on the progress of such calculations in the past few years. Thus in 1977 it was complained "that there is no general formalism available ... to predict the physical properties" of deep centers (Grimmeiss, 1977). However, already the same year, Mircea *et al.* (1977) argued on theoretical grounds against the A level in GaAs (at 0.75–0.8 eV) being due to substitutional oxygen: "Comparison of the calculated activation energies with the experimental ones ... can hardly support our hypothesis that the trap is O at the As site." Two years later Huber *et al.* (1979) proved that, indeed, this trap was not due to O_{As}. A further restriction is that we try to cover the broad picture, applicable to levels in general, and use specific levels primarily as examples. We omit detailed discussion of any one level, since a book reviewing specific deep centers is currently in preparation (Pantelides, 1983).

It must also be kept in mind in connection with both analysis of data and theoretical calculations that most (although certainly not all) treatments assume "simple" levels. By this we mean that a level consists of a single ground state, with excitation to the band represented, if necessary, by a single upper state in the configuration coordinate picture (see Section 10b for a description of this model). Also, in this simple picture, effects of temperature (other than those included via the configuration coordinate model) and field are neglected. It has become apparent that in many cases this is an oversimplified picture. Complications can include excited states (which can also be metastable), multiple charge states, resonant "band" states, a distribution of level energies, and level properties that depend on field and/or temperature. Examples of treatments that include some complications—some of which will again be referred to subsequently—follow. Thus, evidence for an excited state of the one-electron center of GaP(O) has been obtained by Samuelson and Monemar (1978) from photo-cross-section data. Lang and Logan (1977) have postulated metastable states to explain persistent photoconduction in GaAlAs; such states have also been postulated by Bois and Vincent (1977) for a deep oxygen associated level in GaAs, based on photo- and thermal-cross-section results, as well as by Baraff *et al.* (1982) for the two-electron state of GaP(O), from theory. A theoretical treatment of multiphonon recombination preceded by capture at an excited state has been suggested by Gibb *et al.* (1977), and applied to a deep hole trap in GaP. As regards impurities with multiple charge states, a recently very thoroughly studied example is GaAs(Cr). This system, as well as some complications arising from such systems in general, are discussed for instance by Allen (1980), Blakemore (1980), Clark (1980), White (1980), and White *et al.* (1980). In this connection we should also mention the "Anderson negative-U" case, consisting of systems with multiple levels whose relative stability depends on Fermi energy (Baraff *et al.*, 1979); to date levels of this type have been postulated only in Si, although they may well exist also in other semiconductors. Possible effects of resonant states (with energies in the band region but with localized properties) on excitation spectra have been discussed by Pantelides and Grimmeiss (1980).

An excited state of Cr^{2+} lying above the conduction-band edge has since been fairly conclusively established for GaAs by Eaves et al. (1981) and by Hennel et al. (1981). Results from a distribution of levels have been analyzed by Ridley and Leach (1977), and applied to oxygen-doped GaAs by Leach and Ridley (1978) and by Arikan et al. (1980). As regards temperature dependence of level properties, the theory of the influence of entropy on the level ionization energies has been given by Van Vechten and Thurmond (1976). A detailed experimental study of the energy variation with temperature of the one-electron level of GaP(O) has been carried out by Samuelson and Monemar (1978). Results are at some variance with the Van Vechten and Thurmond theory, indicating that other factors may also play a role. There also can be temperature dependences for cross sections; this aspect, for multiphonon recombination, is treated in detailed in Section 10. On field dependences, that of the thermal cross section has recently been studied—experimentally and theoretically—by Vincent et al. (1979), Pons and Makram-Ebeid (1979), Makram-Ebeid (1980, 1981), Makram-Ebeid et al. (1980), Martin et al. (1981), and by Makram-Ebeid and Lannoo (1982a,b). Also, evidence for a field dependence of the optical cross section has been presented by Monemar and Samuelson (1978).

Despite the complications referred to in the above paragraph, many papers, and also in large part this chapter, treat only the simple case. As will become apparent from the subsequent treatment in this review, the simple case can by itself become quite complicated.

As the final part of this Introduction, we list earlier reviews of interest to deep levels and/or III–V semiconductors. There are, of course, the volumes of this treatise. Authored books (i.e., not edited ones consisting of a compilation of articles) include Milnes (1973), of high general relevance, and, since 1975, Bassani and Pastori Parravicini (1975), Stoneham (1975), Englman (1979), Lannoo and Bourgoin (1981), Jaros (1982), all highly relevant to theory, and Watts (1977), combining experimental and theoretical aspects. As regards articles, we shall attempt to mention all of those published since 1977. Fairly general reviews are "Deep Level Impurities in Semiconductors" by Grimmeiss (1977), and "III–V Compound Semiconductors" by Dean (1977). Papers emphasizing experimental methods are by Ikoma et al. (1977), Miller et al. (1977), Lang (1979), Bois and Chantre (1980), Grimmeiss (1980), Mircea et al. (1980), and Kimerling (1981). Theoretical calculations of level energies are covered by Masterov and Samorukov (1978), Pantelides (1978), Jaros (1980), Bernholc et al. (1981), and Vogl (1981). Work on recombination and nonradiative processes in general is reviewed by Stoneham (1977, 1981), Mott (1978), Queisser (1978), and Toyozawa (1978), as well as in the book "Radiationless Processes," edited by Di Bartolo (1981). A listing of various deep levels is given by Ikoma et al. (1977), by Masterov and Samorukov (1978), by Mircea and Bois (1979), and by Martin (1980). Reviews of optical and/or electron paramagnetic resonance characterization are by Kaufmann and Schneider (1980), Cavenett (1981), Corbett *et al.* (1981), Kennedy (1981), Monemar (1981), and Schneider and Kaufmann (1981). Somewhat more specialized aspects are covered by Lang (1977) on "Review of radiation-induced defects in III–V compounds," by Landsberg and Robbins (1978) on "The first 70 semiconductor Auger processes," by Abakumov *et al.* (1978) on the cascade process, "Capture of carriers by attractive centers in semiconductors," by Langer (1980) on "Large defect-lattice relaxation phenomena in solids," as well as by Dean and Choyke (1977) and Kimerling (1978, 1979) on recombination-enhanced defect reactions. Several additional review papers are also included in the 1978 "International Conference on the Physics of Semiconductors," edited by Wilson (1979), as well as in the 1980 one, edited by Tanaka and Toyozawa (1980), as well as in "Semi-Insulating III–V Materials," edited by Rees (1980).

II. Experimental Characterization of Deep Levels

1. GENERAL DISCUSSION

A variety of experimental methods for the characterization of deep levels in semiconductors have been developed in the past decade and are discussed in several excellent review articles. Lang (1979), Mircea *et al.* (1980), Sah (1976), and Sah *et al.* (1970) review the application of space-charge regions to the study of deep levels. Grimmeiss (1974, 1977, 1980) concentrates on the various junction photocurrent and photocapacitance methods for determining the photoionization cross sections of deep states. Miller *et al.* (1977) discuss the capacitance transient methods that have been developed to measure thermal emission and capture rates. Martin (1980) treats, within the context of semi-insulating GaAs, those techniques most suited to highly resistive materials. Kaufmann and Schneider (1980), Kennedy (1981), and Schneider and Kaufmann (1981) review optical spectroscopy and electron spin resonance of deep levels in III–V semiconductors. The book by Milnes (1973) is the standard reference to the work until the early part of the 1970s. It extensively treats the older deep level methods, including photoconductivity and thermally stimulated current.

Despite the success of purely optical methods such as luminescence and absorption in studying shallow energy levels, they are seldom used for characterization of deep states. This is mainly because the deeper levels of interest are usually nonradiative ones or "killer centers" for which such techniques are not applicable. In addition, the deep position within the energy gap means optical experiments must be performed in the infrared where detectors are not as sensitive as in the visible part of the spectrum. Notable instances where optical methods alone have been successfully employed are O (Monemar and Samuelson, 1978), as well as complexes such as Zn–O (Jayson et al., 1972), and some Cu related ones (Monemar et al., 1982; Gislason et al., 1982) in GaP.

Capture and emission processes at a deep center are usually studied by experiments that use either electrical bias or absorbed photons to disturb the free-carrier density. The subsequent thermally or optically induced trapping or emission of carriers is detected as a change in the current or capacitance of a given device, and one is able to deduce the trap parameters from a measurement of these changes.

A description of the emission and capture processes at a trap will be useful before discussing the various experimental methods. Figure 1 depicts the capture and emission processes that can occur at a center with electron energy E_T. The subscripts n and p denote electron and hole transitions, and the superscripts t and o differentiate between thermally and optically stimulated processes. It is assumed here that only thermal capture processes are occurring.

The rate equation for n_T, the density of traps occupied by an electron, is

$$dn_T/dt = (nc_n + e_p^t + e_p^o)(N_T - n_T) \\ - (pc_p + e_n^t + e_n^o)n_T, \quad (6)$$

where N_T is the total density of traps, n and p are the free-electron and free-hole concentrations, and the emission and capture constants are as defined in Fig. 1. If the free-carrier concentrations, n and p, are constant with time and spatially uniform, Eq. (6) is linear with a solution that exponentially approaches the steady state value

$$n_T(t = \infty) = (nc_n + e_p^t + e_p^o)N_T/(nc_n + pc_p + e_n^t + e_p^t + e_n^o + e_p^o). \quad (7)$$

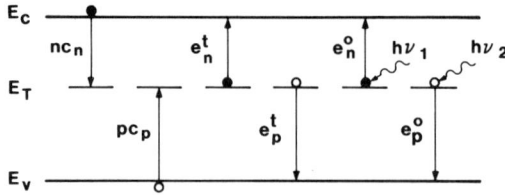

FIG. 1. Emission and capture processes at a deep level of energy E_T. The quantities near the arrows are the rates (in s^{-1}) of the processes. Subscripts indicate whether the transition is by an electron (n) or hole (p), and the superscripts indicate a thermal (t) or optical (o) process. The symbol e is an emission rate, c a capture constant, n and p the concentrations of free electrons and holes in the conduction band (at energy E_c) and valence band (at energy E_v), and hv_1 and hv_2 photon energies.

The time constant of the exponential is given by the sum of all the capture and emission rates:

$$\tau = (nc_n + pc_p + e_n^t + e_p^t + e_n^o + e_p^o)^{-1}. \tag{8}$$

The thermally activated emission rates are proportional to a Boltzmann factor, and by use of the principle of detailed balance can be related to the capture cross section (σ_n^t):

$$e_n^t = (\sigma_n^t \langle v \rangle N_c/g) \exp(-\Delta E_T/kT). \tag{9}$$

The parameter $\langle v \rangle$ is the average thermal velocity of an electron, N_c the density of states in the conduction band, g the degeneracy of the deep level, and $\Delta E_T = E_c - E_T$ the electron transition energy. Equation (9) also relates the capture constant to the emission rate because of the definition

$$c_n = \sigma_n^t \langle v \rangle. \tag{10}$$

Similar equations hold for holes with n replaced by p in the subscripts, N_c by the valence-band density of states N_v, and $\Delta E_T = E_T - E_v$. The optical emission rate is related to the photoionization cross section (σ^o) by

$$e^o = \sigma^o \Phi,$$

where Φ is the incident photon flux density in cm^{-2}s^{-1}.

Equation (9) shows that for σ^t independent of T, measurement of the emission rate as a function of temperature yields the trap energy. The temperature dependence of $\langle v_n \rangle$ ($\propto T^{1/2}$) and $N_c (\propto T^{3/2})$ can be taken into account either by determining the slope of $\ln(e_n^t/T^2)$ as a function of $1/T$ or by subtracting $2kT_m$ from the slope of $\ln e_n^t$ versus $1/T$, where T_m is the average temperature over which the slope of $\ln e_n^t$ was measured (Miller et al., 1977). However, σ^t is often temperature dependent. For instance, Henry and Lang (1977) show that it frequently can be represented by

$$\sigma^t(T) = \sigma_\infty \exp(-E_\infty/kT), \tag{11}$$

where σ_∞ is temperature independent (also see Section 10 for further discussion). This means that the activation energy derived from measurements of the temperature dependence of the emission rate includes a contribution from the capture cross section. Because e^t and σ^t are related by Eq. (10), it is possible to obtain σ^t as a function of temperature by measuring c^t. Once this is done, E_∞ can be evaluated and subtracted from the activation energy of the emission rate to yield ΔE_T.

Deep level experiments can be divided into thermal and optical categories, depending on which constants are being measured. Both categories can be further subdivided into steady state and transient techniques. An example of the former is photoconductivity, whereas deep level transient spectroscopy (DLTS) is in the latter division.

The capture and emission processes at a deep center are most often analyzed by fabricating a device such that the centers under study are in the depletion region of either a p–n or metal–semiconductor junction. This is because the absence of mobile electrons and holes in the space-charge region causes capture to become negligible compared to emission, permitting these two effects to be separated.

It is convenient to adopt the terminology of Miller et al. (1977) and define a center to be an electron trap if $e_n^t \gg e_p^t$ and a hole trap if the reverse is true. In addition, a minority-carrier trap is one for which the emission rate of minority carriers e_{min} is greater than that of majority carriers e_{maj}, whereas for majority-carrier traps $e_{maj} \gg e_{min}$. By these definitions, an electron trap is a majority-carrier trap in an n-type region and a minority-carrier trap in p-type. Note that these definitions are independent of whether the trap is a donor or an acceptor, terms that imply a specific charge state of the center (Pantelides, 1978). Because the emission rates are thermally activated [Eq. (9)], an electron trap usually lies in the upper half of the band gap and a hole trap in the lower.

Before continuing the discussion of deep states, it will be useful to review the p–n junction. Consider a p^+–n step junction reverse biased to V volts and having the n-side uniformly doped with N_D net shallow donors and N_T deep donor traps located in the upper half of the band gap (Fig. 2a). In the abrupt approximation it is assumed that in the region between $x = 0$ and $x = W(V)$ all impurities are ionized and that there are no mobile carriers. As a consequence of this assumption and the uniform distribution of dopants, Poisson's equation can be easily integrated twice to give the electric field, electron potential, and $W(V)$ (Sze, 1969). Because the p-side doping is much greater than that of the n, the width of the space-charge region on the p-side can be neglected, giving

$$W(V) = [2\varepsilon(V_{bi} + V)/q(N_D + N_T)]^{1/2}, \quad (12)$$

where ε is the permittivity of the semiconductor, V_{bi} the built-in potential, and q the elementary charge. Defining the capacitance of the junction as $C \equiv A dQ/dV$, where A is the junction area and dQ the incremental change in charge per unit area on an increment dV in the reverse bias, yields

$$C(V) = \varepsilon A/W(V) = A[q\varepsilon(N_D + N_T)/2(V_{bi} + V)]^{1/2}. \quad (13)$$

To illustrate the basic ideas of deep state experiments, assume that no photons are incident on the junction of Fig. 2a, so that the optical emission rates e_n^o and e_p^o are zero. Under a reverse bias applied sufficiently long to give steady state, because $n = p = 0$, the capture terms of Eq. (7) are also

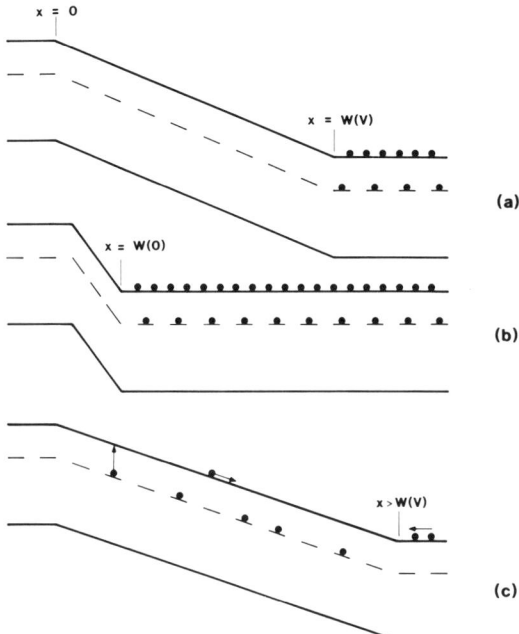

FIG. 2. Schematic energy band diagrams showing the use of a p^+–n junction to study capture and emission processes at a deep level. The junction is shown at steady state with a reverse bias of V volts (a) and with 0 volts (b). The width of the space-charge region under these conditions is $W(V)$ and $W(0)$. Immediately after the reverse bias is switched from 0 to V volts a nonequilibrium condition exists in which electrons occupying traps within the space-charge region are emitted to the conduction band and swept out (c). The shallow donor and acceptor that must be present have been omitted for clarity.

zero, which results in

$$n_T/N_T = e_p^t/(e_n^t + e_p^t) \qquad (14)$$

in the space-charge region between $x = 0$ and $x = W(V)$. Assuming that $E_c - E_T$ is appreciably less than $E_T - E_v$, then, because e_n^t and e_p^t are thermally activated [Eq. (9)], e_p^t is negligible compared with e_n^t, giving $n_T \ll N_T$ and thus empty traps at steady state. If the bias is now switched to zero (Fig. 2b), then in the region between $W(0)$ and $W(V)$ the electron concentration will be restored to the bulk equilibrium value n and traps will become occupied by electrons provided

$$nc_n \gg e_n^t + e_p^t \qquad (15)$$

(pc_p remains negligible because $p \approx 0$). This condition is achieved by choosing a sufficiently low temperature. Upon restoration of the reverse bias at $t = 0$ (Fig. 2c), the time-dependent density of trapped electrons is given by

$$n_T(t) \approx n_T(0)\exp(-e_n^t t), \tag{16}$$

where Eq. (8) has been used to calculate the time constant under the assumption that e_n^t dominates all other terms. For $t > 0$ and with inclusion of the trapped charge, it can be derived from Eq. (12) that constant reverse bias, i.e., constant $(V_{bi} + V)$, requires

$$(N_D + N_T - n_T(t))W^2(t) = \text{const} = (N_D + N_T)W^2(V),$$

which causes the transient space-charge width $W(t)$ to be greater than the steady state value $W(V)$ (Fig. 2c). As trapped electrons are thermally emitted to the conduction band, they are rapidly swept out of the space-charge region by the electric field and therefore not retrapped. In addition, the transient space-charge width relaxes back to the steady state value.

Determination of e_n^t and $n_T(t)$ can be performed by measuring the reverse current through the diode, or the change in space-charge capacitance. For $t > 0$ the total current at the plane $x = W$ can be calculated (Sah et al., 1970) as the sum of the conduction current caused by the emission of electrons from traps

$$i_C(t) = -qA \int_0^W dn_T/dt\, dx \tag{17a}$$

and the displacement current. The latter is due to the movement of conduction-band electrons toward $x = 0$ at the edge of the depletion region as $W(t)$ decays to $W(V)$, and is given by

$$i_d(t) = qA \int_0^W (dn_T/dt)(x/W)\, dx. \tag{17b}$$

Summing (17a) and (17b), the total junction current is

$$i(t) = qAWe_n^t n_T(t)/2. \tag{18}$$

The time dependence of n_T is given by Eq. (16), but the behavior of $i(t)$ is not exponential except for $n_T(t) \ll N_D$ because W is also time dependent through its dependence on n_T. This can be seen by using Eq. (12) and $N_D + N_T - n_T(t)$ in place of $N_D + N_T$ for the positive space-charge in the depletion region, which yields

$$W(t) = [2\varepsilon(V_{bi} + V)/q(N_D + N_T - n_T(t))]^{1/2}. \tag{19}$$

Defining $C(\infty)$ as the steady state capacitance given in Eq. (13) and inserting Eq. (19) into (13), the time-dependent capacitance change is calculated to be

$$\Delta C(t) = C(t) - C(\infty)$$
$$= A[q\varepsilon/2(V_{bi} + V)]^{1/2}[(N_D + N_T - n_T(t))^{1/2} - (N_D + N_T)^{1/2}]. \tag{20}$$

For N_T, $n_T(t) \ll N_D$, this reduces to

$$\Delta C(t)/C(\infty) \approx -n_T(t)/2N_D. \tag{21}$$

Thus measurement of the time constant of either the current or capacitance transient gives e_n^t if $N_T \ll N_D$. Note from Eqs. (18) and (21) that for a current transient $i(t)$ is proportional to $e_n^t n_T(t)$, whereas the capacitance signal is linear in $n_T(t)$. This means that if e_n^t is large (i.e., the time constant of the transient is short), the current measurement is more sensitive. On the other hand, the capacitance transient is independent of e_n^t and therefore more sensitive for slower transients.

As can be seen from Eqs. (18) and (21), the initial amplitude of the current or capacitance transient is proportional to $n_T(0)$. Therefore, if the bias has been kept at zero for a time sufficiently long to fill all traps with electrons, then $n_T(0) = N_T$ and the trap concentration can be determined from the initial amplitude of the transient. The thermal capture rate is measured by restoring the reverse bias before the traps are completely filled by electrons. Adjusting conditions (low temperature) such that inequality (15) holds, then for a width t_f of the filling pulse, the initial amplitude of the trap-emptying transient is given by

$$n_T(t_f) = N_T[1 - \exp(-nc_n t_f)], \tag{22}$$

and the slope of $\ln[N_T - n_T(t_f)]$ gives c_n (n can be separately determined by, for example, a C–V measurement). Similar experiments can be performed by forward biasing the p^+–n junction in order to inject holes into the n region. In this situation the injected holes can be trapped by levels that are occupied by electrons. The preceding analysis is still valid, except that the time constant gives the emission rate of holes to the valence band, and the initial amplitude the number of trapped holes. One difference from the electron case is that the trapping of holes results in an increase of the positive space charge, giving

$$i(t) = qAWe_p^t p_T(t)/2, \tag{23}$$

$$\Delta C(t)/C(\infty) \approx p_T(t)/2N_D. \tag{24}$$

Note that the sign of the current transient is unchanged, whereas the capacitance transient is now positive instead of negative. Therefore, if a minority-carrier injecting, forward-bias pulse is applied to a junction, majority-carrier traps can be distinguished from minority-carrier traps by the sign of their capacitance transients. This is not possible, however, for current transients or for metal–semiconductor junctions where a forward bias cannot inject minority carriers. It is important to note that because both mobile carriers are present during an injection pulse (one that injects minority carriers), recombination can occur through traps. Thus, depending on the relative magnitudes of nc_n, pc_p, e_n^t, and e_p^t, it may be impossible to fill a trap completely or even partially so that an experiment could underestimate the trap concentration or not detect it at all. Also, because of problems in determining the injected minority-carrier concentration, it is difficult to measure minority-carrier capture rates by varying the pulse width [Eq. (22)].

By using photons to ionize traps and cooling the sample to inhibit thermal emission processes, it is possible to measure optical emission rates with techniques similar to those previously discussed. For example, the junction bias can be reduced to zero to fill traps; then, after restoring the bias, illumination with photons of energy $E_c - E_T \leq hv < E_T - E_V$ will give a transient whose time constant is $1/e_n^o(hv)$.

In the previous analysis it was assumed that deep states throughout the space-charge region, $x = 0$ to $x = W$, trap and emit carriers. However, as demonstrated by Braun and Grimmeiss (1973a), there are regions at the edges of the space-charge layer where a trap does not contribute to a current or capacitance transient. Consider again the case of a p^+-n junction with a deep trap in the upper half of the band gap (Fig. 3). The space-charge region is between $x = 0$ and $x = W$, but only the area between x_1 and x_2 gives rise to a trap-emptying transient. This is because p and n are not truly zero in these regions, but exponentially decreasing from the bulk value. As can be seen in Fig. 3, between x_2 and W the deep level lies below the electron quasi-Fermi level and therefore is always filled with electrons. Between $x = 0$ and x_1, pc_p^t is sufficiently large that electrons cannot be trapped. The positions x_1 and x_2 are given by (Braun and Grimmeiss, 1973a)

$$p(x_1)c_p^t = e_n^t + e_p^t + e_n^o + e_p^o = n(x_2)c_n^t.$$

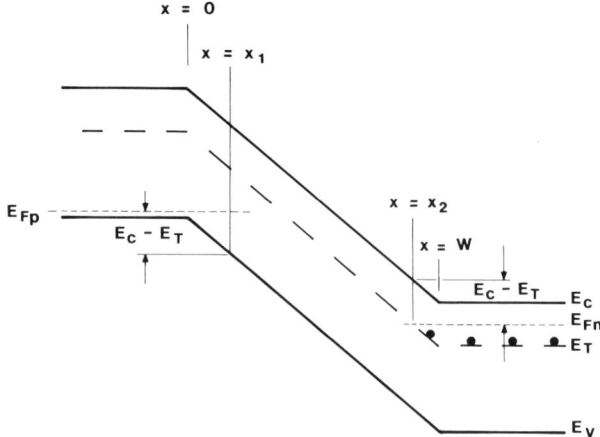

FIG. 3. Schematic energy level diagram of a p^+-n junction showing the edges of the space-charge region, $x = 0$ to x_1 and $x = x_2$ to W, within which a deep trap does not trap and emit carriers. E_{Fp} and E_{Fn} are the quasi-Fermi levels for holes and electrons.

For the case where $e_n^t \gg e_p^t + e_n^o + e_p^o$ and $c_n^t/c_p^t = 1$, these conditions reduce to $p(x_1) = n(x_2) = n_1/g$ where $n_1 = N_c \exp[-(E_c - E_T)/kT]$ and Eqs. (9) and (10) have been used to relate e_n^t to c_n^t. If $N_C = N_V$ and $g = 1$, then x_1 and x_2 are the positions where the quasi-Fermi levels for holes and electrons are $E_c - E_T$ from the valence and conduction bands, respectively (Fig. 3). Thus, for deep levels it is possible that the region $x_2 - x_1$ in which they are effective as traps is substantially less than the total depletion layer thickness W. For the case of a current transient the effect of the edge region is to change the expression given by Eq. (18) to (Grimmeiss, 1977)

$$i(t) = qA(x_2 - x_1)e_n^t n_T(t)/D, \tag{25}$$

where D can assume any value between 1 and 2 and is given by (Lang, 1979)

$$1/D = \int_{x_1}^{x_2} (1 - x/W)\,dx/(x_2 - x_1) = 1 - (x_1 + x_2)/2W.$$

For a p^+–n junction where the edge region is negligible, $x_1 = 0$ and $x_2 = W$ so that $D = 2$ and Eq. (25) reduces to Eq. (18). Analogous corrections must also be considered for capacitance transients; results for the Schottky diode case have, for instance, been given by Noras and Szawelska (1982).

2. Experimental Techniques

Because of the previously mentioned reviews (Sah *et al.*, 1970; Milnes, 1973; Grimmeiss, 1974, 1977, 1980; Sah, 1976; Miller *et al.*, 1977; Lang, 1979; Martin, 1980), this section will be brief and mainly discuss techniques not previously reviewed. However, in view of the widespread use of deep level transient spectroscopy (DLTS) and the number of variations it has inspired, it is appropriate to describe briefly the technique. The reader is referred to the original article by Lang (1974b) and the review papers by Miller *et al.* (1977) and Lang (1979) for details.

The sample, a reverse-biased p–n or metal–semiconductor junction, is placed in a capacitance bridge and the quiescent capacitance signal nulled out. The diode is then repetitively pulsed, either to lower reverse bias or into forward bias, and the transient due to the emission of trapped carriers is analyzed. As discussed in the preceding section, for a single deep state with $N_T \ll N_D$ the transient is exponential with an initial amplitude that gives the trap concentration, and a time constant, its emission rate. The capacitance signal is processed by a "rate window" whose output peaks when the time constant of the input transient matches a preset value. The temperature of the sample is then scanned (usually from ~ 77 to $\sim 450°K$) and the output of the rate window plotted as a function of the temperature. This produces a "trap spectrum" that peaks when the emission rate of carriers equals the value determined by the window and is zero otherwise. If there are several traps present, the transient will be a sum of exponentials, each having a time constant equal to the value set by the rate window at some characteristic temperature. The rate window separates the exponentials so that scanning the temperature produces a series of peaks, one for each trap contributing to the transient, whose amplitudes are proportional to the corresponding trap concentrations. This is the basic DLTS mode and, because of the spectrumlike data that it produces, it is useful as a survey technique. DLTS is also capable of measuring the activation energy of traps, their capture coefficients, and their spatial distribution in the direction perpendicular to the junction. If the sample is a p^+-n junction and the amplitude of the pulse is sufficient to cause hole injection, negative DLTS peaks are due to electron emission and positive peaks to hole emission. Finally, because the capacitance transients are repetitively generated, the rate window function can be simultaneously performed with signal averaging, making DLTS very sensitive. The window function was originally performed with a dual-channel boxcar integrator, but a lock-in amplifier or exponential correlator (Miller *et al.*, 1975) is also used. The rate window concept of DLTS has also been applied to current transients (Wessels, 1976; Hurtes *et al.*, 1978; Fairman *et al.*, 1979; Borsuk and Swanson, 1980).

Miller *et al.* (1975) generalized the rate window concept by showing that it is equivalent to correlation of the unknown transient with a weighting function. Thus the boxcar weighting function is a pair of positive and negative rectangular pulses of unit amplitude. In the limit of very wide boxcar gates, this function becomes a square wave, which is the correlation function for a lock-in amplifier. Miller *et al.* (1975) calculate that an exponential weighting function gives the best signal-to-noise ratio, which is a factor of 2 greater than for the lock-in. Other weighting functions have been examined by Hodgart (1978, 1979) and by Crowell and Alipanahi (1981).

Lang (1974a) also developed a fast transient-response capacitance bridge for use in the DLTS apparatus, but for many routine applications a commercial capacitance meter is employed (Miller *et al.*, 1977). Guldberg (1977) and Jansson *et al.* (1981) describe circuits that perform the dual sampling that is usually done by a two-channel boxcar integrator (Lang 1974b). The effects of the capacitance meter time constant on the measured trap emission rates are also discussed. Miller and Patterson (1977) give a circuit, to be used with a lock-in amplifier implementation, for synchronization and processing of the capacitance signal to remove the large transient caused by pulsing the sample bias. Day *et al.* (1979a) describe the effects of the boxcar gate widths being large compared to the delay times, as well as the effects of lock-in amplifier phasing and gate-off period, on the measured emission rate and activation energy. They also discuss the relative merits of the two methods of performing the emission rate analysis. In another paper Day *et al.* (1979b) show how Lang's original capacitance bridge can be modified by inclusion of a dummy diode to keep the bridge balanced as the DLTS spectrum of a leaky diode is being taken. Other authors describe a calibration circuit (Troxell and Watkins, 1980), a circuit that enables a single-channel boxcar integrator to be used in place of a dual-channel one (Atanasov, 1980), and use of a two-phase lock-in amplifier to emulate a dual-channel boxcar integrator (Pons et al., 1980). Okushi and Tokumaru (1981) use an analog squaring device to analyze the square of the capacitance transient and thus extend DLTS to the case where $n_T(t)$ is of the order of N_D [cf. Eq. (20)]. Goto et al. (1979) examine the shape of a DLTS peak in order to estimate the energy and capture cross section of a deep level in a single temperature scan. Their method has been extended by Le Bloa et al. (1981). Kirov and Radev (1981) describe a DLTS method based on measurement of the charge released as traps thermally empty.

A number of computer-based DLTS systems have been described. The chief attraction of such systems is reduced data acquisition time because only one temperature scan is performed. Jack et al. (1980) process the transient in the usual manner with a computer-controlled boxcar integrator. Wagner et al. (1980) employ a fast, triggered, digital voltmeter to digitize the transient at selected times, then perform the rate window function mathematically in the computer. Okushi and Tokumaru (1980) and Kirchner et al. (1981) digitize the capacitance transient and use the computer to extract the time constant. Algorithmically, the implementation by the latter is the more ambitious. The authors have two boxcar simulation routines, as well as programs employing the Fourier transform and the method of moments. The Fourier transform method is shown to be best for single, exponential transients, whereas the method of moments is superior for cases where there are multiple exponential transients.

A number of refinements and variations of the original DLTS method have been developed. Mitonneau et al. (1977a) use a pulse of sub-band-gap energy photons to perturb the occupation of deep states in the depletion region of a Schottky diode in steady state reverse bias. This permits minority-carrier traps to be studied, something that cannot be done using only electrical methods. In a slight modification of this method, Brunwin et al. (1979) generate minority carriers outside the depletion region with a pulse of intrinsic illumination. Carriers within a diffusion length of the depletion region are collected, trapped, and reemitted, giving rise to a DLTS signal. The advantage of this method over using extrinsic illumination is that centers that are not optically active can be detected because the capture process is thermal. Minority-carrier transients generated by this method can also be directly analyzed instead of being processed into a DLTS spectrum (Hamilton, 1974; Hamilton et al., 1979). Takikawa and Ikoma (1980) use successive electrical and light pulses to change the population of trapped carriers. They also point out that the amplitude of an optically induced, minority-carrier transient is temperature dependent, and correct for this effect.

Lefèvre and Schulz (1977) developed a technique in which two reduced bias pulses, the second of slightly smaller amplitude, are alternately applied to the sample. The transients due to the two pulses are separately processed in the usual DLTS fashion, then these two signals are subtracted. This defines a spatial window in the depletion region of the sample that can be moved by varying the quiescent sample bias in order to obtain a high resolution spatial profile of the deep level concentration. This method has also been used to measure the effects of electric field on the emission rate of a charged carrier (Makram-Ebeid, 1980). Johnson et al. (1979) describe an apparatus where the technique of Lefèvre and Schulz is combined with feedback that maintains the sample capacitance constant by varying the applied bias. Processing the transients in the bias permits trap concentrations to be measured accurately even at high densities.

Petroff and Lang (1977) combined DLTS with scanning electron microscopy in a method that allows spatial imaging of deep states in the plane of a junction. The scanned electron beam is pulsed on and off and the resulting thermally stimulated current or capacitance transient is analyzed using the usual DLTS methods.

Several methods not based on DLTS have also been described. White et al. (1976) present a two-light source, scanned photocapacitance technique that yields a spectrum of the deep states in the depletion region of a junction. The method is fast and sensitive, but most useful as a survey technique because knowledge of the dependence of the photoionization cross section on photon energy is required to obtain accurate trap depths.

Chantre et al. (1981) have automated and extended the transient photocapacitance methods described by Grimmeiss (1977). Initial trap populations are established by electrically pulsing the diode bias, by choice of sample temperature or illumination, and the time constant of the photocapacitance transient is determined by measuring its derivative at $t = 0$. Grimmeiss and Kullendorf (1980) extended the dual-light source, steady photocapacitance technique (Grimmeiss, 1977) to the case of large trap densities by use of feedback to maintain the sample capacitance constant. The high trap density gives increased signal and the use of two light sources decreases the measurement time.

Majerfeld and Bhattacharya (1978) and Zylbersztejn (1978) both describe methods that derive the energy of a trap relative to the bulk equilibrium Fermi level by measuring the amount of band bending between the edge of the depletion region and the position where the quasi-Fermi level crosses the energy level of the trap. Majerfeld and Bhattacharya obtain this value from the difference between $C-V$ measurements with the trap empty and filled. However, Grimmeiss et al. (1980) have shown that the filling of traps in the edge of the depletion region can be extremely slow and can cause a serious misinterpretation of the results. Zylbersztejn (1978) estimates the band bending from the time dependence of the trap-refilling process at various reverse biases. The trap refilling is expected to be a single exponential below a certain voltage, V_{sco}, and two above, and the band bending can be calculated from V_{sco}. The effects of the transition zone at the edge of the depletion region have also been analyzed by Pons (1980), Vincent (1980), Noras (1981a,c) and Noras and Szawelska (1982) for capacitance transients, and by Borsuk and Swanson (1981) for current transients.

A disadvantage of methods that utilize a junction to analyze deep states is the effect of the electric field. Extrinsic photoconductivity can be performed at low fields and high sensitivity, but because capture processes are no longer negligible, the spectral dependence of the photoionization cross section cannot be determined except in special circumstances. Grimmeiss and Ledebo (1975a) have shown that if the light intensity is adjusted to maintain a constant photocurrent as the photon energy is varied, the photoionization cross section is proportional to the inverse of the photon flux. The method is also useful for materials, such as II–VI compounds, where fabrication of a p–n junction is not possible.

By optically creating carriers with a pulse of above band-gap illumination, then monitoring the subsequent current transient due to thermal detrapping, Hurtes et al. (1978) and Fairman et al. (1979) were able to apply the DLTS method to bulk, high-resistivity materials. This method, however, is unable to distinguish between electron and hole traps, and the calculation of trap densities is difficult.

3. IDENTIFICATION OF DEEP LEVELS

Although it is usually possible to show that an impurity introduced into a semiconductor gives rise to one or more deep levels, it is often difficult to identify the defect responsible for the observed states. For example, at least two donor and eight acceptor levels have been associated with the presence of Cu in GaAs (Milnes, 1973). A donor level at $E_c - 0.07$ eV has been attributed to interstitial Cu, and states at $E_v + 0.15$ and $E_v + 0.45$ eV to Cu_{Ga} acting as a double acceptor. Other work, however, shows that the two acceptor levels have different symmetries and, therefore, are not caused by the same center (Willmann et al., 1973). Another case is the dominant electron trap at $E_c - 0.75$ eV in GaAs, which was assumed to be associated with O, but the identification has been disproven (Huber et al., 1979). Still another example is provided by Cr-doped GaAs. For those interested in the difficulties of clear-cut center identifications, a good summary of this case is given by several papers in "Semi-Insulating III–V Materials" (Rees, 1980). As a final note we mention that even identification of the relevant impurity is sometimes difficult; thus Kumar and Ledebo (1981) argue that Cu is in fact responsible for the level at $E_v + 0.4$ eV in GaAs, which had earlier been attributed to Ni.

Deep state experiments measure carrier capture or emission rates, processes that are not sensitive to the microscopic structure (such as chemical composition, symmetry, or spin) of the defect. Therefore, the various techniques for analysis of deep states can at best only show a correlation with a particular impurity when used in conjunction with doping experiments. A definitive, unambiguous assignment is impossible without the aid of other experiments, such as high-resolution absorption or luminescence spectroscopy, or electron paramagnetic resonance (EPR). Unfortunately, these techniques are usually inapplicable to most deep levels. However, when absorption or luminescence lines are detectable and sharp, the symmetry of a defect can be deduced from Zeeman or stress experiments (see, for example, Ozeki et al. 1979b). In certain cases the energy of a transition is sensitive to the isotopic mass of an impurity, and use of isotopically enriched dopants can yield a positive chemical identification of a level.

EPR is an extremely powerful tool that is capable of determining the symmetry of a center. If hyperfine splittings can be resolved, isotope experiments can chemically identify the impurity. Unfortunately, the III–V compounds are composed of elements having large nuclear spins that broaden resonance lines, thereby rendering the technique ineffective in most cases (Watts, 1977). Nevertheless, EPR has recently been used with notable success for several systems, among the most important being GaAs(Cr) and anti-site defects; this work has been reviewed, for instance, by Corbett et al. (1981), Kennedy (1981), and Schneider and Kaufmann (1981).

A principal obstacle to identification of defects is the difficulty of comparing the results from EPR, luminescence, absorption, and deep state experiments. Probably the least ambiguous is that between EPR and luminescence when, as for transition metal impurities, it is possible for optical Zeeman measurements of a sharp luminescence line to determine the ground state g factor. If the optical and EPR measurements give the same value, then the correlation is made (Watts, 1977). In some cases, when optical excitation enhances or quenches the EPR signal, there may be a similar response in the photoconductivity or luminescence excitation spectrum.

III. Observation and Intercomparison of Deep Levels. Results for GaAs and GaP

4. INTRODUCTION

Any listing of deep levels and/or intercomparisons between levels detected by different techniques must take cognizance of several problems. Thus one hindrance to comparing deep state energies derived from optical and thermal measurements is caused by a fundamental difference between the two types of experiments. As discussed by Miller et al. (1977) and by Engström and Alm (1978), this is due to the fact that the slope of e^t/T^2 versus $1/T$ yields the enthalpy H, whereas optical methods measure the free energy G. In general, these two quantities are temperature dependent, vary differently with temperature, and are not equal except at $T = 0$. A second problem in comparing optical and thermal values is that lattice relaxation can contribute a nonelectronic energy difference (Franck–Condon shift). In order to characterize completely a center using thermal methods, the temperature dependence of the thermal capture cross section must be determined. This can be done, for example, by measuring the filling rate of a center, as described in Section 1; alternatively, the free energy can be obtained by a determination of the distribution of occupied centers in the space-charge region (see, for example, Pons, 1980). Finally, in optical work, to extract energy values from photoionization cross sections, one must ultimately fit the data to a model (Section 12). To take cognizance of these difficulties, we use three means of representing literature data.

Our tabulation consists, first (e.g., Figs. 4 and 5), of energies reported in the literature (mainly those reported since 1974, although a few important levels that were well characterized previously are also included). The method used in the characterization is also included. A listing here does not necessarily imply that the free energy of the level [ΔE_T of Eq. (9)] has been reliably determined. Indeed, in view of the difficulties in comparing data taken by different researchers and the relatively few centers that have been completely characterized using thermal methods, it has become common to use the so-called signature of a level, consisting of a plot of e^t/T^2 (or its reciprocal) versus $1/T$, rather than an energy and a capture cross section to identify a center (Ikoma et al., 1977; Martin et al., 1977; Mitonneau et al., 1977b; Mircea and Bois, 1979). Tabulations of e^t/T^2 versus $1/T$ for electron traps (Martin et al., 1977) and hole traps (Mitonneau et al., 1977b) in GaAs, as well as electron and hole traps in both GaAs and GaP (Mircea and Bois, 1979), have been published, and we present such plots as our second tabulation (e.g., Figs. 6 and 7). However, despite our use of these "signature" plots, two cautionary notes are still in order. First, if the emission rate is field dependent, then the signature will depend on the field strength during the measurement and is thus not unique to a level (e.g., Pons and Makram-Ebeid, 1979; Pautrat et al., 1980). Second, DLTS techniques give true emission rates only if the carrier concentration changes sufficiently abruptly at the edge of the depletion region; otherwise the effective emission rate can depend on carrier concentration (Noras, 1981c).

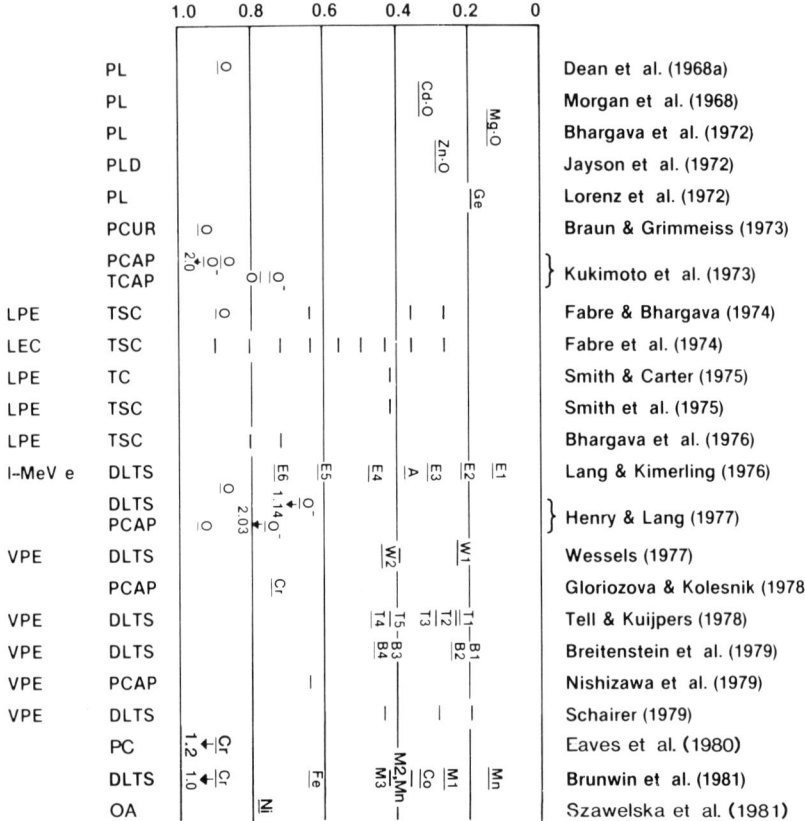

FIG. 4. Energy below the conduction band of levels reported in the literature for GaP. States are arranged from top to bottom chronologically, then by author. At the left is an indication of the method of sample growth or preparation: liquid phase epitaxy (LPE), liquid encapsulated Czochralski (LEC), irradiated with 1-MeV electrons (1-MeV e), and vapor phase epitaxy (VPE). Next to this the experimental method is listed: photoluminescence (PL), photoluminescence decay time (PLD), junction photocurrent (PCUR), photocapacitance (PCAP), transient capacitance (TCAP), thermally stimulated current (TSC), transient junction dark current (TC), deep level transient spectroscopy (DLTS), photoconductivity (PC), and optical absorption (OA).

A further desirable aspect in the tabulation of levels is the evaluation of correspondence between optical and thermal energies, i.e., values for the Franck–Condon shift ($\equiv d_{F-C}$). We list in Table II such values for several centers (as well as the corresponding thermal energies), where the evaluation

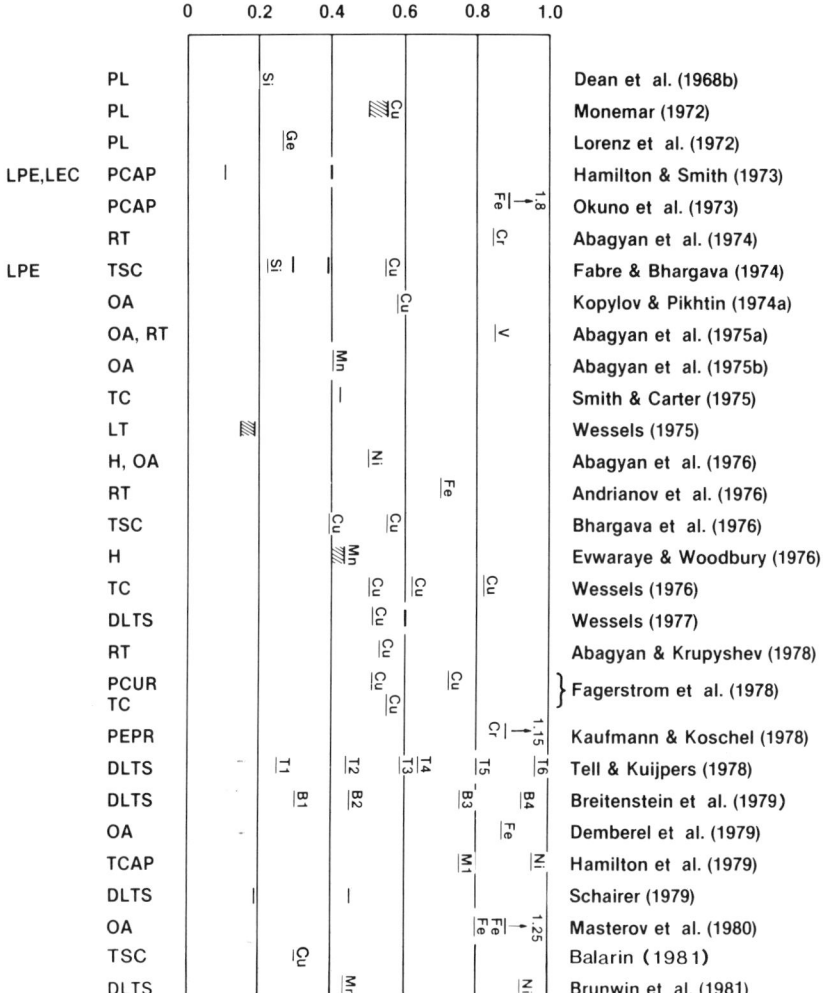

FIG. 5. Energy above the valence band of levels reported in the literature for GaP. Arrangement and notations are the same as in Fig. 4. Abbreviations for experimental methods not defined in Fig. 4. are temperature dependence of resistivity (RT), temperature dependence of minority-carrier lifetime (LT), Hall effect (H), and photostimulated electron paramagnetic resonance (PEPR).

of d_{F-C} has been carried out at least in part by optical means; the procedures used to obtain these values are described in Section 11. An alternate method for evaluating d_{F-C} has also been suggested, namely use of multiphonon

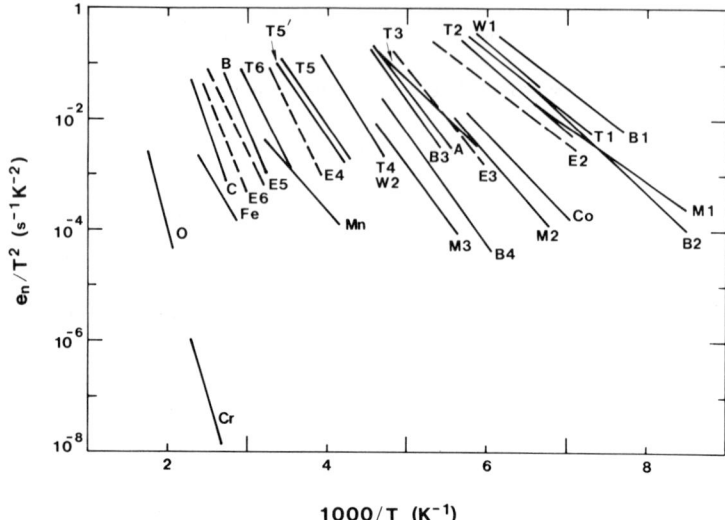

FIG. 6. Plots of e_n/T^2 as a function of $1000/T$ for electron levels in GaP. Levels B1–B4 are from Breitenstein et al. (1979) and M1–M3, Co, Mn, Fe and Cr from Brunwin et al. (1981). The rest are from the compilation by Mircea and Bois (1979). The dashed lines indicate levels induced by electron irradiation. Inserting $\langle v \rangle = (3kT/m_e^*)^{1/2}$ and $N_c = 2(2\pi m_{de}^* kT/h^2)^{3/2}$ with $m_{de}^* = 6^{2/3} m_e^*$ (i.e., assuming six valleys) into Eq. (9) gives $e_n^t/T^2 = 6\sigma_n^t (m_e^*/m_0)(1/g)(3.26 \times 10^{21})\exp(-\Delta E_T/kT)$, where m_e^* is the effective mass of an electron in one valley. Using $m_e^* = (m_\perp^2 m_\parallel)^{1/3}$, $m_\perp = 0.25 m_0$, $m_\parallel \cong 2 m_0$ (Suzuki and Miura, 1976) and assuming a degeneracy of 12, the same as for shallow donors, gives $e_n^t/T^2 = \sigma_n^t (8.2 \times 10^{20}) \exp(-\Delta E_T/kT)$. Thus dividing the value of e_n^t/T^2 at $1000/T = 0$ by 8.2×10^{20} allows an estimate of $\sigma_n^t (T = \infty)$ to be made for the various traps. (Note that the above evaluation neglects rigorous use of the conductivity mass in $\langle v \rangle$.)

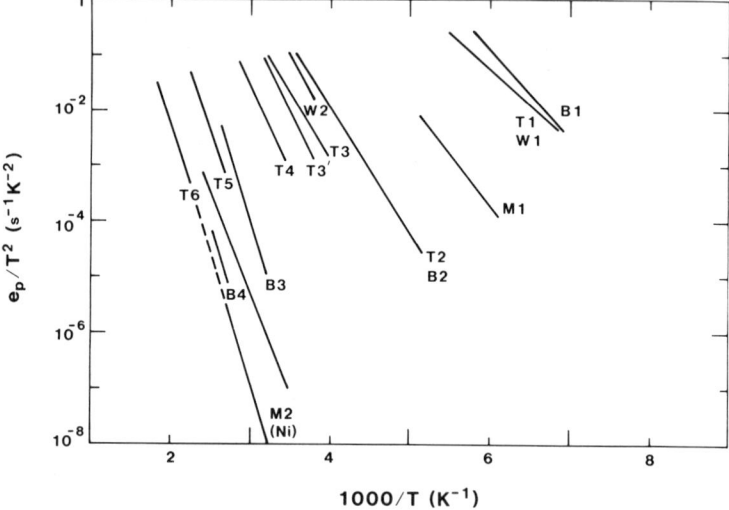

FIG. 7. Plots of e_p/T^2 as a function of $1000/T$ for hole levels in GaP. Levels B1–B4 are from Breitenstein et al. (1979) and Mn is from Brunwin et al. (1981). The rest were compiled by Mircea and Bois (1979). Using $\langle v \rangle$ and N_c as given in the caption of Fig. 6, $m_{dh}^* = m_h^*$ and $m_h^* = 0.6 m_0$ (Schwerdtfeger, 1972; Bradley et al., 1973), $g = 4$, and proceeding analogously, one obtains $e_p^t/T^2 = \sigma_p^t (4.88 \times 10^{20}) \exp(-\Delta E_T/kT)$.

TABLE II

VALUES OF FRANCK–CONDON SHIFT (d_{D-C}) FOR VARIOUS CENTERS

Center	Energy (eV)	d_{F-C} (eV)	Temp. to which energy is referred (°K)	Reference
GaP (Cu),	$E_v + 0.58$	0.34	Not given, probably $\simeq 90$	Kopylov and Pikhtin (1974a)
A center	$E_v + 0.5$	0.105	0	Grimmeiss et al. (1978)[a]
GaP (O),	$E_c - 0.89$	0.09	293	Kopylov and Pikhtin (1974b)
one-electron center	$E_c - 0.896$	0.09	0	Henry and Lang (1977)
	$E_c - 0.896$	0.08	0	Jaros (1977)[b]
	$E_c - 0.899$	0.085	0	Monemar and Samuelson (1978)[a]
GaP (Zn–O)	$E_c - 0.282$	0.19	300	Henry and Lang (1977)
GaAs ("O"), EL2 center	$E_c - 0.69$ $E_c - 0.64$	0.14	0 190	Tyler et al. (1977)
	$E_c - 0.74$	0.12	300	Chantre and Bois (1980)[c] and Chantre et al. (1981)[c]
GaAs, O doped	$E_c - 0.3$	0.11	0	Arikan et al. (1980)
GaAs (Cr)	$E_c - 0.8$	0.2	Fitting between 77 and 300	Leyral et al. (1981)
	$E_c - 0.66$	0.09	296	Ridley and Amato (1981)
GaAs (Cr)	$E_v + 0.7$	0.2	Fitting between 77 and 300	Leyral et al. (1981)
	$E_v + 0.74$	0.18	Fitting between 4 and 300	Martinez et al. (1981)
GaAs (Cu)	$E_v + 0.4$	~ 0	≈ 150	
GaAs (E3)	$E_c - 0.24$	0.2	≈ 150	Chantre et al. (1981)[c]
GaAs (EL6)	$E_c - 0.3$	~ 0.6	≈ 150	
GaAs (EL3)	$E_c - 0.44$	~ 0.05	≈ 200	

[a] Determined by convolution procedure—see Section 11.
[b] Based on an analysis of data of Kukimoto et al. (1973) and of Braun and Grimmeiss (1973b), taken in a field region.
[c] Measurements in a field region.

recombination theory (Section 10) together with measured capture cross-section data, and applied to the B center in GaAs (e.g., Burt, 1979, 1981; Robbins 1980). However, Markvart (1981) is able to fit the same data without use of a Franck–Condon shift (i.e., without lattice relaxation) by use of a different model for the vibrational behavior (specifically, with use of a non-linear electron–phonon interaction and of anharmonic terms). In view of this discrepancy, we omit tabulation of d_{F-C} values obtained via analyses of only thermal data. A similar caution against purely thermal determinations of d_{F-C} has also been given by Bois and Chantre (1980). It should still be noted that the data of Table II, unless otherwise indicated, were taken in "bulk" material, i.e., in a field-free (or low field) case.

5. GaP

The energies of deep electron and hole traps, arranged chronologically and according to author, are shown for GaP in Figs. 4 and 5, and plots of e^t/T^2 are given in Figs. 6 and 7. Comparison of the energies reported in different papers for the same level, sometimes derived using the same measurement technique, illustrates the difficulty in identifying a state by only its energy. As an example Mircea and Bois (1979) have observed that the emission rate plots are identical (Fig. 6) for the electron trap W2 reported by Wessels (1977) and T4 by Tell and Kuijpers (1978). Thus it was concluded that these levels are caused by the same defect, whereas it cannot be determined whether either T4 or T5 is the same as W2 by examination of the energies alone in Fig. 4.

Hamilton et al. (1979) have shown that a level at 0.75 eV above the valence band (M1 in Figs. 5 and 7) is the dominant recombination center in epitaxial layers and controls the minority-carrier lifetime in n-type GaP. Another state at $E_v + 0.92$ eV has been shown to be caused by the persistent presence of Ni in vapor phase epitaxial (VPE) GaP (Dean et al., 1977).

6. GaAs

Energy levels of deep centers in GaAs are indicated in Figs. 8 and 9, and plots of e^t/T^2 are given in Figs. 10 and 11. The dominant trap in bulk and VPE GaAs is the electron state EL2 (also referred to as O, L2 and A in various papers), which was attributed to oxygen until Huber et al. (1979) showed that there was no correlation between the concentrations of oxygen and EL2. This has been corroborated by Parsey et al. (1981), who have demonstrated that the careful control of As pressure and the addition of Ga_2O_3 to the melt result in horizontal Bridgman-grown GaAs with no detectable electron traps.

1. DEEP LEVELS IN WIDE BAND-GAP III–V

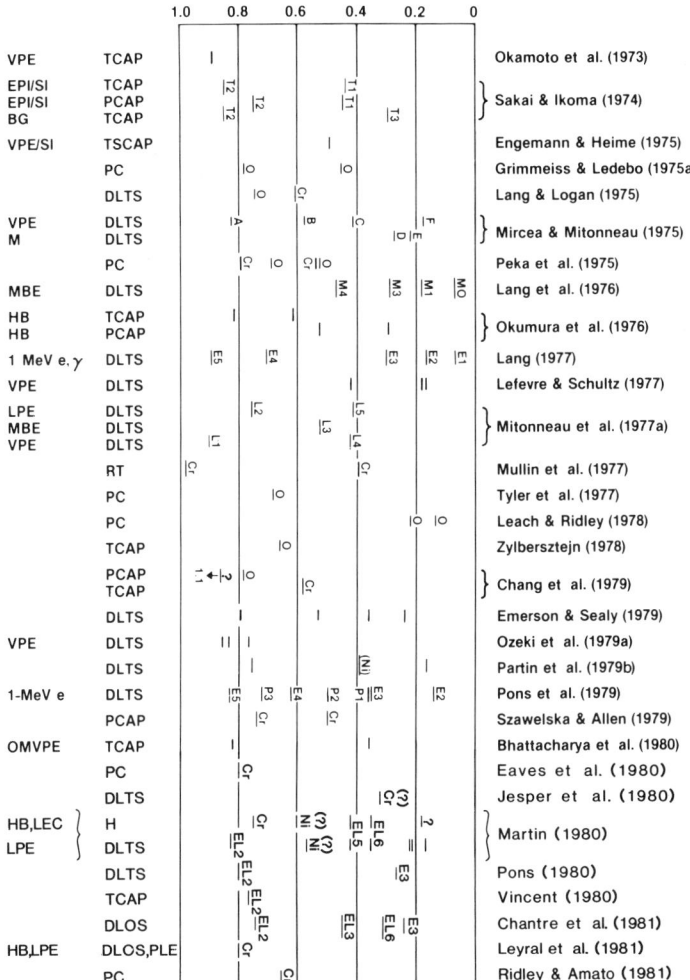

FIG. 8. Energy below the conduction band of levels reported in the literature for GaAs. Arrangement and notations are the same as for Figs. 4 and 5. Notations not defined there are epitaxial layer on semi-insulating substrate (EPI/SI), boat-grown (BG), vapor phase epitaxial layer on semi-insulating substrate (VPE/SI), melt-grown (M), molecular beam epitaxy (MBE), horizontal Bridgman (HB), irradiated with 1-MeV electrons or γ rays (1-MeV e, γ), thermally stimulated capacitance (TSCAP), photoluminescence excitation (PLE), and deep level optical spectroscopy (DLOS).

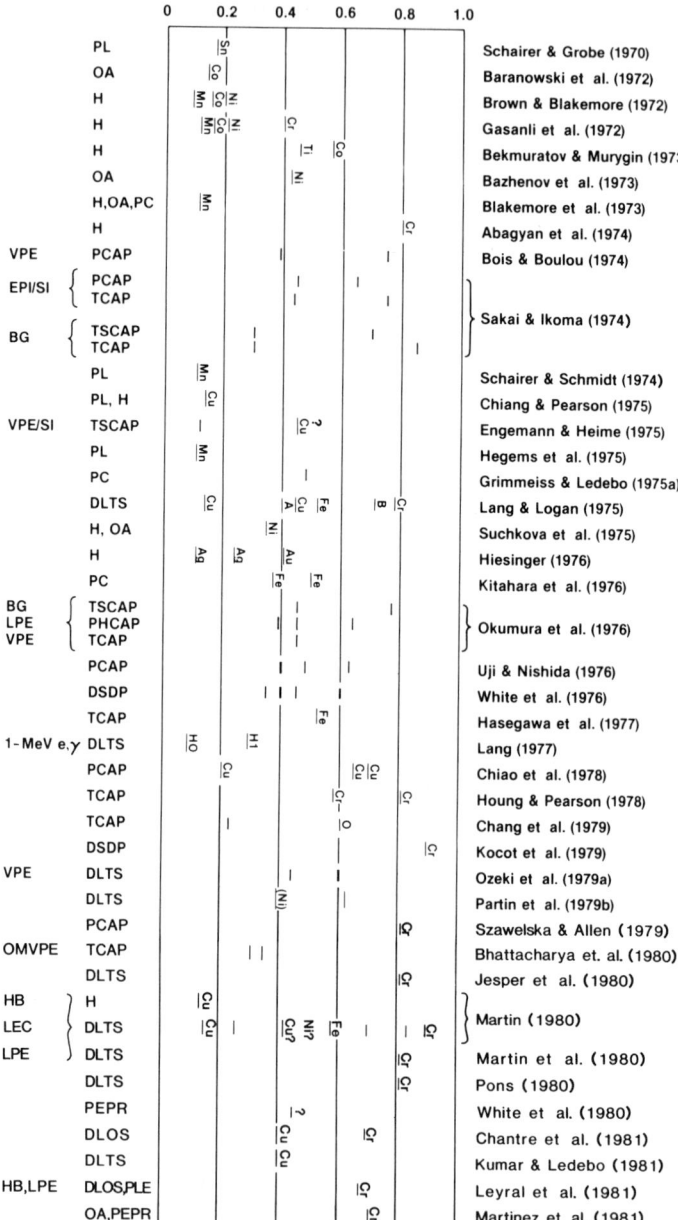

FIG. 9. Energy above the valence band of levels reported in the literature for GaAs. Arrangement and notations are as in Figs. 4, 5, and 8. DSDP indicates double source differentiated photocapacitance.

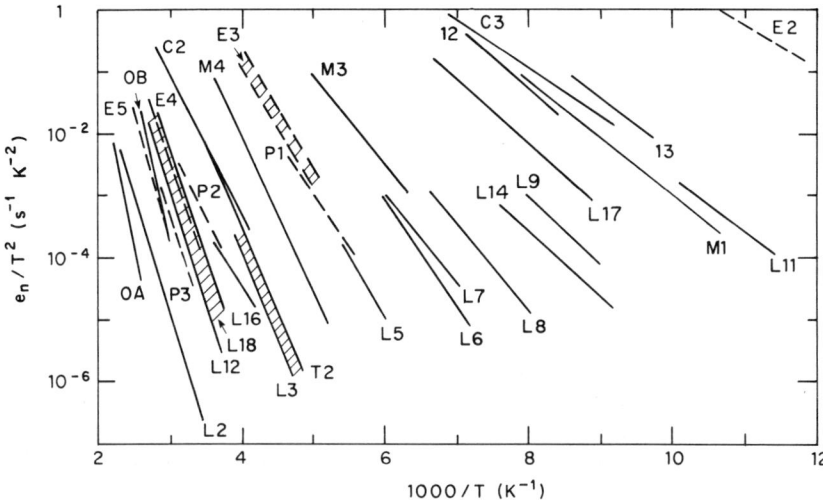

FIG. 10. Plots of e_n/T^2 as a function of $1000/T$ for electron levels in GaAs. Levels OA and OB are from Ozeki et al. (1979a), and C2 (Ni) and C3 from Partin et al. (1979b). All of the others are from the tabulation by Mircea and Bois (1979). Dashed lines represent levels caused by electron irradiation. Proceeding as in the caption of Fig. 7, using $m_e^* = 0.067 m_0$ (Stillman et al., 1977) and $g = 2$, one obtains $e_n^t/T^2 = \sigma_n^t (1.09 \times 10^{20}) \exp(-\Delta E_T/kT)$.

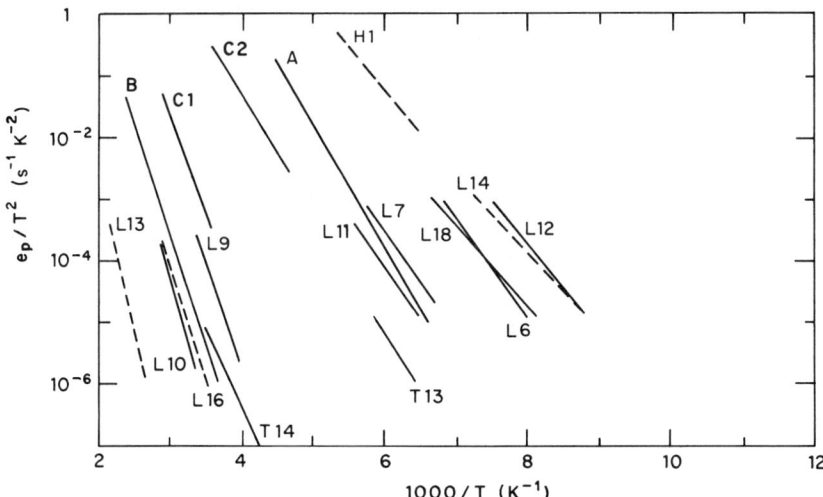

FIG. 11. Plots of e_p/T^2 as a function of $1000/T$ for hole levels in GaAs. Levels C1(Ni) and C2 are from Partin et al. (1979b) and the remainder were compiled by Mircea and Bois (1979). Dashed lines are states produced by electron irradiation. Proceeding as in the caption to Fig. 7, using $m_h^* \cong 0.5$ (Skolnick et al., 1976) and $g = 4$, one obtains $e_p^t/T^2 = \sigma_p^t (4.07 \times 10^{20}) \exp(-\Delta E_T/kT)$.

IV. Theory of Recombination

7. RADIATIVE RECOMBINATION

The subject of radiative recombination (luminescence) at deep centers will be covered only very briefly here. One reason is that transition probabilities for photon emission at deep centers are in principle no different from those at shallow centers, and the latter have been extensively treated in the luminescence literature. Reviews are given, for example, by Bergh and Dean (1972), Frova (1973), Queisser and Heim (1974), and Curie (1978). Another reason lies in the type of data that is usually obtained. One type consists of broad bands, due to large phonon coupling; although limited use has been made of such bands to aid in evaluation of the Franck–Condon shift (Mircea-Roussel and Makram-Ebeid, 1981), overall it is difficult to obtain a unique fit from data of this type. The second type, sharp lines, generally results from impurities, such as the transition metals, with localized atomic-like levels (i.e., "structured" impurities; see, for example, Robbins and Dean, 1978). Here, the main interest is in the ordering of the states (which is also used to determine between which levels the transitions take place). This aspect is more in the realm of level calculations and has, for instance, been reviewed by Masterov and Samorukov (1978); more recent work in this area is by Fleurov and Kikoin (1979), Hemstreet and Dimmock (1979a,b), Partin et al. (1979a,b), Fazzio and Leite (1980), and Hemstreet (1980). Note that interesting data can be, and has been, obtained on impurity complexes, but we omit consideration of this type here; recent examples are the detailed analyses done on GaP(Cu) by Monemar et al. (1982) and Gislason et al. (1982).

8. NONRADIATIVE RECOMBINATION: GENERAL ASPECTS

Nonradiative transitions at defect levels in semiconductors can be classified into three "primary" categories. In alphabetical order they are (1) via the Auger effect, (2) via cascade capture through excited states, and (3) via multiphonon emission (MPE). The first two are easily visualized in a simple electronic energy diagram—an example of an Auger process, that at a one-electron trap, being shown in Fig. 12 and a schematic of cascade capture in Fig. 13. Multiphonon emission must include representation of vibrational states as well as the electronic ones; this is conventionally done via a configuration coordinate diagram, discussed in detail in Section 10b. The energy loss mechanisms for these transitions are as follows: In the Auger effect there is interaction between two electrons and one hole (or, correspondingly, between two holes and one electron). Recombination occurs between one of the electrons and the hole, and the resultant energy excites the second electron. This second electron then returns to the ground state (usually the band

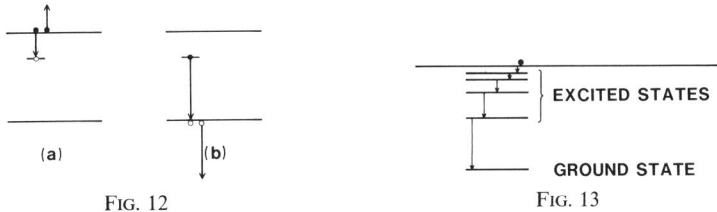

FIG. 12. Schematic of Auger capture at a one-electron trap (a) and of Auger recombination from there (b).

FIG. 13. Schematic of cascade capture.

edge) by sequential phonon emission. In the cascade process, a carrier is captured into progressively deeper excited (electronic) states of the center. A phonon is emitted at each transition. For MPE, there is a transition from an upper electronic state to an excited vibrational state of a lower electronic level. Phonons are emitted by relaxation of the vibrational excited state. To distinguish among these three processes the following two characteristics are helpful: (1) Auger processes involving free carriers show a dependence on carrier concentration; and (2) for cascade, the cross section is expected to increase with decreasing temperature. In addition to the mentioned three processes, a novel type has recently been suggested by Queisser (1978). This is recombination induced by elastically stressed regions in piezoelectric semiconductors. Charge motion in such regions results in energy release by elastic deformation. However, it has not yet been established whether this mechanism is of importance in practice. In this chapter we emphasize primarily the MPE process (Section 10). This area has a strong concentration of recent advances. Moreover, most recent reviews (e.g., Haug, 1972; Stoneham, 1977; Mott, 1977, 1978; Toyozawa, 1978) and books (e.g., Stoneham, 1975; Englman, 1979) cover only relatively briefly literature slanted towards semiconductor aspects of MPE theory; only one recent review (Stoneham, 1981) does emphasize this topic—it was written concurrently with this chapter and turned out to complement ours nicely.

Some relatively new analyses in the theory of nonradiative transitions have followed from the fact that there is no basic reason why our three "primary" processes cannot also take place in combination. Thus Gibb *et al.* (1977) propose a process of cascade capture into an excited electronic state and subsequent multiphonon emission from there. The results of this model were applied to capture and emission properties of the 0.75-eV trap in GaP. A more detailed analysis has since been given by Rees *et al.* (1980). Similarly, cascade capture followed by an Auger process with a free carrier seems a quite likely process. However, we are not aware that such a model has as yet been suggested. The third possible combination of processes, namely Auger with multiphonon, has been examined by Rebsch (1979) and by Chernysh *et al.* (1980). They find that one can obtain an enhancement of the nonphonon Auger rate; for a particular set of reasonable parameter values Rebsch (1979) obtains an increase by a factor of 4, and Chernysh *et al.* (1980), for a different case, by two orders of magnitude. In addition to there being an interaction between two processes, as just discussed, two or more processes can also affect, independently, the properties of a particular center; in principle, given proper circumstances such as the requirement of an excited electronic state for cascade, they must, although in practice one will generally dominate. Thus it has been suggested by Robbins (1980) that data on the B center in GaAs can be explained by assuming the Auger process to dominate at low temperature, the MPE at high temperature, and a contribution of a radiative process at the same center has been proposed by Pässler (1980a) [although as regards this B center, alternate models have also been suggested—see Table 5 in Stoneham (1981)].

9. Nonradiative Recombination: Auger and Cascade

a. *Auger*

The Auger effect at a center can take place in various ways, since the three particles required for the process can be located on the particular center, on another center, and/or in the bands. The case of all three particles located at the center has been well treated fairly recently by Robbins and Dean (1978) and will therefore not be considered further. The remaining cases are those where either one or two particles are free (in one of the bands). For both these situations, the transition probability p depends on the free-carrier concentration; it is customarily defined in terms of Auger coefficients B and C, respectively, for one and two free carriers. For the case of one free hole,

$$P = Bp, \qquad (26)$$

whereas for two free holes (illustrated in Fig. 12b)

$$P = Cp^2. \qquad (27)$$

The latter case (two free carriers) has been well summarized in several prior reviews (for example, Landsberg, 1970; Stoneham, 1975; Landsberg and Robbins, 1978); we therefore here mention only two very recent papers and then proceed to the former type i.e., one free carrier. Papers by both Haug (1980) and by Robbins and Landsberg (1980) conclude that the magnitude of the Auger coefficient is not greatly influenced by details of the wave function. It is the depth that has the primary influence (Haug, 1980).

The Auger effect involving one free carrier, by definition, must include two trapped carriers. These two trapped carriers can be either on the same defect or on nearby defects. A further subdivision is that the two trapped carriers can

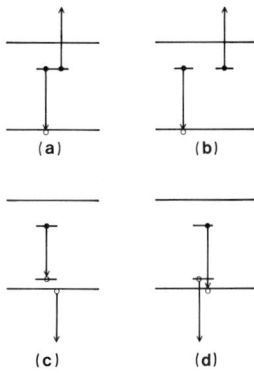

FIG. 14. Some Auger processes involving one-free carrier (holes as illustrated): The case of two trapped electrons on the same center is shown in (a), and the situation for trapping on nearby centers is shown in (b). The case of an exciton (isoelectronic) type center, with electron recombination to the trapped hole is shown in (c), and recombination with a free hole in (d) [note that in practice these two processes have to be considered in parallel (see, for example, Neumark, 1973)].

either be the same or they can be different; these two types are illustrated in Fig. 14—for both the same in Fig. 14a (two-electron case) and for the two different (exciton case) in Fig. 14b.

Regarding centers that trap two electrons, these have been analyzed by Neumark (1973) on the basis of He type wave functions and, more recently, by Jaros (1978) and by Riddoch and Jaros (1980). One conclusion from this work is that quantitative values of Auger cross sections depend strongly on the band structure, with the transition matrix element (M_A) given by Neumark (1973) as

$$M_A = F_{ac}F_{bd}J,$$

where F_{ac} is an interband overlap integral, F_{bd} is an intraband one, J is a transition integral including the impurity wave function, and where $B \sim M_A^2$. Also, a preliminary calculation by Jaros (1978) gave $B \approx 10^{-7}$–10^{-9} cm^3 s^{-1}, whereas with more detailed inclusion of the band structure Riddoch and Jaros (1980) obtained $B \approx 10^{-13}$ cm^3 s^{-1}. A further point made by Riddoch and Jaros (1980) is that use of effective-mass theory with He-type wave functions may not be suitable for such centers. Two comments are in order on this. First, Riddoch and Jaros (1980) estimate that for the He case the use of $F_{ac}F_{bd} \approx 1$ leads to $B \approx 10^{-8}$ cm^3 s^{-1}, but also estimate $F_{ac}F_{bd} \approx 0.01$, so that inclusion of the F factors gives $B \approx 10^{-12}$ cm^3 s^{-1} (since $B \sim |F_{ac}F_{bd}|^2$), i.e., a value within an order of magnitude of the value of their more detailed calculation; this emphasizes the importance of the $F_{ac}F_{bd}$ factors if the more approximate treatment of Neumark (1973) is used, as already emphasized in that earlier paper. Second, Riddoch and Jaros (1980) conclude that the wave function "localization parameter," which relates to the radius in an effective-mass type treatment, does not greatly depend on level depth. This correlates with earlier results of Jaros (1977) and implies that in a He-type treatment the quantity to be established is the impurity radius and not the impurity energy. Since Neumark's (1973) calculation used radii as primary parameters, we thus feel some interesting conclusions of this work are still valid; namely, that results do not depend greatly on the type of He wave function (three different functions gave Auger coefficients agreeing to 35%), and also that Auger rates are not strongly influenced by correlation (since correlation was included in one of the three wave functions).

The exciton-type Auger effect has been treated in several papers. Robbins and Dean (1978) have examined the situation for structured impurities and conclude that Auger recombination can be significant here. Sinha and DiDomenico (1970) treated the "nonstructured" case with use of effective-mass wave functions of p symmetry for the hole, and applied this to the Zn–O nearest-neighbor center in GaP. After being corrected for exchange by Neumark (1973), this approach predicted an Auger coefficient (B) of $\sim 10^{-10}$ cm^3 s^{-1}, for this center. Since recent experimental determinations of this coefficient (Neumark et al., 1977; van der Does de Bye, 1976) yield values in the range of $2-5 \times 10^{-11}$ cm^3 s^{-1}, it is not certain that use of such p-type functions is appropriate. An alternate, more standard, effective-mass treatment with s-type envelope functions for both electron and hole can, however, yield estimates in agreement with the experimental values (Neumark, 1973; Neumark et al., 1977). As regards the Auger effect at donor–acceptor pairs—which is the analog of the exciton case but with the electron and hole on different centers—this was treated by Landsberg and Adams (1973), who concluded that for usual impurity concentrations the radiative process will dominate over the Auger effect. However, a recent treatment by Chernysh et al. (1980) points out that if there is a strong electron–phonon interaction, the phonon-aided Auger effect for this case can be several orders of magnitude larger than the no-phonon one. Since Landsberg and Adams did not include phonon effects, perhaps their conclusion on the dominance of the radiative path should be reinvestigated.

b. Cascade

This process (Fig. 13) was first suggested by Lax (1959) as an explanation for large capture cross sections at some centers, the so-called giant traps. It has been adequately covered in previous reviews such as Landsberg (1970), Milnes (1973), Stoneham (1975), and Abakumov et al. (1978). The latter paper also includes a revised analysis of the process. Of more recent work, one should mention that of Beleznay and Andor (1978); they have carried out Monte Carlo calculations of this process and obtain improved agreement with observed cross sections (for GaAs as well as for Ge). Also, Pickin (1978, 1979, 1980) has analyzed the consequences of the cascade process on the kinetics of capture; however, his quantitative evaluation has been restricted to Ge.

In connection with the cascade process it should also be mentioned that cascade, by itself, could explain carrier *recombination* (i.e., not merely capture, but, e.g., the recombination of a trapped electron with a valence-band hole) *only* if a center provides a continuum of levels in the gap separated in energy by less than the maximum phonon energy. Such a scheme of levels appears very unlikely.

10. Nonradiative Recombination: Multiphonon Emission

a. Introduction

The most interesting recent advances in recombination theory have been in the area of multiphonon emission (MPE). This process has been relatively neglected for several years—probably in view of the calculations of Gummel and Lax (1957), which gave theoretical cross-section values several orders of magnitude lower than many of the experimental values in Si and Ge (see also Lax, 1959). Henry and Lang (1977) (see also Lang and Henry, 1975) have now revitalized it: First, they observed carrier capture cross sections—in both GaAs and GaP—which increased with increasing temperature, a behavior that is incompatible with the cascade process. Second, from the dependence (or lack of it) of the cross section on the concentration of free and trapped carriers they were able to rule out an Auger process (quite definitely for at least three centers). This leaves only MPE for these centers.

In this chapter, we first present (Section 10b) some qualitative aspects of MPE processes based on their visualization via the configuration coordinate model. We next discuss the problem more quantitatively (Section 10c). It is also to be noted that although the configuration coordinate model (i.e., visualization in terms of a single "effective" frequency) follows from the quantitative treatment only in certain approximations, in practice most of the literature does use such approximations in one form or another. Finally (Section 10d), we review the literature.

b. Qualitative Considerations

A *qualitative* overview of multiphonon processes, their origin and consequences, can be given in terms of the configuration coordinate model (see, for example, Klick and Schulman, 1957). It is here assumed that the ionic motion of relevance to the transition can be represented in terms of a single (effective) lattice coordinate. This is an approximation, but this model provides the best (and probably almost the only) means of visualizing the situation. A next approximation that is often used, and that we also use here, is that this motion can be represented in terms of a parabola. Considering one ground state and one excited state, some resultant possibilities are shown in Fig. 15, with energy E plotted versus lattice coordinate Q. Figure 15a shows two equal parabolas, the one for the excited (electron) state simply displaced upward in energy. In our next example, Fig. 15b, the excited state parabola is again

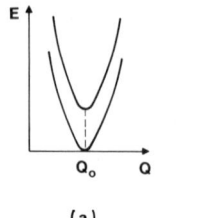
(a)

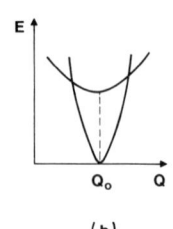
(b)

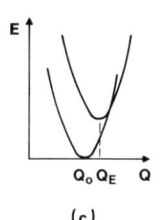
(c)

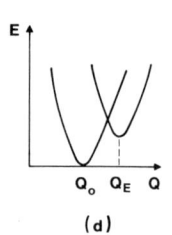
(d)

FIG. 15. The configuration coordinate model illustrated for various cases: (a) two equal parabolas, not displaced, (b) two unequal parabolas, not displaced, (c) two equal parabolas, with the origin displaced but with the minimum of the upper parabola inside the lower, and (d) two displaced parabolas, with the minimum of the upper outside the lower.

displaced straight upward, but it has a lower frequency so that the two curves intersect. Figure 15c shows two equal parabolas, but with the excited state displaced to a different equilibrium lattice coordinate (from Q_0 to Q_E). Note that in Fig. 15c the excited state minimum is still inside the lower parabola. This latter aspect is changed in Fig. 15d, where the upper minimum is now outside the lower parabola. Of course, cases of different lattice frequencies together with different equilibrium coordinates can also occur.

To analyze differences among the four "basic" diagrams of Fig. 15a–15d, we also have to consider types of possible transitions. For the optical transitions, we assume Franck–Condon processes. Resultant types are shown, schematically, on Fig. 16, which is an enlarged diagram of Fig. 15c. The transition of energy E_a is the (optical) absorption, whereas the thermal excitation has energy E_0. Thus the thermal and optical energies are not equal, differing by an energy ΔE,

$$E_a = E_0 + \Delta E, \tag{28}$$

where ΔE is related to the curvature and displacement of the parabolas. This energy ΔE is referred to as the Franck–Condon shift d_{F-C} (see Table II and Section 11). Optical emission, again assuming a Franck–Condon process, is at an energy E_e. For parabolas of equal curvature,

$$E_e = E_a - 2\Delta E. \tag{29}$$

The difference in absorption and emission energies ($= 2\Delta E$ for the present model) is the so-called Stokes shift. Also, note that ΔE increases with

Fig. 16. The configuration coordinate model of Fig. 15c on an expanded scale. Various relevant energies (see text) have been defined. Also illustrated are the three "primary" types of return to the electronic ground state: (1) the direct transition from the upper minimum, (2) the tunneling (horizontal) transition to an excited vibrational state of the electronic ground state, and (3) the thermally activated transition to the intersection of the two parabolas.

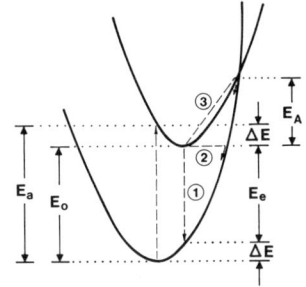

increasing displacement of the parabolas, i.e., with increasing lattice relaxation. We next consider the numbered transitions, which correspond to possible recombination paths. Transition 1 can be radiative (as just discussed, of energy E_e), but, of more interest in the present context of nonradiative transitions, it can in principle also be multiphonon (although unlikely in practice except for small values of E_0—see, for example, Haug, 1972). Transition 2 is of the tunneling type, with subsequent "gradual" phonon loss by vibrational deexcitation to the ground state. And, transition 3 takes place by vibrational excitation of the excited state to the crossing point, with a subsequent "gradual" phonon loss analogous to that of transition 2. However, for this process the capture cross section shows an activation E_A. Of course, in quantitative treatments (Sections 10c and 10d), all these transitions should be included; after all, tunneling can take place anywhere between the parabola minimum and the crossover, but with the occupancy of the required higher vibrational levels determined by an appropriate activation energy. As is well known, "pure" tunneling will dominate at low temperatures and the activated process at high temperatures. However, as has recently been emphasized for instance by Ridley (1978a), these two limiting cases can give appreciably different results at intermediate temperatures. In connection with activated capture cross sections, it should still be pointed out that since both Figs. 15b and 15c can lead to transition 3, both can give activated capture; thus occurrence of this process does not prove the presence of lattice relaxation. Additionally, we wish to point out that not all transitions 1–3 are possible for the various situations of Figs. 15a–15d. Namely, for the case of Fig. 15a, only transitions 1 and 2 are possible. For Figs. 15b and 15c, all transitions are possible, but for Fig. 15d transition 1 cannot take place. This has an interesting consequence: If the transition probabilities for transitions 2 and 3 are low (note that this can be accomplished for 3 by lowering the temperature), the excited state will be *metastable*. Models of this type have, for instance, been proposed by Lang and Logan (1977) for persistent photoconductivity in GaAlAs and by Vincent and Bois (1978) for persistent photocapacitance quenching in GaAs.

As a further point, we mention that the vibrational energy loss ΔE (Fig. 16) is also often expressed in terms of the strength of the electron–phonon coupling or of the Huang–Rhys factor S (see, for example, Stoneham, 1975, 1977, 1981), where

$$S = \Delta E/\hbar\omega_0, \qquad (30)$$

with ω_0 the vibration frequency. Here strong coupling refers to $S \gg 1$ and weak coupling to $S \ll 1$. In most treatments the factor S is considered an empirical parameter, but there has also been some recent renewed interest in evaluating it from first principles (Ridley, 1978b; Halperin and Englman, 1979; Stoneham, 1979).

Also of interest in connection with Fig. 16 is a process that has been labeled "hot transfer" (see, for example, Toyozawa, 1978; Kayanuma and Nasu, 1978; Jortner, 1979). Here it is suggested that in the case $\Delta E > E_A$ the transition to the ground state can take place during the lattice relaxation, i.e., before the excited state has reached its equilibrium position. This effect was first suggested by Dexter et al. (1955) and further analyzed by Bartram and Stoneham (1975) for F centers. A recent slight modification, in terms of fast capture of a majority carrier subsequent to that of a minority carrier, has been suggested by Sumi (1981); he points out that this process may be active in recombination-enhanced defect reactions.

Last but not least we must still emphasize that in practice there may well be complications on the (relatively) simple cases discussed so far. One is that usually more than two parabolas are active in real physical systems of present interest. Henry and Lang (1977) already represented the valence–conduction band and center system by three parabolas; recent reviews of this case, along with summaries of types of resultant recombination behavior (analogous to our discussion of Fig. 15) have been given by Bois and Chantre (1980) and Lang (1980). The next step, a fourth parabola (for an excited state), has been used by Vincent and Bois (1978) to explain data obtained on the "oxygen" center (at $\sim E_c - 0.75$ eV) of GaAs. Moreover, the number of parabolas can go from there on up, for instance if degenerate states are considered [as for Cr in GaAs—see Hennel et al. (1981)] or if the conduction band is considered in more detail (e.g., Sumi, 1980b, 1981; Chantre et al., 1981). A second complication is that the visualization via parabolas uses a one-dimensional representation and the simplest one at that. In general, one may have to deal with multidimensional potential energy surfaces and/or the nonparabolic (anharmonic) case. Such situations have been treated, for instance, by Englman (1979), Englman and Ranfagni (1980), Markvart (1981), Martinez et al. (1981), and Stoneham (1981). We generally neglect these various complications in our following quantitative analysis of MPE (Section 10c), but include some further discussion in our literature review (Section 10d).

c. *Quantitative Aspects*

We now consider problems of a quantitative analysis of multiphonon transitions. Here an exact treatment seems hopeless at the present time, and to make headway at all a fair number of approximations are required. We shall give an overview of the general difficulties, discuss some (unfortunate) confusion on "Born–Oppenheimer" terminology, and then illustrate some quantitative problems using the "adiabatic" formulation (see below). The present discussion will also be used as a basis for subdividing the various papers, to be discussed in Section 10d, into various (perhaps somewhat arbitrary) categories.

The multiphonon problem involves a complex many-body (both many-ion and many-electron) Hamiltonian and, in order to induce transitions, some perturbation. Essentially, headway on this problem has been possible only by (1) approximating the unperturbed Hamiltonian, (2) assuming some sensible wave functions for this unperturbed Hamiltonian, and (3) assuming some perturbation.

Most multiphonon treatments are based on an approximation technique suggested by Born and Oppenheimer (1927). An essential point here is that the total wave function is expressed as a product of electronic and vibrational wave functions. The resultant use of different basis sets for the electronic wave functions has however led to problems. One difficulty has been confusion in terminology, which was already noted by Markham (1956), has recently been reviewed by Englman (1979) and Stoneham (1981), and will also be discussed below. Another, more serious, problem has been that such different methods have led to orders-of-mangitude differences in values of absolute cross sections. The dilemma has been resolved only recently, in two steps. First (chronologically) it was realized that the different basis sets must be carried to equivalent orders in perturbation theory (Huang, 1981; Gutsche, 1981, 1982; Burt, 1982a). Second, a penetrating and elegant treatment by Peuker *et al.* (1982) has shown that, in addition, proper consideration must be given to the perturbation Hamiltonian, and that many previous treatments did not properly include the full nondiagonal part of the total Hamiltonian.

Before we can discuss the recent developments further, we must discuss terminology. This will also provide a guide to earlier literature as well as to the classification we use in our review of recent papers (Section 10d). As mentioned, the total wave function in the Born–Oppenheimer method is expressed as a product of electronic and vibrational wave functions. What has resulted is that different types of electronic wave functions have been used (which is not necessarily confusing), *and* that in many cases a particular selection has been called *the* Born–Oppenheimer approximation (which *has* led to confusion). We discuss here only the two predominant choices of functions (see Englman, 1979 or Stoneham, 1981, for some others), which are referred to as the static and adiabatic approximations. Using ψ as the (zero-order) total wave function, ϕ for the corresponding electronic, and χ for the vibrational ones, then the *static* approximation is given by

$$\psi_{ln}(r, Q) = \phi_l(r, Q^0)\chi_{n(l)}(Q), \tag{31}$$

where r and l are the electron coordinates and quantum numbers, respectively, and Q and n are the corresponding quantities for the nuclei. The quantity Q^0 refers to the *equilibrium* position of the ions; hence the name *static* approximation. For the *adiabatic* approximation

$$\psi_{ln}(r, Q) = \phi_l(r, Q)\chi_{n(l)}(Q). \tag{32}$$

In this case the zero-order electronic wave functions are, in principle, referred to a Hamiltonian that contains the potential from the ions at their actual positions, i.e., the electrons follow the ionic motion *adiabatically*. Since both these approximations are sometimes referred to as *the* Born–Oppenheimer approximation, this has led to confusion in terminology: for example, Mott (1977) refers to the Born–Oppenheimer approximation, but gives wave functions of the adiabatic type, whereas Englman (1972) differentiates between the two forms, but specifically calls the static form the Born–Oppenheimer method. [We note that, historically, the adiabatic form was first suggested by Seitz (1940)—see, for example, Markham (1956) or Haug and Sauermann (1958)]. In this chapter, we shall preferentially use the terminology "static" and "adiabatic." [Note that the term "crude adiabatic" is also sometimes used for the static approximation, mainly in the chemical literature—see, for example, Englman (1972, 1979).]

We can now state the conclusions of the recent comparisons of these methods. One conclusion is that results from the adiabatic method in the non-Condon approximation (see further below) and those from the static approach are fully equivalent to a first-order perturbation in the linear electron–phonon interaction (Huang, 1981; Gutsche, 1981, 1982; Burt, 1982a; Peuker et al., 1982). A further conclusion, reached by Peuker et al. (1982), is that if in the adiabatic treatment the "full non-diagonal part of the total Hamiltonian is taken into account rather than only its non-adiabatic contribution", then the adiabatic treatment in the Condon approximation also gives the same result (again to first order); in the non-Condon treatment the terms introduced by this correction are negligible. A corollary conclusion by Peuker et al. (1982) is that use of this correct treatment shows that "the static approach is the lowest order approximation to the adiabatic result." Further, it can be proven (e.g., Peuker et al., 1982) that the old, incorrect adiabatic method gives results that are too low by a factor $(E_0/\hbar\omega_0)^2$, where ω_0 is the vibrational frequency and E_0 is the energy difference of the electronic states (defined in Fig. 16). This factor can amount to several orders of magnitude, as had been noted in the earlier comparisons of the various treatments.

We next present a brief description of the mathematical manipulations based on the framework of the adiabatic approximation, just indicating briefly the step that Peuker et al. (1982) have shown to be missing. Our aim is not so much to present the overall method [which is presented in great detail by Peuker et al. (1982), and where the "old" adiabatic analysis has been given in many earlier comprehensive reviews—e.g., Born and Huang (1954), Englman (1972, 1979), Haug (1964, 1972), Markham (1959), or Perlin (1963)], but rather to provide a means to catalogue and understand the various "recent" papers on this subject. (We define "recent" in Section 10d). For this purpose, emphasis on the adiabatic method seems sensible: most treatments to date have employed it, we can discuss the Condon approximation within the context of this approach (the Condon approximation holds by definition in the static case), and, since the error is a function of $E_0/\hbar\omega_0$ (Peuker et al., 1982), all the extensive old work on temperature dependences—a topic of prime interest—remains valid.

The total Hamiltonian (H) of the system can be put into the form

$$H = T_E(r) + T_N(Q) + V_E(r) + V_N(Q) + V_{EN}(r, Q). \tag{33}$$

Here T refers to the kinetic energy terms and V to the potential energy, with the subscripts E for the electronic and N for the nuclear parts, respectively. One then considers the electronic energy E_{El} of the state l as given by

$$[T_E(r) + V_E(r) + V_{EN}(r, Q)]\phi_l(r, Q) = E_{El}(Q)\phi_l(r, Q). \tag{34}$$

By use of Eq. (32) and the standard relation $H\psi = E\psi$, and if the dependence of $\phi_l(r, Q)$ on Q is neglected, one obtains for the energy E_{nl} of the vibrational state (nl)

$$[T_N(Q) + V_N(Q) + E_{El}(Q)]\chi_{nl}(Q) = E_{nl}\chi_{nl}(Q), \tag{35}$$

where $\chi_{n(l)}$ of Eq. (32) is now written as χ_{nl}. Since the χ_{nl} are now approximate (the Q dependence of ϕ has been neglected), the difference term H', where

$$H' = H - E_{nl}, \tag{36a}$$

provides a perturbation and induces transitions. However, since the ϕ_l are also approximate, there is a corresponding term

$$H'' = T_E + V_E + V_{EN}, \tag{36b}$$

which can also induce transitions via nondiagonal matrix elements (Peuker et al., 1982). We henceforth consider only the H' term. This leads to a result that, for a given ψ_{ln}, can be expressed in the form

$$H'\psi_{ln} = T_N(Q)\phi_l(r, Q)\chi_{nl}(Q) - \phi_l(r, Q)T_N(Q)\chi_{nl}(Q). \tag{37}$$

If one uses the (normalized) expression for T_N, namely

$$T_N = -\frac{1}{2}\sum_k \frac{\partial^2}{\partial Q_k^2}, \tag{38}$$

where the sum k is usually over the normal modes (which can be either appropriate local modes or lattice modes), then the matrix element for a transition from state ln to sm can be expressed as

$$\langle \phi_s \chi_{ms} | H' | \phi_l \chi_{nl} \rangle = -\sum_k \iint dr\, dQ \left[\phi_s^* \chi_{ms}^* \frac{\partial \phi_l}{\partial Q_k} \frac{\partial \chi_{nl}}{\partial Q_k} + \frac{1}{2} \phi_s^* \chi_{ms}^* \chi_{nl} \frac{\partial^2 \phi_l}{\partial Q_k^2} \right]. \tag{39}$$

Within the old adiabatic approximation, Eq. (39) is the basic "starting point." However, from here on, the various approximations diverge. For ease of discussion, we shall first still make the *Condon* approximation, and then give the further approximations. However, it must be kept in mind that many similar approximations are also made in papers that use a "non-Condon" approach. The basic premise of the Condon approximation is that the electronic part of the matrix element varies sufficiently slowly with Q so that it can be taken out of the integration over dQ. The matrix element then reduces to products of electronic and vibrational integrals. In Dirac notation

$$\langle \phi_s \chi_{ms} | H' | \phi_l \chi_{nl} \rangle = -\sum_k \left[\left\langle \phi_s \left| \frac{\partial}{\partial Q_k} \right| \phi_l \right\rangle \left\langle \chi_{ms} \left| \frac{\partial}{\partial Q_k} \right| \chi_{nl} \right\rangle \right.$$
$$\left. + \frac{1}{2} \left\langle \phi_s \left| \frac{\partial^2}{\partial Q_k^2} \right| \phi_l \right\rangle \left\langle \chi_{ms} | \chi_{nl} \right\rangle \right]. \tag{40}$$

Specifically, the various papers working within both the adiabatic and the Condon approximations, and using the (frequent) assumption of harmonic vibrations, can still differ in how many and what type (optical, acoustic, or local) modes they consider and in how they approximate the four separate integrals on the right-hand side of Eq. (40). And the choice of modes applies to both the ground and the excited states (so does the choice of electronic wave functions, but this choice is implicit in the evaluation of the electronic integrals.) It is this choice regarding the two states that was emphasized in connection with Fig. 15 (Section 10b). It can be seen that even within the stated approximations (adiabatic, Condon, harmonic) there is an appreciable number of permutations and combinations.

Before discussing specific papers (Section 10d), it seems worthwhile to still make three further comments on Eq. (40). First, the term involving $\partial^2/\partial Q^2$ is often disregarded: If the variation of ϕ with Q is slight, it is assumed that $\partial^2/\partial Q^2$ can be neglected. In some models, for instance if ϕ_s and ϕ_l have opposite parity, this neglect is rigorously correct (see Stoneham, 1975). However, in other cases it is "arbitrary" as expressed by Stoneham (1975), who also points out that if ϕ_s and ϕ_l have the same parity, then the $\partial^2/\partial Q^2$ term is the only one which remains. Second, if one assumes a large number (N) of modes (e.g., lattice modes), each with wave function X, and uses the relation

$$\chi_{nl}(Q) = \prod_{k=1}^{N} X_{n(k)l}(Q_k), \qquad (41)$$

then with the further assumption that addition of a single vibrational overlap factor is unimportant, one obtains from Eq. (40)

$$\langle \phi_s \chi_{ms} | H' | \phi_l \chi_{nl} \rangle = C \prod_k \langle X_{m(k)s} | X_{n(k)l} \rangle, \qquad (42)$$

where

$$C = -\sum_k \left[\left\langle \phi_s \left| \frac{\partial}{\partial Q_k} \right| \phi_l \right\rangle \left\langle X_{m(k)s} \left| \frac{\partial}{\partial Q_k} \right| X_{n(k)l} \right\rangle + \frac{1}{2} \left\langle \phi_s \left| \frac{\partial^2}{\partial Q_k^2} \right| \phi_l \right\rangle \right] \qquad (43)$$

(for details of this derivation, see, for example, Brailsford and Chang, 1970). Thus, the Frank–Condon overlap factor $\prod_k \langle X_{m(k)s} | X_{n(k)l} \rangle$ is involved in both terms on the right-hand side of Eq. (40). It is probably partly because of this aspect that this overlap factor is emphasized in many papers. For example, Englman and Jortner (1970), Freed and Jortner (1970), and Brailsford and Chang (1970) all express their results in terms of the parameter C of Eq. (43), but do not evaluate this parameter (not even the $\langle X_{m(k)s} | \partial/\partial Q | X_{n(k)l} \rangle$ part). Another reason why this overlap term is usually emphasized, and this is also the third of our comments, is that due to the change in level occupancy with temperature (i.e., effectively change of n) a temperature dependence of the transition probabilities can result from this term. Indeed, this is the primary cause of the temperature dependence of capture cross sections. [But note that, analogously, a temperature dependence can also result from the $\langle X_{m(k)s} | \partial/\partial Q_k | X_{n(k)l} \rangle$ term—although as discussed in Section 10d(1) this is probably not too serious. Specific results on the temperature dependence are given in our review of papers—Section 10d(1).]

The evaluation of the electronic wave functions of Eq. (34) has been neglected in this section; this aspect was not required for our classification of papers in Section 10d. Information on this topic can be obtained from reviews (such as, Stoneham, 1975; Pantelides, 1978; Jaros, 1980), or from some of the papers discussed in Section 10d. Additionally, we discuss some parts of this problem in Section 12 in connection with evaluation of optical cross sections.

d. Review of Recent Papers

The main reviews of multiphonon transition theory that include a good list of references to papers relevant to semiconductors are the ones by Markham (1959), by Perlin (1963), and by Stoneham (1981). Since this chapter was written concurrently with Stoneham's, we had decided to emphasize papers in this area from 1963 on; in view of differences in emphasis between Stoneham

(1981) and ourselves, we retain papers from 1963 on as "recent." We shall here omit most of a multitude of papers (including those by Jortner and by Brailsford and co-workers referred to in Section 10c) that apply this theory primarily to molecular systems. However, we do include several papers that consider impurities or defects (e.g., F centers) in solids in general, even without specific application to semiconductors. We shall first discuss work using the adiabatic method, where the work that uses the Condon approximation is discussed in Section 10d(1), with Section 10d(2) for the "non-Condon" approaches. The approaches in the static method are reviewed in Section 10d(3). As discussed in Section 10c., the static and the non-Condon adiabatic methods have very recently been shown to be equivalent (Huang, 1981; Gutsche 1981, 1982; Peuker et al., 1982), but we nevertheless separate the two treatments for convenience. Some "other" approaches are given in Section 10d(4).

(1) *Adiabatic and "Condon."* Since the Condon approximation separates the electronic and the vibrational parts of the problem (see Section 10c), most of the papers using this approximation emphasize the latter aspect. Specifically, they analyze the so-called Franck–Condon overlap integrals [the $\langle X_m | X_n \rangle$ of Eq. (42)] and the occupancy factors of these vibrational levels. Such analyses include (1) the influence of mode type and/or number as well as (2) for various specific mode cases, the temperature dependence of the transition probability. It is worth noting that since the Condon adiabatic treatment has been shown to be in error by the temperature-*independent* factor $(E_0/\hbar\omega_0)^2$ (Peuker et al., 1982; and already discussed in Section 10c), the temperature dependence should be properly given by such work (at least for single-frequency models). Nevertheless, it must also be remembered that the temperature dependence of the $\langle X_m | \partial/\partial Q | X_n \rangle$ term [see Eq. (43) and its discussion], is usually neglected. However, it is known that this term can be expressed as a linear combination of $\langle X_m | X_{n+1} \rangle$ and of $\langle X_m | X_{n-1} \rangle$ (for details, see Löffler, 1969), and it is thus unlikely that this $\partial/\partial Q$ term would greatly alter the temperature dependence given by the (usually considered) $\langle X_m | X_n \rangle$ terms (although we are not aware of any detailed analysis of this point).

Regarding the temperature dependence of the transition probabilities for the single configuration coordinate (SCC) model with displaced parabolas (Fig. 15c), quite detailed analyses have been given by Struck and Fonger (1975) and by Robertson and Friedman (1976). [Similiar work, but within a non-Condon framework, was also done, for example, by Löffler (1969), by Ridley (1978a), by Goto et al. (1980), and, in the static case, by Pässler (1978, 1980b, 1981)]. Struck and Fonger (SF) include numerical results for the general case, including that of unequal frequencies of the two parabolas. In addition, they give an analytical expression for equal frequencies ($\equiv \omega_0$) and for weak coupling [small S—see Section 10b and Eq. (30)]. Robertson and Friedman (RF) confirm the results of SF for this analytic case and, in addition, give analytic results for the high temperature and strong coupling situation (for equal frequencies). The analytic expressions are

$$W_T = W_0[1 + n(\omega_0)]^p \tag{44}$$

for weak coupling, low temperature, and

$$W_T \sim (kT)^{-1/2} \exp(-E_A/kT) \tag{45}$$

for strong coupling, high temperature. Here W_T is the transition probability, W_0 is that at $T = 0°K$, E_A is the activation energy to crossover (defined in Fig. 16), $n(\omega_0)$ is the phonon occupation of the mode,

$$n(\omega_0) = [\exp(\hbar\omega_0/kT) - 1]^{-1}, \tag{46}$$

and p is the number of phonons involved in the transition,

$$p = E_0/\hbar\omega_0, \tag{47}$$

where E_0 is the energy difference of the electronic states (defined in Fig. 16). It should be noted that Eq. (45), other than the $(kT)^{1/2}$ factor, is equivalent to the frequently used Eq. (11).

It is interesting to note that the temperature dependence, as given by Eqs. (44) and (45), can, basically, be split into three different types of behavior. One is the simple activated one given by Eq. (45). The second is an essentially temperature-independent behavior of Eq. (44) at *low* temperatures [$n(\omega_0) \ll 1$]. This corresponds to the tunneling transition (2) indicated in Fig. 16. The third corresponds to a temperature dependence in Eq. (44) at higher temperatures, but where such temperatures are still low enough so that Eq. (45) does not yet apply. The behavior in this range is *not* activated; quantitative comparisons to illustrate this in several specific cases have been given by SF (see, for example, their Figs. 5 and 7). The same point has also been emphasized by Ridley (1978a). It remains to discuss, more quantitatively, the validity ranges of Eqs. (44) and (45). Both SF and RF express these conditions in terms of requirements on *both* the temperature and on the parabola displacements. Ridley (1978a), on the other hand, gives a criterion for deep traps, say $p > 10$, which combines both these parameters. In terms of the Huang–Rhys parameter S, which for the simple case is given by Eq. (30), the low temperature condition is

$$(p/2S)^2[n(n+1)]^{-1} \gg 1, \tag{48}$$

with the corresponding high temperature condition of

$$(p/2S)^2[n(n+1)]^{-1} \ll 1, \tag{49}$$

where the temperature dependence results from that of n [Eq. (46)]. Ridley also makes the point that in the transition region the situation is not well described by either Eq. (44) or Eq. (45). Thus, at an appropriate intermediate temperature, the two results differ by $\approx 4p \exp(p/5)$ [see Ridley, 1978a, Eq. (66)]; for deep levels, this is a large number. Of additional particular interest, independent of the detailed behavior, is the fact that the temperature dependence of the cross section may thus *not* be exponential in the range of experimental results. Consequently, any analyses that rely on such an exponential dependence [Eq. (11)] to extract additional information must be viewed with caution. [An example here is Mircea and Mitonneau (1977), who extract the temperature dependence of the level in this fashion.]

The next areas to be discussed are the effects of equal versus unequal frequencies for the two parabolas (Fig. 15b), and those of mode type and number (i.e., abandonment of the SCC model). The overall conclusion is that these factors do not greatly change the *qualitative* results obtained by the SCC model with equal frequencies. However, depending on the parameters, there can be appreciable *quantitative* differences. A review of specific papers follows.

The problem of unequal frequencies for the ground and the excited electronic states has been included in the paper by Struck and Fonger (1975) referred to above. They conclude that, in general, with such unequal frequencies "no new types of quenching would be uncovered," i.e., there would be no qualitative changes in the possible temperature dependencies obtained with equal frequencies. This same problem is also addressed in a second paper by Robertson and Friedman (1977), and analytic expressions are given. These are, in general, quite complicated, and Robertson and Friedman therefore analyze only the $T = 0$ case in detail. The result for weak coupling, emission of a large number of phonons ($p \gg 1$) and a large difference in frequencies ($\omega_0 \gg \omega_1$, where ω_0 is the frequency of the ground state and ω_1 that of the excited state) is relatively tractable and can be compared to the result for displaced equal frequency parabolas (given for instance by Robertson and Friedman, 1976). The main terms of the two transition probabilities (W_0) are

$$W_0(\omega_0 = \omega_1 = \omega) \sim \exp\left[-\left(\frac{E_0}{\hbar\omega}\right)\left(\ln\frac{E_0}{\Delta E} - 1\right)\right] \exp\left(-\frac{\Delta E}{\hbar\omega}\right) \quad (50)$$

and

$$W_0(\omega_0 \gg \omega_1) \sim \exp\left[-\left(\frac{E_0}{\hbar\omega_0}\right)\left(-\ln\frac{\omega_0 - \omega_1}{\omega_0 + \omega_1}\right)\right]$$

$$\times \exp\left[-\left(\frac{\Delta E}{\hbar\omega_0}\right)\left(\frac{2\omega_1}{\omega_0 + \omega_1}\right)\right]. \quad (51)$$

Since $\omega_0 \gg \omega_1$, one can use

$$(\omega_0 - \omega_1)/(\omega_0 + \omega_1) \approx [1 - (2\omega_1/\omega_0)] \quad \text{and} \quad \ln(1 - x) \approx -x$$

to simplify Eq. (51)

$$W_0(\omega_0 \gg \omega_1) \sim \exp\left[-\left(\frac{E_0}{\hbar\omega_0}\right)\left(\frac{2\omega_1}{\omega_0}\right)\right]\exp\left[-\left(\frac{\Delta E}{\hbar\omega_0}\right)\left(\frac{2\omega_1}{\omega_0}\right)\right]. \quad (51a)$$

It is apparent that quantitative values can be appreciably different in the two cases.

The next problem, that of multiple oscillators for a given (or both) parabola, means abandonment of the SCC model, which we (and much literature) have emphasized. Nevertheless, it has been shown by Fonger and Struck (1974), and reemphasized for MPE by Struck and Fonger (1975), that if the multiple (lattice) oscillators all have the same frequency, the results are the same as for the SCC model. The only (trivial) difference is that the lattice frequency replaces the local frequency of the SCC case. The more general case of a frequency distribution of the lattice modes has been treated by Englman (1977). He finds that for strong coupling, "for a wide set of circumstances" one can utilize a description either in terms of a single-phonon mode (or of a few modes), or of a "continuum" model (i.e., one with a broad frequency spectrum). Nevertheless, this situation does not always apply. Whether or not it does can, however, be checked by appropriate criteria given by Englman (1977). A further treatment of this problem is given by Weissman and Jortner (1978), who develop an improved approximation scheme for the case with phonon dispersion. The conclusion from this work is essentially the same as from Englman (1977): For strong dispersion, the single-mode approximation breaks down. Moreover, Weissman and Jortner (1978) show that in this case the single-mode approximation does poorly both as regards the magnitude and the temperature dependence of the transition probability. To show this, we reproduce part of their Fig. 3b as our Fig. 17. Again, quantitative results can be different from those of the single-mode case.

It should still be pointed out that with increasing complexity of vibrational behavior the standard Born–Oppenheimer treatments become increasingly difficult. Some alternate approaches will be mentioned in Section 10d(4).

(2) *Adiabatic and "Non-Condon."* The "non-Condon" approaches, as mentioned in Section 10c, retain some interaction between the electronic and vibrational matrix elements. As a general conclusion, two primary results emerge from all these treatments. First, the temperature dependence of the transition probability is still mainly determined by the $\langle X_m | X_n \rangle$ part [see Eq. (42)] through the occupancy of the vibrational levels. This follows since the

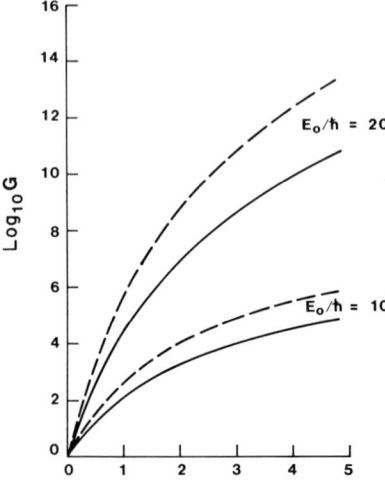

FIG. 17. A comparision of the temperature dependence of the "line-shape function" (G) of the transition probability for the multimode case (solid line) as against a single mode approximation (dashed line). Here the phonon frequency spectrum (A) is assumed to be of Gaussian form, $A(\omega) = L(2\pi\sigma^2)^{-1/2}\exp[-(\omega - \omega_0)^2/2\sigma^2]$, where L is the coupling strength and is related to a generalized (multifrequency) Huang–Rhys factor. The temperature dependence is expressed by the phonon occupation [n_0, see Eq. (46)] of the central mode. $L = 0.5$, $\sigma = 0.3$. [After Weissman and Jortner (1978, Fig. 3b).]

probability is still expressed in terms of combinations of the $\langle X_m | X_n \rangle$ and of their derivatives with respect to Q (i.e., of $\langle X_m | \partial/\partial Q | X_n \rangle$, etc.); and, as already mentioned in Section 10d(1), these derivatives can be expressed as linear combinations of various $\langle X_m | X_{n \pm 1} \rangle$ integrals (see, for example, Löffler, 1969). The consequence, of course, is that the resultant temperature dependences are essentially the same as those obtained in the Condon treatments, and discussed in Section 10d(1). The second and important point to emerge from these "non-Condon" treatments is that the absolute transition probability for deep levels is, however, several orders of magnitude higher than from equivalent Condon approaches. As previously mentioned (Section 10c), these two conclusions, of course, follow automatically from the recent work of Peuker et al. (1982).

Regarding the specific papers in this area, probably the first treatments were by Kovarskii (1962), Kovarskii and Sinyavskii (1962, 1964), Sinyavskii (1966), and Sinyavskii and Kovarskii (1967), with the increase in transition probability for deep traps emphasized already in the first paper. Specific results for the F center in Cu_2O are given in the 1966 paper, and for thermal trapping in Ge, Si, and CdS in the 1967 paper. A comparison to experimental values given in the 1967 paper shows reasonably good agreement.

Another series of papers, in the late 1960s and early 1970s by Rampacher (1968), Löffler (1969), and Stumpf (1969, 1971), gave a quite detailed exposition and applied it to F centers in alkali halides.

More recent work in this area has been largely catalyzed by the papers of Lang and Henry (1975) and of Henry and Lang (1977) (which were already mentioned in the Introduction to this MPE section). These recent papers are applied specifically to the III–V semiconductors. Included here is a general theory by Ridley (1978a) together with a specific comparison to Henry and Lang's (1977) data on the B center in GaAs, and a related investigation by Burt (1979, 1981). A somewhat different general treatment has been given by Goto et al. (1980), together with specific application to GaAs(Cr). Regarding the temperature dependence, Ridley obtains the same low and high temperature limits as given in the Condon approximation [Section 10d(1), Eqs. (44) and (45)]; Goto et al. consider the whole range and find overall similar behavior but with some quantitative differences (see their Fig. 7). Ridley concludes that the high temperature (exponential) regime, Eq. (45), is essentially of the same form as that obtained by Henry and Lang (1977), Eq. (11), and used by them to fit their data. However, Ridley notes that Huang–Rhys factors (S) of sufficient magnitude [see Eq. (49)] such that this high temperature regime is indeed reached at practical temperatures (less than the melting point of the material) are very unlikely in semiconductors. Ridley is able to fit, with $S \approx 1$, the temperature dependence of the Henry and Lang data on the B center with Eq. (44), i.e., with the low temperature case. The change in phonon occupation [see Eq. (46)] thus introduces sufficient temperature dependence to approximate an exponential over the measurement range. Nevertheless, the subsequent work by Burt (1979) shows that a *quantitative* fit to the absolute low temperature cross section cannot be obtained from the Ridley theory with $S \approx 1$ but requires, instead, $S \approx 3$; indeed the $S \approx 1$ value leads to an absolute cross section several orders of magnitude lower than the observed one (Burt, 1979). The problem with $S \approx 3$ is that although it gives good agreement of theory and experiment at low temperatures, the room temperature theoretical value is now about a factor of 5 too small (Burt, 1979). To complicate the situation even further, Pässler (1980b) and Morante et al. (1982) estimate $S \approx 7.7$ and $S \approx 5$, respectively, by fitting over the whole temperature range. The latter two papers use the static method versus the non-Condon adiabatic one of Burt (1979) (based on Ridley, 1978a), but as mentioned (Section 10c), these methods are equivalent. The problem is partly that even once the basic approach is selected, there are still many possible variations regarding the electron–phonon interaction, the vibrational modes, and the electronic wave functions [on this point, see, for example, Sections 10b and 10c.; also of interest is Burt (1982b)]. Moreover, as emphasized by Burt (1981), any fit is primarily sensitive only to S as a function of p [or, equivalently, of $\hbar\omega_0$—see Eq. (47)], i.e., S and p are not readily independently determined. A useful summary of some of these difficulties is given by Burt (1982b). Of course, all the above attempts at fitting the temperature dependence assume the MPE process to dominate over the entire temperature range and also assume the SCC model; other interpretations of the data on the B center suggest a contribution from radiative (Pässler, 1980a) or Auger (Robbins, 1980) recombination [although these two interpretations have already been questioned—see Burt (1981) and Morante et al. (1982)], or a nonlinear electron–phonon interaction with anharmonic effects [Markvart, 1981; also see Section 10d(4)]. The overall conclusion from this work is that, given a theory, one can predict the temperature dependence; however, given a temperature-dependent cross section, one may not be able to decide unambiguously on the appropriate theoretical model. Similar conclusions have been reached by Stoneham (1981).

(3) *Static Approach.* There appears to be only one author, namely Pässler (1974a,b, 1975a,b, 1976a,b, 1977a,b, 1978, 1980a,b, 1981), who has extensively worked with the static approximation, Eq. (31), in application to semiconductors, with a very recent additional such treatment by Morante et al. (1982). Other recent work, on rare earth ions in ionic crystals, is, for instance, given by Pukhov and Sakun (1979).

Pässler considers both the single-frequency case (1974a) and that with phonon dispersion (1974a, 1975a), and then carries out most subsequent work in terms of a "mean phonon energy" (1975a, 1978). He has also considered the temperature dependence (1978, 1980b, 1981) and applied it (1980b) to the B center in GaAs [as just discussed above—Section 10d(2)]. Regarding the magnitude of the cross section, he concludes (1976a) that his approach can, indeed, account for (large) observed cross-section values. *However*, to obtain this agreement requires an impurity potential of a short-range (δ-function) type, such as that suggested by Lucovsky (1965). Agreement for coulombic (hydrogenic) impurities is poor. Unfortunately, Pässler does not give any physical insight as to why this agreement with experiment depends relatively critically on the nature of the electronic wave functions. Morante et al. (1982) apply their results to the A and B centers in GaAs, where aspects relating to the latter center have already been discussed above [Section 10d(2)].

(4) *"Other" Approaches.* We here summarize various treatments of the MPE problem that do not fit into our "main" categories of Sections 10d(1)–10d(3).

One step in such a different direction can be referred to as a "mixed" method, and is used in the previously mentioned paper of Henry and Lang (1977), with an extension by Jaros and Brand (1976b). Thus, Henry and Lang use the usual adiabatic product function [Eq. (32)] up to a critical point $Q = Q^1$ close to the intersection of the two parabolas, and assume that beyond this point the wave function no longer changes, i.e.,

$$\phi_l(r, Q) = \phi_l(r, Q^1), \qquad Q > Q^1 \tag{52}$$

This latter equation does not correspond to the static approximation of Eq. (31), just referred to Q^1 instead of Q^0, since $\phi(Q > Q^1)$ is now considered constant, and we therefore classify it as a "mixed" approach. The results agree with the magnitudes of observed cross sections in GaAs and GaP. The temperature dependence is predicted exponential at high temperatures [as given by Eq. (11)]. The paper by Jaros and Brand (1976b) uses better electronic wave functions, and employs Henry and Lang's experimentally determined cross-section activation energies to fix Q^1. This yields values of cross section. These calculations were done for O and for the Zn–O complex in GaP, and the theoretical cross-section values are factors of 2–3 lower than experimental. In line with our discussion in Section 10d(1), if the criticism of the exponential dependence is correct, an appropriately revised treatment would be useful. A further modification of the Henry–Lang theory has been suggested recently (Abram, 1980) and involves an alternate method of solving their Hamiltonian.

A second approach is semiclassical and calculates crossings from one parabola to the other by a Landau–Zener (see Zener, 1932 or Landau and Lifshitz, 1958) approach, introducing a (random) time dependence in the energy splitting and the coupling. This method was used as a second approach by Henry and Lang (1977), and at high temperatures (it is not valid at low temperatures) it agrees with their quantum results. Further extensions and more detailed mathematical treatments have been given by Lax and Shugard (1977) and by Sumi (1980a,b). In his second paper Sumi (1980b) explicitly introduces a conduction band of a *finite* width, i.e., multiple levels, rather than the single level usually assumed in earlier MPE work; indeed, he questions whether the usual Born–Oppenheimer perturbation treatment applies to most semiconductors, which have a relatively wide bandwidth. In a subsequent paper, Sumi (1981) further considers the case of fast capture of a majority carrier subsequent to that of a minority carrier; the excited state here does not have time to relax to its equilibrium configuration, and this case is thus analogous to the "hot" transfer mentioned earlier (Section 10b). Another interesting use of Landau–Zener theory has been by Markvart (1981), who considers a nonlinear electron–phonon interaction along with anharmonic effects. With this approach, he can fit the temperature-dependent cross section of the B center, which the SCC theory could not do very well [as already mentioned in Sect. 10d(2)].

Still another method, namely use of the WKB (Wentzel–Kramers–Brillouin) theory, has been suggested by Englman and Ranfagni (1980) for use with multidimensional or complex energy surfaces.

Finally, we should still mention work by Makram-Ebeid and Lannoo (1982a,b) that calculates (in the adiabatic approach) the effect of an electric field in enhancing, with phonon assistance, the tunneling rate. Very good agreement is obtained with experimental data (Makram-Ebeid et al., 1980; Makram-Ebeid, 1981). Moreover, this approach appears to be useful for the evaluation of S and $\hbar\omega_0$, as claimed [however, we are not certain whether or not there are similar problems as with the evaluation of these parameters from simple thermal data—see the discussion in connection with the GaAs B center in Section 10d(2)].

V. Analysis of Optical Cross Sections

11. GENERAL DISCUSSION. COMPLICATIONS

One of the areas of research that has lately seen considerable activity in III–V compounds, as well as in semiconductors in general, is analysis of optical cross sections. The reasons for this are not hard to fathom—such studies can, in principle, give information about the parity of the impurity wave function and thus about the symmetry of a center. Moreover, by various extensions of these studies—such as study of the temperature dependence or of a comparison with thermal energies—one also obtains information on the extent of lattice relaxation [see, for example, Monemar and Samuelson (1978) and the review by Grimmeiss (1977)]. In addition, once the magnitude of phonon broadening has been established, temperature studies can give the shift of the level with respect to the band edges (see, for example, Samuelson and Monemar, 1978). Thus, various useful characterizations can be obtained.

In this section, we shall review various factors that must be considered in accurate evaluation of center parameters from the "raw" data. First, the extent of the (already mentioned) lattice relaxation must be considered (see below). Second, there can be additional complications, such as excited states or a field dependence of the cross section. In any case, one tries to separate out such complications and thus obtain an "electronic" cross section. This latter can then be compared to appropriate theory (Section 12).

A possible influence of electric fields on cross sections has in the past perhaps been emphasized more in connection with thermal cross sections than with optical ones. However, as pointed out, for instance, by Monemar and Samuelson (1978), Pantelides and Grimmeiss (1980), and Szawelska et al. (1981), it can also influence optical ones. Figure 18 (Fig. 3 of Monemar and Samuelson, 1978) compares the hole cross section for GaP(O) in the bulk and in a space-charge region. Thus, measurements in bulk appear preferable. The alternate approach of calculating this influence of a field has not yet, to our knowledge, been carried out.

A probable role of an excited metastable state for the "oxygen" center of GaAs has already been mentioned in connection with our discussion of the configuration coordinate model (Section 10b). This was suggested by Vincent and Bois (1978) to explain a slow decrease of photocapacitance observed after

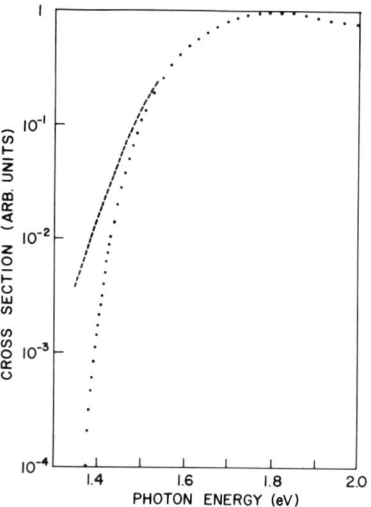

FIG 18. Comparision between the spectral dependence of the photoneutralization cross section (σ_{p1}^0) for O in GaP without field (dotted line, taken in bulk by photoluminescence excitation) and with field (dashed line, taken in a junction by photocapacitance). [After Monemar and Samuelson (1978, Fig. 3), with the photocapacitance data provided by C. H. Henry, and data taken at 190°K.]

the normal increase (see also Bois and Vincent, 1977). Similarly, Samuelson and Monemar (1978) require a (formally) temperature-dependent electronic optical cross section to explain their data on the one-electron state of GaP(O). They interpret this temperature dependence as arising from a population shift between a ground and excited state (with the cross section from each state now being temperature independent, as expected). We are not aware of any theoretical analysis that has included such excited states a priori for either of these two systems. Interestingly, a very recent theoretical analysis (Baraff et al., 1982) has predicted a metastable state for the two-electron center of GaP(O).

The next important aspect to be considered is the electron–phonon interaction (lattice relaxation). Here, the effect of momentum conserving phonons, or "promoting" modes, can in principle be included in the electronic cross section; this is discussed, for instance, by Monemar and Samuelson (1976) and Stoneham (1977). However, the configuration coordinate (CC) phonons (or "accepting" modes) are treated separately. The effect of these CC modes is usually expressed by the Franck–Condon factor d_{F-C}, where this factor is the same as the ΔE defined in our Fig. 16. Thus assuming a single mode,

$$d_{F-C} = S\hbar\omega_0, \tag{53}$$

where S and ω_0 have been previously defined (Section 10b). It has recently been amply established—as shown in Table II—that the magnitude of d_{F-C} can be appreciable for deep levels. This is important, since d_{F-C} shifts the threshold of the optical cross section relative to the thermal energy (see Fig. 16), even at 0°K. In addition, at higher temperatures there is also an edge broadening (see,

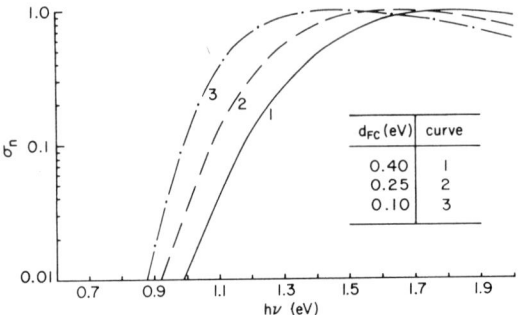

FIG. 19. Predicted dependence of the photoionization spectral dependence on the Franck–Condon factor [d_{F-C}—see Eq. (53)]. The parameter values are appropriate for the electron cross section (σ_n) for O in GaP. The level depth is $E_i = 0.9$ eV, the band gap is $E_g = 2.2$ eV, the average optical gap (the "Penn" gap) is $E_p = 5.8$ eV, and the temperature is 400°K. [After Jaros (1977, Fig. 5e).]

for example, Monemar and Samuelson, 1978; Samuelson and Monemar, 1978; Ridley and Amato, 1981). To illustrate the effect of d_{F-C}, comparative curves are shown in Fig. 19 for different values of d_{F-C} (this is a reproduction of Fig. 5e of Jaros, 1977); similar curves are also given, for instance, by Chantre et al. (1981). It can be noted that a change in d_{F-C} causes an appreciable shift in the curves, i.e., the effect of d_{F-C} cannot be neglected!

Treatments that separate out the d_{F-C} contribution generally use the adiabatic and Condon approximations and assume a linear electron–phonon interaction. [Note that the Condon approximation should generally be satisfactory for optical cross sections—see, for example, Stoneham (1975)—in contrast to the situation for the thermal case (Section 10).] Within this framework, two distinct approaches have evolved. One approach, introduced by Monemar and Samuelson (1976), uses a deconvolution technique to separate the phonon part from the (raw) data, yielding the electronic cross section. This is a neat technique, but due to a required high sensitivity it has to date been used only with luminescent centers. The second technique incorporates the phonon interaction in the theory. In practice, d_{F-C} is here usually introduced as an empirical parameter to obtain a fit between theory and data, although there has been recent renewed interest in the evaluation of the Huang–Rhys factor S from first principles (Ridley, 1978b; Halperin and Englman, 1979; Stoneham, 1979). In this fitting procedure d_{F-C} can in principle be obtained in various ways: for instance, by comparison to ionization cross-section data at two (or more) temperatures (e.g., Kopylov and Pikhtin, 1974a; Jaros, 1977; Ridley and Amato, 1981), by comparison of the thermal and the optical activation energy (e.g., Bois et al., 1979; Chantre et al., 1981), or by analysis of the "tail" of the cross section (e.g.,

Ridley and Amato, 1981). In practice, since all these methods include fitting to observed cross-section data, the question is whether resultant fits are reasonably unique. This point is definitely nontrivial. Thus Chantre et al. (1981) feel that with an adjustable d_{F-C} almost any model can fit up to the maximum of the cross section; in line with this, Noras (1980, 1981b) argues that the best way to obtain a reliable optical energy is to analyze data well above threshold since thermal broadening effects are negligible in that range. On the other hand, Ridley and Amato (1981) and Ridley (1981) argue in favor of fitting near threshold by approximating the thermal effects, since further away there is too much sensitivity to factors such as details of the band structure and of symmetry. An alternate, somewhat intermediate approach, is used by Chantre et al. (1981); they essentially use the extent of the bound state wave function as a fitting parameter to match curves over a relatively wide energy range. A further complication is that all these models assume a single-phonon frequency, which may not be realistic; thus Monemar and Samuelson (1978) require two phonons in their deconvolution for GaP(O). To assist in this regard, a two-phonon fitting model has recently been developed by Martinez et al. (1981), and applied to GaAs(Cr).

It remains to judge, in view of the just-mentioned difficulties, whether one can obtain reliable d_{F-C} values by optical means [regarding difficulties by use of thermal cross sections, see Sections 4 and 10d(2)]. To help with such evaluations, we are aware of two suggestions. One is to measure the cross sections for a given center to and from *both* bands; in this case the sum of the optical threshold energies gives the band gap plus the Stokes shift, i.e., plus $2d_{F-C}$ for parabolas with equal frequencies, and thus provides an internal check (Noras, 1981b). The other suggestion is to measure as a function of pressure. This has been proposed by Blow and Inkson (1980b) as a means to check on the role of higher minima. However, Hennel et al. (1980a,b, 1981) and Martinez et al. (1981) checked the effect of pressure on GaAs(Cr) and interpret the results in terms of a shift of the level with respect to the band edge. The extent to which these two effects (role of higher bands versus shift of level) can be separately determined does not seem to have been investigated as yet [one possibility here would be to use theoretical estimates for the pressure shift of deep levels—such estimates have, for instance, been recently given by Ren et al. (1982)]. A further guide to the reliability of d_{F-C} values—although an empirical one—can be obtained by comparing such values for a given center as determined by different methods. Referring to Table II for cases where two or more optical estimates are available, the agreement overall is good. The exceptions are GaP(Cu) and the donor in GaAs(Cr). We are not certain of the reasons for these two discrepancies, but we suggest the following. For GaP(Cu), Kopylov and Pikhtin (1974a) used absorption data and moreover fitted to it at low absorption, so that experimental difficulties could be of importance here. For GaAs(Cr) it was noted by Amato et al. (1980), Hennel et al. (1981), and Martinez et al. (1981) that the observed band appears to be due to a sum of transitions from two levels, and this could easily render interpretation ambiguous. A further intercomparison of d_{F-C} values can be obtained by use of the results from the field emission analysis of Makram-Ebeid and Lannoo (1982a,b); the agreement with the values of Table II is here good for three out of four centers (the exception being the E3 center for GaAs, for which Makram-Ebeid and Lannoo obtain $d_{F-C} = 0.125$ eV). [Note that we do not list the field emission values in our Table II since they are obtained by a basically thermal method—see discussion in Sections 4 and 10d(2)] Overall, in view of the degree of agreement, we thus tend to conclude that despite difference in the details of the fitting procedures, optical fitting for d_{F-C} does work.

12. Electronic Cross Section

The electronic optical cross section $\sigma(hv)$ can be expressed in the dipole approximation as

$$\sigma(hv) \sim (1/hv) \sum_{n\mathbf{k}} [\langle \psi_i | \mathbf{\varepsilon}_\lambda \cdot \mathbf{p} | \phi_{n\mathbf{k}} \rangle]^2 \delta(E_i + E_{n\mathbf{k}} - hv), \qquad (54)$$

where v is the frequency of the incident radiation, ψ_i the initial (impurity) wave function, $\phi_{n\mathbf{k}}$ the final (band) one, with n denoting the bands (or band extrema) and $\mathbf{k}$ the reduced wave vector, $\mathbf{p}$ the momentum operator, E_i the impurity ionization energy (and corresponds to E_0 of Fig. 16), $E_{n\mathbf{k}}$ the electron energy after ionization, and $\mathbf{\varepsilon}_\lambda$ the polarization vector of the radiation (see, for example, Stoneham, 1975). An average over degenerate initial and final states is implicit in the matrix element in Eq. (54). Moreover, it is usually assumed that the matrix element is a scalar function of energy [for a fuller discussion of this aspect see, for example, Jaros (1977), especially the discussion leading to his Eq. (22)]. Then, with neglect of $\mathbf{\varepsilon}_\lambda$ (immaterial in our subsequent treatment), the "Golden Rule" leads to

$$\sigma(hv) \sim (1/hv) \sum_n [\mathbf{M}_n(k)]^2 \rho_n(E), \qquad (55)$$

where $\rho_n(E)$ is the density of states in the nth band, and where the matrix element $\mathbf{M}_n$ is defined as

$$\mathbf{M}_n(k) \equiv \langle \psi_i | \mathbf{p} | \phi_{n\mathbf{k}} \rangle. \qquad (56)$$

Equation (55) is the usual starting point for calculations of the cross section. It can be seen to depend on the matrix element, on the various bands that must be considered as contributing to the transition, and on the density of states in such bands. In Tables III and IV we list various recent (1974 on) papers that both develop a particular approximation and apply it to GaAs (Table III) or GaP (Table IV), together with a sorting of the approximations in terms of these

TABLE III

Theoretical Analyses of σ_{opt} for GaAs

Reference	Center	d_{F-C} included	Wave functions			Density of states
			Band	Impurity		
Rynne et al. (1976)	Mn, σ_p Cu, σ_p	No	Plane waves	Calculated from inversion of data		Parabolic
Grimmeiss and Ledebo (1975b)	"O", σ_n	No	Plane waves	From δ potential; $m^* = m_0$		Parabolic
Tyler et al. (1977)[a]	"O", σ_n	Yes	Bloch functions, details to be published	To be published		Parabolic, but with discussion of validity range
Chantre and Bois (1980) and Chantre et al. (1981)	"O" $\{\sigma_n \atop \sigma_p$ "E3," σ_n	Yes	Dipole matrix element assumed constant or $\sim k$	Effective-mass type, with radius used as fitting parameter		Parabolic, but with inclusion of higher minima for σ_n
Ridley and Amato (1981)	Cr, σ_n	Yes	Bloch functions, including Coulomb correction	Hard-core envelope modulated by periodic part ("Billiard ball" model)		Nonparabolicity included

[a] Also, see comments regarding their interpretation of the data in Chantre et al. (1981).

TABLE IV

THEORETICAL ANALYSES OF σ_{opt} FOR GaP

Reference	Center	$d_{\text{F-C}}$ included	Wave functions Band	Wave functions Impurity	Density of states
Kopylov and Pikhtin (1974a)	Cu, σ_p	Yes	Dipole matrix element assumed constant or $\sim k$	Expanded in band functions, via Koster–Slater model	Parabolic
Kopylov and Pikhtin (1974b)	O $\{\sigma_{n1}, \sigma_{p1}\}$				
Grimmeiss et al. (1974) and Morgan (1975)	O $\{\sigma_{n1}, \sigma_{n2}\}$	No	Dipole matrix element assumed constant or $\sim k$	Matrix elements approximated, with emphasis on symmetry aspects	Parabolic
Jaros (1975a)	O $\{\sigma_{n1}, \sigma_{n2}\}$	Yes	Approximated via Penn model	Expanded in band functions, with subsequent use of approximate pseudopotential approach	Parabolic
Jaros (1975b)	O, σ_{n2}	Yes	Detailed pseudopotential, multiband, calculations		Calculated
Jaros (1977)	O $\{\sigma_{n1}, \sigma_{n2}, \sigma_{p1}, \sigma_{p2}\}$	Yes	Matrix elements approximated via pseudopotential approach and Penn model	Matrix elements approximated	Calculated
Monemar and Samuelson (1978)	O $\{\sigma_{n1}, \sigma_{p1}\}$	Yes	Dipole matrix element assumed constant or $\sim k$	Matrix elements approximated	Calculated
Blow and Inkson (1980a)	O, σ_{n1}	No	Approximated via Penn model	Core function matched to complex band functions	Parabolic
Blow and Inkson (1982)			Free-electron-like, but with proper symmetry		Parabolic, but with inclusion of higher minima

various contributions. The approximations to the matrix element are discussed in terms of the wave functions for the initial (impurity) and final (band) states. (Note that the "O" center in Table III is now usually referred to as EL2—see Section 6.)

We shall now discuss this matrix element in more detail. Several approximations—namely the quantum defect model (e.g., Bebb, 1969), the use of a δ function (Lucovsky, 1965), or of a Coulomb potential for the impurity— have long been well known and have been reviewed earlier (e.g., Milnes, 1973; Stoneham, 1975). Here we emphasize various recent developments. First, we present some work that analyzes the cross section in terms of the contributing matrix elements and of their symmetry. Results from this are important, since they show certain aspects to be relatively independent of the details of the deep state. Subsequently we give other recent papers, some of which consider various modifications to the impurity wave functions, others to the band wave functions.

The importance of symmetry aspects of the matrix element $\mathbf{M}_n$ defined in Eq. (56) has been variously emphasized, for instance, by Grimmeiss et al. (1974), Morgan (1975), Jaros (1977), Monemar and Samuelson (1978)—see their Appendix A, Banks et al. (1980a,b), Blow and Inkson (1980a,b, 1982), and Inkson (1981). To proceed, we first expand the impurity wave function in terms of the band functions, i.e.,

$$\psi_i = \sum_{n'\mathbf{k}'} A_{n'\mathbf{k}'} \phi_{n'\mathbf{k}'}, \tag{57}$$

Thus Eq. (56) can be put in the form

$$\mathbf{M}_n(k) = \sum_{n'\mathbf{k}'} A_{n'\mathbf{k}'} \langle \phi_{n'\mathbf{k}'} | \mathbf{p} | \phi_{n\mathbf{k}} \rangle. \tag{58}$$

The evaluation of the A's is the next step. These are obtained from the one-electron impurity Schrödinger equation. Using an impurity potential h, with an impurity energy E_i,

$$(H_0 + h)\psi_i = E_i \psi_i, \tag{59}$$

which gives

$$-A_{n'\mathbf{k}'} = \langle \phi_{n'\mathbf{k}'} | h | \psi_i \rangle / (E_{n'\mathbf{k}'} - E_i). \tag{60}$$

From the treatment so far, one can see several points:

(1) The overall transition probability can be expressed in terms of two types of matrix elements, namely $\langle \phi_{n'\mathbf{k}'} | \mathbf{p} | \phi_{n\mathbf{k}} \rangle$ and $\langle \phi_{n'\mathbf{k}'} | h | \psi_i \rangle$ and will thus depend on whether each of these two is allowed or approximately forbidden. [Note that in a real system transitions are frequently not completely forbidden (see e.g., Jaros 1977)]. This point has, for instance, been emphasized by Grimmeiss et al. (1974) and Morgan (1975), who analyze photoconductivity

data [on GaP(O)] in terms of allowed and forbidden type transitions to several band minima.

(2) In many cases, higher bands or band extrema must be included in the analysis, i.e., one must consider $\sum_n$ in Eq. (55). This point has been made in the same papers by Grimmeiss et al. (1974) and Morgan (1975), especially if transitions to the lower band(s) are symmetry forbidden. It has also been emphasized for derivative spectroscopy by White et al. (1976).

(3) For most deep levels, one must include the contributions of several bands for a good impurity wave function. This is the $\sum_{n'}$ in Eq. (57). This aspect has, for instance, been emphasized by Jaros and Brand (1976a).

(4) Although we have listed the above points (1)–(3) separately for emphasis, in any particular situation these factors may well interact. Thus impurity symmetry will determine both whether a transition is allowed and which bands are important, etc. There have been several recent papers discussing this overall problem, quite relevant ones being by Banks et al. (1980a,b), Blow and Inkson (1980a,b, 1982), Chantre et al. (1981), and Inkson (1981).

It appears difficult to obtain much in the way of general results beyond the ones given so far. Nevertheless, it seems worthwhile to still highlight some aspects obtained by emphasizing the range close to the threshold of transitions to a particular band (or band extremum). Such extrema consist of states with particular symmetries. Thus for a neutral impurity state that has a well-defined symmetry (which is not necessarily the case, as deep states are in general composed of contributions of many bands—see e.g., Jaros and Brand, 1976a), some quantitative conclusions about the matrix elements can be stated. (For charged states modifications are required—see further below.) For forbidden transitions to a band edge at $k = k_0$, these elements are expected to be proportional to $(k - k_0)$ (e.g., Grimmeiss et al., 1974). For allowed transitions, on the other hand, the matrix element $\langle \phi_{nk} | h | \psi_i \rangle$ is frequently constant, independent of k. This was already a consequence of the early Lucovsky (1965) model. Jaros (1977) has now emphasized that, for deep states, this result can follow also for more general potentials than the δ-function one. Specifically, using $\phi_{nk} \sim \exp(ikr)$, $\psi_i \sim (1/r)\exp(-\alpha r)$, $h \sim r\exp(-r/r_c)$, he obtains

$$\langle \phi_{nk} | h | \psi_i \rangle \sim u/(k^2 + u^2)^2, \tag{61}$$

where $u = \alpha + 1/r_c$. Since for deep states $\alpha \approx 0.5$ a.u., $r_c \approx 2$ a.u., he concludes that in this case the condition $k \ll u$ usually holds. By inverting this reasoning, it is obvious that as long as $k \ll u$, the matrix element will be a constant regardless of the detailed form of the potential. For example, $r_c = \infty$ is satisfactory provided α is sufficiently large. [Also, any form of h

and ψ_i that leads to $h\psi_i \sim \exp(-ur)$ is obviously equivalent.] Moreover, an extension of the Jaros treatment shows that a similar result also holds for $h\psi_i \sim (1/r)\exp(-ur)$ [retaining $\phi_{nk} \sim \exp(ikr)$], for which

$$\langle \phi_{nk}|h|\psi_i \rangle \sim 1/(k^2 + u^2). \tag{62}$$

It thus follows that for deep levels and small k (i.e., close to threshold) the simplified analysis should be valid for a relatively wide range of realistic potentials and wave functions. Note that as long as the matrix elements are indeed reasonably constant for allowed transitions and proportional to k for forbidden transitions, it is a good approximation to follow the approach of Grimmeiss et al. (1974): near a band edge, the cross section to this band (σ_n) is given by

$$\sigma_n \sim (h\nu)^{-3}(h\nu - E_i)^f \rho_n(h\nu - E_i), \tag{63}$$

where $f = 0, 1$, or 2 depending on whether both matrix elements are allowed, one is forbidden and one allowed, or both are forbidden. Of course, the various approximations are expected to break down at higher k. In general, this breakdown is expected near or beyond the maximum in the cross section (e.g., Jaros, 1977). However, in some cases, for instance with strong mixing between bands, the breakdown can be close to threshold. An example here is the cross section to the conduction band of GaP; due to mixing between the X_1 and X_3 bands the momentum matrix element changes from proportional to k to approximately constant at ≈ 10 meV from threshold (Kopylov and Pikhtin, 1977). Another breakdown of the above conclusion, Eq. (63), occurs for nonvertical transitions; this aspect has recently been treated by Inkson (1981), who shows that the $(h\nu)^{-3}$ factor in Eq. (63) is now replaced by $(h\nu)^{-1}$.

In order to proceed beyond the qualitative and semiquantitative points presented above, improved impurity wave functions are required. Steps in this direction have recently been taken by several groups. A quite detailed treatment via a Green's function method [an extension of the Koster-Slater (1954) approach] has been given by Banks et al. (1980a,b), and applied to the transitions of centers with a_1 (s-like) and t_2 (p-like) symmetry to both bands in GaAs and in GaP. Another detailed treatment, with an LCAO formalism, is by Peña and Mattis (1981), but has to date been applied only to Si. Another school of thought has tried to keep the essential physics, but otherwise simplify the treatment in various ways. Noteworthy here is the work by Inkson and coworkers, which uses various evanescent wave techniques (Inkson, 1980, 1981; Blow and Inkson, 1980a,b, 1982; Burt, 1980; Dzwig et al., 1982). Included are applications to levels of various depths in GaAs (Dzwig et al., 1982), as well as to indirect materials such as GaP (Blow and Inkson, 1980a, 1982). The latter workers show particularly strikingly that in this case for *deep* states the maximum of σ is shifted drastically to lower energies compared to the

Lucovsky model; we show this result in Fig. 20, which is part of Fig. 5 of Blow and Inkson (1980a). Note that the Lucovsky δ-function approach had been assumed to be good for *deep* states. Of course, as already mentioned further above, the Lucovsky model corresponds to an allowed transition, and thus can apply only if the wave functions are of appropriate symmetry. Other work is by Bois and Chantre (1980), Chantre and Bois (1980), and Chantre *et al.* (1981), who set up a semiempirical model, using as a fitting parameter the spatial extent of an effective-mass-type impurity wave function. Detailed fits have been obtained for the cross sections to *both* bands for several centers in GaAs, namely the "O" (Chantre and Bois, 1980), and Cr centers (Leyral *et al.*, 1981). Still another approach by Ridley and co-workers, uses a model that "is probably the simplest one which incorporates all the basic features of a deep-level centre" (Ridley and Amato, 1981). Here Ridley (1980) has given analytic solutions for various cases (also see below) by use of a wave function that is constant within a core region and zero outside. This approach is referred to (Ridley, 1980) as the billiard-ball or BB model. A subsequent model, by Amato and Ridley (1980) "bridges" between the BB model and a normalized quantum defect model. This bridging model is expected to be generally valid and is thus useful to determine the validity ranges of other models. The drawback of the bridging model is that it can be treated only numerically. The conclusion (Amato and Ridley, 1980) regarding the BB model is that it is good for repulsive centers, but otherwise only for very deep ($E_i \gtrsim 100 E_{\text{hydrogenic}}$). This model has been applied to GaAs(Cr) by Amato *et al.* (1980) and Ridley and Amato (1981).

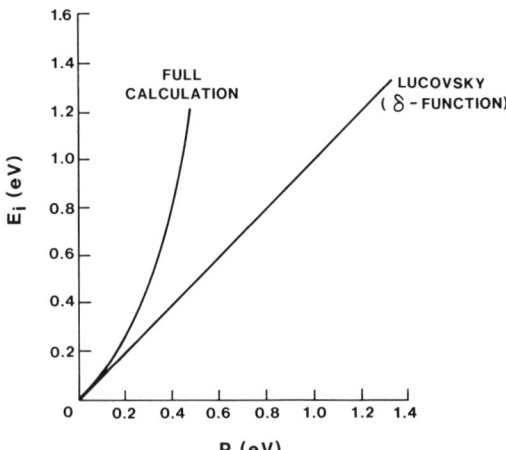

FIG. 20. A plot showing the dependence of the peak of the photoionization cross section (P) on the level energy (E_i) for the Lucovsky (δ-function) model and for a more accurate ("full calculation") evaluation of impurity wave functions. [After Blow and Inkson (1980a, Fig. 5).]

An alternate, quite interesting, approach to impurity wave functions has been developed by Rynne et al. (1976). These authors invert the experimental data to yield the shape of the impurity wave function. Unfortunately, it is not obvious whether the wave functions obtained by these authors are indeed meaningful—in view of various remaining assumptions such as use of a parabolic density of states and of plane waves for the band functions.

The next factor to be considered in the calculation of the cross section consists of the band wave functions. Here three points should be mentioned. First, recent work has generally used Bloch waves (Jaros, 1977; Monemar and Samuelson, 1978; Amato and Ridley, 1980; Banks et al., 1980a,b; Blow and Inkson, 1980a,b, 1982; Burt, 1980; Chantre and Bois, 1980; Ridley, 1980, 1981; Inkson, 1981; Peña and Mattis, 1981) rather than the earlier plane waves. Second, Pantelides and Grimmeiss (1980) have pointed out that a strong short-range potential can modify the continuum states giving resonant states with localized wave functions. This model has been used to explain data for ZnSe(Cu), but may well also be of influence for deep levels in the III–V materials. Third, the band states are influenced by the charge of the impurity center. This effect, the so-called final state scattering or Coulomb correction, should also be included. Unfortunately, this aspect has not been specifically considered in most of the papers of Tables III and IV. Treatments incorporating Coulomb corrections are given by Baltenkov and Grinberg (1976), Amato and Ridley (1980), Burt (1980), Ridley (1980), and by Ridley and Amato (1981). However, application to a specific center is carried out in only one instance, for GaAs(Cr) by Ridley and Amato (1981), and for this case they conclude that the results are not greatly influenced by charge. Nevertheless, Ridley's (1980) paper and its BB model are useful in this connection: analytic solutions can be obtained, at least near threshold, for attractive (σ_A) and repulsive (σ_R) centers as well as for the neutral case (σ_0):

$$(\sigma_A/\sigma_0) = (x-1)^{-1/2}, \qquad (64)$$

$$(\sigma_R/\sigma_0) = (x-1)^{-1/2} \exp[-2\pi\eta/(x-1)^{1/2}], \qquad (65)$$

where $x = h\nu/E_i$ and $\eta = |Z|(E_{\text{hydrogenic}}/E_i)^{1/2}$. Overall, these papers show that the Coulomb effect might be appreciable, but details still appear ambiguous. Thus Amato and Ridley (1980) conclude that for a relatively shallow attractive center, with $\eta = 0.7$ (i.e., $E_i/E_{\text{hydrogenic}} \approx 2$), the maximum in σ shifts from $h\nu \approx 2E_i$ to $h\nu \approx 1.3E_i$ (their Figs. 3, 5), but that the shift decreases for deeper states [e.g., GaAs(Cr)]. On the other hand, Burt (1980) illustrates a sizable shift for a deep state (his Fig. 1), and states "for the shallower levels the deviations are not so severe and the essential features ... are not lost when unscattered final states are used."

There still remains one part of the electronic cross section to be discussed: the density of states. It has been increasingly realized lately that the density of states is by now relatively well known for most semiconductors. It can thus be incorporated properly in the cross section, although possibly only by a numerical analysis. As can be seen from Tables III and IV, most recent papers that treat a specific center do indeed either include the proper density of states or show that a parabolic density is appropriate in the range of their analysis. That this density of states can be very important in the *shape* of the cross section has also been recently emphasized (Nazareno and Amato, 1982).

As an overall summary regarding the cross section and its potential for information: With no a priori information on a center or the band structure of the host, it is very difficult (or impossible) to extract information. To emphasize this, we show in Table V a summary of the $(hv - E_i)$ dependencies of cross sections from Eq. (63) for neutral and attractive centers using the Ridley (1980) result for the latter [Eq. (64)], and for parabolic and linear densities of states [an approximately linear density is, for instance, obtained for the valence band of GaP from $\sim 0.01-0.1$ eV above threshold—see, for example, Monemar and Samuelson (1978), their Fig. 15]. It can be seen that if f is not known, then very little information can be extracted; and even with known f the neutral, parabolic case gives the same result as the attractive, linear case. Nevertheless, emphasis on difficulties as represented by Table V is by no means the whole story; indeed, based on the various analyses reviewed here as well as earlier ones, considerable progress is being made. Helpful techniques here include measurements as a function of pressure (e.g., Blow and Inkson, 1980b; Hennel *et al.*, 1980a,b, 1981; Martinez *et al.*, 1981), as well as from the same center to *both* bands (Bois and Chantre, 1980; Chantre and Bois, 1980; Chantre *et al.*, 1981; Noras, 1981b). Also, with awareness of the various "nonsymmetry" factors of importance (the main ones probably being the vibrational relaxation, the Coulomb correction, and the density of states), it *should* be possible to obtain information on f values. As discussed in Section 11, d_{F-C} can

TABLE V

FUNCTIONAL DEPENDENCE OF σ_{opt}[a]

Center	Density of states	
	Parabolic	Linear
Neutral	$f + \frac{1}{2}$	$f + 1$
Attractive	f	$f + \frac{1}{2}$

[a] See Eq. (63).

probably be adequately corrected for, given enough data. Similarly, the density of states for most of the materials of interest can be reasonably well evaluated (although it may be laborious). Thus, if the charge state of the center is known, as it frequently is, good f values should be obtainable. Consequently our expectation is that there will soon be more and better f values and resultant information on center symmetries.

ACKNOWLEDGMENTS

We are especially grateful to R.N. Bhargava for his guidance and helpful comments during the preparation of this paper, and would also like to acknowledge helpful discussions with P. J. Dean, T. N. Morgan, and A. M. Stoneham.

REFERENCES

Abagyan, S. A., and Krupyshev, R. S. (1978). *Fiz. Tekh. Poluprovodn.* (*Leningrad*) **12**, 2360 [*Sov. Phys.—Semicond.* (*Engl. Transl.*) **12**, 1403].
Abagyan, S. A., Ivanov, G. A., Kuznetsov, Yu. N., Okunev, Yu. A., and Shanurin, Yu. E. (1974). *Fiz. Tekh. Poluprovodn.* (*Leningrad*) **7**, 1474 [*Sov. Phys.—Semicond.* (*Engl. Transl.*) **7**, 989].
Abagyan, S. A., Ivanov, G. A., Kuznetsov, Yu. N., and Okunev. Yu. A. (1975a). *Fiz. Tekh. Poluprovodn.* (*Leningrad*) **8**, 1691 [*Sov. Phys.—Semicond.* (*Engl. Transl.*) **8**, 1096].
Abagyan, S. A., Ivanov, G. A., Koroleva, G. A., Kuznetsov, Yu. N., and Okunev, Yu. A. (1975b). *Fiz. Tekh. Poluprovodn.* (*Leningrad*) **9**, 369 [*Sov. Phys.—Semicond.* (*Engl. Transl.*) **9**, 243].
Abagyan, S. A., Ivanov, G. A., and Koroleva, G. A. (1976). *Fiz. Tekh. Poluprovodn.* (*Leningrad*) **10**, 1773 [*Sov. Phys.—Semicond.* (*Engl. Transl.*) **10**, 1056 (1977)].
Abakumov, V. N., Perel', V. I., and Yassievich, I. N. (1978). *Fiz. Tekh. Poluprovodn.* (*Leningrad*) **12**, 3 [*Sov. Phys.—Semicond.* (*Engl. Transl.*) **12**, 1].
Abram, R. A. (1980). *J. Phys. C* **13**, L, 753.
Allen, J. W. (1980). *In* "Semi-Insulating III–V Materials, Nottingham 1980" (G. J. Rees, ed.), p. 261. Shiva Publ., Orpington, UK.
Amato, M. A., and Ridley, B. K. (1980). *J. Phys. C* **13**, 2027.
Amato, M. A., Arikan, M. C., and Ridley, B. K. (1980). *In* "Semi-Insulating III–V Materials, Nottingham 1980" (G. J. Rees, ed.), p. 249. Shiva Publ., Orpington, UK.
Andrianov, D. G., Grinshtein, P. M., Ippolitova, G. K., Omel'yanovskii, É. M., Suchkova, N. I., and Fistul', V. I. (1976). *Fiz. Tekh. Poluprovodn.* (*Leningrad*) **10**, 1173 [*Sov. Phys.—Semicond.* (*Engl. Transl.*) **10**, 696].
Arikan, M. C., Hatch, C. B., and Ridley, B. K. (1980). *J. Phys. C* **13**, 635.
Atanasov, R. D. (1980). *Rev. Sci. Instrum.* **51**, 1277.
Balarin, M. (1981). *Solid State Commun.* **39**, 1277.
Baltenkov, A. S., and Grinberg, A. A. (1976). *Fiz. Tekh. Poluprovodn.* (*Leningrad*) **10**, 1159 [*Sov. Phys.—Semicond.* (*Engl. Transl.*) **10**, 688].
Banks, P. W., Brand, S., and Jaros, M. (1980a). *J. Phys. Soc. Jpn., Suppl. A* **49**, 243.
Banks, P. W., Brand, S., and Jaros, M. (1980b). *J. Phys. C* **13**, 6167.
Baraff, G. A., Kane, E. O., and Schlüter, M. (1979). *Phys. Rev. Lett.* **43**, 956.
Baraff, G. A., Kane. E. O., and Schlüter, M. (1982). *Phys. Rev. B* **25**, 548.
Baranowski, J. M., Grynberg, M., and Magerramov, E. M. (1972). *Phys. Status Solidi B* **50**, 433.
Bartram, R. H., and Stoneham, A. M. (1975). *Solid State Commun.* **17**, 1593.

Bassani, F., and Pastori Parravicini, G. (1975). "Electronic States and Optical Transitions in Solids." Pergamon, Oxford.
Bazhenov, V. K., Rashevskaya, E. P., Solov'ev, N. N., and Foigel', M. G. (1973). *Fiz. Tekh. Poluprovodn. (Leningrad)* **7**, 1601 [*Sov. Phys.—Semicond. (Engl. Transl.)* **7**, 1067 (1974)].
Bebb, H. B. (1969). *Phys. Rev.* **185**, 1116.
Bekmuratov, M. F., and Murygin, V. I. (1973). *Fiz. Tekh. Poluprovodn. (Leningrad)* **7**, 83 [*Sov. Phys.—Semicond. (Engl. Transl.)* **7**, 55].
Beleznay, F., and Andor, L. (1978). *Solid State Electron.* **21**, 1305.
Bergh, A. A., and Dean, P. J. (1972). *Proc. IEEE* **60**, 156.
Bernholc, J., Lipari, N. O., Pantelides, S. T., and Scheffler, M. (1981). *Conf. Ser. Inst. Phys.* No. 59, p. 1.
Bhargava, R. N., Michel, C., Lupatkin, W. L., Bronnes, R. L., and Kurtz, S. K. (1972). *Appl. Phys. Lett.* **20**, 227.
Bhargava, R. N., Harnack, P. M., Herko, S. P., Mürau, P. C., and Seymour, R. J. (1976). *J. Lumin.* **12/13**, 515.
Bhattacharya, P. K., Ku, J. W., Owen, S. J. T., Aebi, V., Cooper, C. B., III, and Moon, R. L. (1980). *Appl. Phys. Lett.* **36**, 304.
Blakemore, J. S. (1962). "Semiconductor Statistics." Pergamon, New York.
Blakemore, J. S. (1980). *In* "Semi-Insulating III–V Materials, Nottingham, 1980" (G. J. Rees, ed.), p. 29. Shiva Publ., Orpington, UK.
Blakemore, J. S., Brown, W. J., Jr., Stass, M. L., and Woodbury, D. A. (1973). *J. Appl. Phys.* **44**, 3352.
Blow, K. J., and Inkson, J. C. (1980a). *J. Phys. C* **13**, 359.
Blow, K. J., and Inkson, J. C. (1980b). *In* "Semi-Insulating III–V Materials, Nottingham, 1980" (G. J. Rees, ed.), p. 274. Shiva Publ. Orpington, UK.
Blow, K. J., and Inkson, J. C. (1982). *J. Phys. C* **15**, 3711.
Bois, D., and Boulou, M. (1974). *Phys. Status Solidi A* **22**, 671.
Bois, D., and Chantre, A. (1980). *Rev. Phys. Appl.* **15**, 631.
Bois, D., and Vincent, G. (1977). *J. Phys. Lett. (Paris)* **38**, L-351.
Bois, D., Chantre, A., Vincent, G., and Nouailhat, A. (1979). *Conf. Ser. Inst. Phys.* No. 43, p. 295.
Born, M., and Huang, K. (1954). "Dynamical Theory of Crystal Lattices" Oxford, London.
Born, M., and Oppenheimer, J. R. (1927). *Ann. Phys. (Leipzig)* **84**, 457.
Borsuk, J. A., and Swanson, R. M. (1980). *IEEE Trans. Electron Devices* **ED-27**, 2217.
Borsuk, J. A., and Swanson, R. M. (1981). *J. Appl. Phys.* **52**, 6704.
Bradley, C. C., Simmonds, P. E., Stockton, J. R., and Stradling, R. A. (1973). *Solid State Commun.* **12**, 413.
Brailsford, A. D., and Chang, T. Y. (1970). *J. Chem. Phys.* **53**, 3108.
Braun, S., and Grimmeiss, H. G. (1973a). *J. Appl. Phys.* **44**, 2789.
Braun, S., and Grimmeiss, H. G. (1973b). *Solid State Commun.* **12**, 657.
Breitenstein, O., Rheinlander, B., and Bindemann, R. (1979). *Phys. Status Solidi A* **51**, 70.
Brown, W. J., Jr., and Blakemore, J. S. (1972). *J. Appl. Phys.* **43**, 2242.
Brunwin, R., Hamilton, B., Jordan, R., and Peaker, A. R. (1979). *Electron Lett.* **15**, 350.
Brunwin, R. F., Hamilton, B., Hodgkinson, J., Peaker, A. R., and Dean, P. J. (1981). *Solid State Electron.* **24**, 249.
Bube, R. H. (1960). "Photoconductivity of Solids." Wiley, New York.
Burt, M. G. (1979). *J. Phys. C* **12**, 4827.
Burt, M. G., (1980). *J. Phys. C* **13**, 1825.
Burt, M. G. (1981). *J. Phys. C* **14**, L845.
Burt, M. G. (1982a). *J. Phys. C* **15**, L381.
Burt, M. G. (1982b). *J. Phys. C* **15**, L965.

Cavenett, B. C. (1981). *Conf. Ser. Inst. Phys.* No. 59, p. 69.
Chang, C. D., Damestani, A., and Forbes, L. (1979). *Solid State Electron.* **22**, 1053.
Chantre, A., and Bois, D. (1980). *J. Phys. Soc. Jpn., Suppl. A.* **49**, 247.
Chantre, A., Vincent, G., and Bois, D. (1981). *Phys. Rev. B* **23**, 5335.
Chernysh, L. V., Sheinkman, M. K., Sinyavskii, E. P., and Kovarskii, V. A. (1980). *Phys. Status Solidi B* **100**, K149.
Chiang, S. Y., and Pearson, G. L. (1975). *J. Lumin.* **10**, 313.
Chiao, Sun-hai, S., Mattes B. L., and Bube, R. H. (1978). *J. Appl. Phys.* **49**, 261.
Clark, M. G. (1980). *J. Phys. C* **13**, 2311.
Corbett, J. W., Kleinhenz, R. L., and Wilsey, N. D. (1981). In "Defects in Semiconductors" (J. Narayan and T. Y. Tan, eds.), Materials Research Society Proc., Vol. 2. p. 1, North-Holland, New York.
Crowell, C. R., and Alipanahi, S. (1981). *Solid State Electron.* **24**, 25.
Curie, D. (1978). In "Luminescence of Inorganic Solids" (B. Di Bartolo, ed.), p. 337. Plenum, New York.
Day, D. S., Tsai, M. Y., Streetman, B. G., and Lang, D. V. (1979a). *J. Appl. Phys.* **50**, 5093.
Day, D. S., Helix, M. J., Hess, K., and Streetman, B. G. (1979b). *Rev. Sci. Instrum.* **50**, 1571.
Dean, P. J. (1977). *Top. Appl. Phys.* **17**, 63.
Dean, P. J., and Choyke, W. J. (1977). *Adv. Phys.* **26**, 1.
Dean, P. J., Henry, C. H., and Frosch, C. J. (1968a). *Phys. Rev.* **168**, 812.
Dean, P. J., Frosch, C. J., and Henry, C. H. (1968b). *J. Appl. Phys.* **39**, 5631.
Dean, P. J., White, A. M., Hamilton, B., Peaker, A. R., and Gibb, R. M. (1977). *J. Phys. D* **10**, 2545.
Demberel, L. A., Popov, A. S., and Kushev, D. B. (1979). *Phys. Status Solidi A* **52**, 653.
Dexter, D. L., Klick, C. C., and Russell, G. A. (1955). *Phys. Rev.* **100**, 603.
Di Bartolo, B., ed. (1981). "Radiationless Processes" (NATO Advanced Study Institute Series, Vol. B62). Plenum, New York.
Dishman, J. M., and DiDomenico, M., Jr., (1970). *Phys Rev. B* **1**, 3381.
Dzwig, P., Burt, M. G., Inkson, J. C., and Crum, V. (1982). *J. Phys. C* **15**, 1187.
Eaves, L., Williams, P. J., and Uihlein, C. (1981). *J. Phys. C* **14**, L693.
Eaves, L., Englert, T., Instone, T., Uihlein, C., Williams, P. J., and Wright, H. C. (1980). In "Semi-Insulating III–V Materials, Nottingham 1980" (G. J. Rees, ed.), p. 145. Shiva Publ., Orpington, UK.
Emerson, N. G., and Sealy, B. J. (1979). *Electron. Lett.* **15**, 554.
Engemann, J., and Heime, K. (1975). *CRC Crit. Rev. Solid State Sci.* **5**, 485.
Englman R. (1972). "The Jahn–Teller Effect in Molecules and Crystals," Chap. II. Wiley, London.
Englman, R. (1977). *J. Chem. Phys.* **66**, 2212.
Englman, R. (1979). "Non-Radiative Decay of Ions and Molecules in Solids." North-Holland, Amsterdam.
Englman, R., and Jortner, J. (1970). *Mol. Phys.* **18**, 145.
Englman, R., and Ranfagni, A. (1980). *Physica B + C (Amsterdam)* **98B**, 151.
Engström, O., and Alm, A. (1978). *Solid State Electron.* **21**, 1571.
Evwaraye, A. O., and Woodbury, H. H. (1976). *J. Appl. Phys.* **47**, 1595.
Fabre, E., and Bhargava, R. N. (1974). *Appl. Phys. Lett.* **24**, 322.
Fabre, E., Bhargava, R. N., and Zwicker, W. K. (1974). *J. Electron. Mater.* **3**, 409.
Fagerström, P. O., Grimmeiss, H. G., and Titze, H. (1978). *J. Appl. Phys.* **49**, 3341.
Fairman, R. D., Morin, F. J., and Oliver, J. R. (1979). *Conf. Ser. Inst. Phys.* No. 45, p. 134.
Fazzio, A., and Leite, J. R. (1980). *Phys. Rev. B* **21**, 4710.
Fleurov, V. N., and Kikoin, K. A. (1979). *J. Phys. C* **12**, 61.
Fonger, W. H., and Struck, C. W. (1974). *J. Lumin.* **8**, 452.
Freed, K. F., and Jortner, J. (1970). *J. Chem. Phys.* **52**, 6272.

Frova, A., ed. (1973). *J. Lumin.* **7**.
Gasanli, Sh. M., Emel'yanenko, O. V., Nasledov, D. N., and Talalakin, G. N. (1972). *Fiz. Tekh. Poluprovodn.* (*Leningrad*) **6**, 2053 [*Sov. Phys.—Semicond.* (*Engl. Transl.*) **6**, 1747 (1973)].
Gibb, R. M., Rees, G. J., Thomas, B. W., Wilson, B. L. H., Hamilton, B., Wight, D. R., and Mott, N. F. (1977). *Philos. Mag.* **36**, 1021.
Gislason, H. P., Monemar, B., Dean, P. J., Herbert, D. C., Depinna, S., Cavenett, B. C., and Killoran, N. (1982). *Phys. Rev. B* **26**, 827.
Gloriozova, R. I., and Kolesnik, L. I. (1978). *Fiz. Tekh. Poluprovodn.* (*Leningrad*) **12**, 117 [*Sov. Phys.—Semicond.* (*Engl. Transl.*) **12**, 66].
Goto, H., Adachi, Y., and Ikoma, T. (1979). *Jpn. J. Appl. Phys.* **18**, 1979.
Goto, H., Adachi, Y., and Ikoma, T. (1980). *Phys. Rev. B* **22**, 782.
Grimmeiss, H. G. (1974). *Conf. Ser. Inst. Phys.* No. 22, p. 187.
Grimmeiss, H. G. (1977). *Annu. Rev. Mater. Sci.* **7**, 341.
Grimmeiss, H. G. (1980). *Lect. Notes Phys.* **122**, 50.
Grimmeiss, H. G., and Kullendorff, N. (1980). *J. Appl. Phys.* **51**, 5852.
Grimmeiss, H. G., and Ledebo, L.-Å. (1975a). *J. Appl. Phys.* **46**, 2155.
Grimmeiss, H. G., and Ledebo, L.-Å. (1975b). *J. Phys. C* **8**, 2615.
Grimmeiss, H. G., Ledebo, L-Å., Ovren, C., and Morgan, T. N. (1974). In "Proceedings of the International Conference on the Physics of Semiconductors, 12th" (M. H. Pilkuhn, ed.), p. 368. Teubner, Stuttgart.
Grimmeiss, H. G., Ledebo, L-Å., and Meijer, E. (1980). *Appl. Phys. Lett.* **36**, 307.
Grimmeiss, H. G., Monemar, B., and Samuelson, L. (1978). *Solid State Electron.* **21**, 1505.
Guldberg, J. (1977). *J. Phys. E* **10**, 1016.
Gummel, H., and Lax, M. (1957). *Ann. Phys.* (*N. Y.*) **2**, 28.
Gutsche, E. (1981). *J. Lumin.* **24/25**, 689.
Gutsche, E. (1982). *Phys. Status Solidi B* **109**, 583.
Hall, R. N., (1952). *Phys. Rev.* **87**, 387.
Halperin, B., and Englman, R. (1979). *J. Lumin.* **20**, 329.
Hamilton, B. (1974). *Conf. Ser. Inst. Phys.* No. 22, p. 218.
Hamilton, B., and Smith, B. L. (1973). *Appl. Phys. Lett.* **22**, 674.
Hamilton, B., Peaker, A. R., and Wight, D. R. (1979). *J. Appl. Phys.* **50**, 6373.
Hasegawa, H., Kojima, K., and Sakai, T. (1977). *Jpn. J. Appl. Phys.* **16**, 1251.
Haug, A. (1964). "Theoretische Festkörperphysik," I. Franz Deuticke, Wien.
Haug, A. (1972). In "Festkörperprobleme" (O. Madelung, ed.), XII, p. 411. Pergamon/Vieweg, Braunschweig.
Haug, A. (1980). *Phys. Status Solidi B* **97**, 481.
Haug, A., and Sauermann, G. (1958). *Z. Phys.* **153**, 269.
Hegems, M., Dingle, R., Rupp, L. W., Jr., (1975). *J. Appl. Phys.* **46**, 3059.
Hemstreet, L. A. (1980). *Phys. Rev. B* **22**, 4590.
Hemstreet, L. A., and Dimmock, J. O. (1979a). *Phys. Rev. B* **20**, 1527.
Hemstreet, L. A., and Dimmock, J. O. (1979b). *Solid State Commun.* **31**, 461.
Hennel, A. M., Szuszkiewicz, W., and Martinez, G. (1980a). *Rev. Phys. Appl.* **15**, 697.
Hennel, A. M., Szuszkiewicz, W., Martinez, G., Clerjaud, B., Huber, A. M., Morillot, G., and Merenda, P. (1980b). In "Semi-Insulating III–V Materials, Nottingham, 1980" (G. J. Rees, ed.), p. 228. Shiva Publ., Orpington, UK.
Hennel, A. M., Szuszkiewicz, W., Balkanski, M., Martinez, G., and Clerjaud, B. (1981). *Phys. Rev. B* **23**, 3933.
Henry, C. H., and Lang, D. V. (1977). *Phys. Rev. B* **15**, 989.
Hiesinger, P. (1976). *Phys. Status Solidi A* **33**, K39.
Hodgart, M. S. (1978). *Electron. Lett.* **14**, 388.

Hodgart, M. S. (1979). *Electron. Lett.* **15**, 724.
Houng, Y. M., and Pearson, G. L. (1978). *J. Appl. Phys.* **49**, 3348.
Huang, Kun (1981). *Sci. Sin. (Engl. Ed.)* **24**, 27.
Huber, A. M., Linh, N. T., Valladon, M., Debrun, J. L., Martin, G. M., Mitonneau, A., and Mircea, A. (1979). *J. Appl. Phys.* **50**, 4022.
Hurtes, Ch., Boulou, M., Mitonneau, A., and Bois, D. (1978). *Appl. Phys. Lett.* **32**, 821.
Ikoma, T., Takikawa, M., and Okumura, T. (1977). *Jpn. J. Appl. Phys.* **16**, *Suppl.* **16-1**, 223.
Inkson, J. C. (1980). *J. Phys. C* **13**, 369.
Inkson, J. C. (1981). *J. Phys. C* **14**, 1093.
Jack, M. D., Pack, R. C., and Henriksen, J. (1980). *IEEE Trans. Electron Devices* **ED-27**, 2226.
Jansson, L., Kumar, V., Ledebo, L.-Å., and Nideborn, K. (1981). *J. Phys. E* **14**, 464.
Jaros, M. (1975a). *J. Phys. C* **8**, L264.
Jaros, M. (1975b). *J. Phys. C* **8**, 2455.
Jaros, M. (1977). *Phys. Rev. B* **16**, 3694.
Jaros, M. (1978). *Solid State Commun.* **25**, 1071.
Jaros, M. (1980). *Adv. Phys.* **29**, 409.
Jaros, M. (1982). "Deep Levels in Semiconductors." Adam Hilger, London.
Jaros, M., and Brand, S. (1976a). *Phys. Rev. B* **14**, 4494.
Jaros, M., and Brand, S. (1976b). *Int. Conf. Phys. Semiconduct., Proc., 13th* p. 1090.
Jayson, J. S., Bhargava, R. N., and Dixon, R. W. (1970). *J. Appl. Phys.* **41**, 4972.
Jayson, J. S., Bachrach, R. Z., Dapkus, P. D., and Schumaker, N. E. (1972). *Phys. Rev. B* **6**, 2357.
Jesper, T., Hamilton, B., and Peaker, A. R. (1980). *In* "Semi-Insulating III–V Materials, Nottingham, 1980" (G. J. Rees, ed.), p. 233. Shiva Publ., Orpington, UK.
Johnson, N. M., Bartelink, D. J., Gold, R. B., and Gibbons, J. F. (1979). *J. Appl. Phys.* **50**, 4828.
Jortner, J. (1979). *Philos. Mag. B* **40**, 317.
Kaufmann, U., and Koschel, W. H. (1978). *Phys. Rev. B* **17**, 2081.
Kaufmann, U., and Schneider, J. (1980). *In* "Festköperprobleme" (J. Treusch, ed.), Vol. 20, p. 87. Vieweg, Braunschweig.
Kayanuma, Y., and Nasu, K. (1978). *Solid State Commun.* **27**, 1371.
Kennedy, T. A. (1981). *In* "Nuclear and Electron Resonance Spectroscopies Applied to Materials Science" (E. N. Kaufmann and G. K. Shenoy, eds.), Materials Research Society Pioc., Vol. 3, p. 95. North-Holland, New York.
Kimerling, L. C. (1978). *Solid State Electron.* **21**, 1391,
Kimerling, L. C. (1979). *Conf. Ser. Inst. Phys.* No. 46, p. 56.
Kimerling, L. C. (1981). *In* "Defects in Semiconductors" (J. Narayan and T. Y. Tan, eds.), Materials Research Society Proc., Vol. 2, p. 85. North-Holland, New York.
Kirchner, P. D., Schaff, W. J., Maracas, G. N., Eastman, L. F., Chappell, T. I., and Ranson, C. M. (1981). *J. Appl. Phys.* **52**, 6462.
Kirov, K. I., and Radev, K. B. (1981). *Phys. Status Solidi A* **63**, 711.
Kitahara, K., Nakai, K., Ozeki, M., Shibatomi, A., and Dazai, K. (1976). *Jpn. J. Appl. Phys.* **15**, 2275.
Klick, C. C., and Schulman, J. H. (1957). *Solid State Phys.* **5**, 97.
Kocot, K., Rao, R. A., and Pearson, G. L. (1979). *Phys. Rev. B* **19**, 2059.
Kopylov, A. A., and Pikhtin, A. N. (1974a). *Fiz. Tverd. Tela* **16**, 1837 [*Sov. Phys.—Solid State (Engl. Transl.)* **16**, 1200 (1975)].
Kopylov, A. A. and Pikhtin, A. N. (1974b). *Fiz. Tekh. Poluprovodn. (Leningrad)* **8**, 2398 [*Sov. Phys.—Semicond. (Engl. Transl.)* **8**, 1563 (1975)].
Kopylov, A. A., and Pikhtin, A. N. (1976). *Fiz. Tekh. Poluprovodn. (Leningrad)* **10**, 15 [*Sov. Phys.—Semicond. (Engl. Transl.)* **10**, 7].

Kopylov, A. A., and Pikhtin, A. N. (1977). *Fiz. Tekh. Poluprovodn. (Leningrad)* **11**, 867 [*Sov. Phys—Semicond. (Engl. Transl.)* **11**, 510].
Koster, G. F., and Slater, J. C. (1954). *Phys. Rev.* **95**, 1167; **96**, 1208.
Kovarskii, V. A. (1962). *Fiz. Tverd. Tela* **4**, 1636 [*Sov. Phys.—Solid State (Engl. Transl.)* **4**, 1200]
Kovarskii, V. A., and Sinyavskii, É. P. (1962). *Fiz. Tverd. Tela* **4**, 3202 [*Sov. Phys.—Solid State (Engl. Transl.)* **4**, 2345].
Kovarskii, V. A., and Sinyavskii, É. P. (1964). *Fiz. Tverd. Tela* **6**, 636 [*Sov. Phys.—Solid State (Engl. Transl.)* **6**, 498].
Kukimoto, H., Henry, C. H., and Merritt, F. R. (1973). *Phys. Rev. B* **7**, 2486.
Kumar, V., and Ledebo, L.-Å. (1981). *J. Appl. Phys.* **52**, 4866.
Landau, L. D., and Lifshitz, E. M. (1958). "Quantum Mechanics," p. 310ff. Pergamon, London.
Landsberg, P. T. (1970). *Phys. Status Solidi* **41**, 457.
Landsberg, P. T., and Adams, M. J. (1973). *Proc. R. Soc. London, Ser. A* **334**, 523.
Landsberg, P. T., and Robbins, D. J. (1978). *Solid State Electron.* **21**, 1289.
Lang, D. V. (1974a). *J. Appl. Phys.* **45**, 3014.
Lang, D. V. (1974b). *J. Appl. Phys.* **45**, 3023.
Lang, D. V. (1977). *Conf. Ser. Inst. Phys.* No. 31, p. 70.
Lang, D. V. (1979). *Top. Appl. Phys.* **37**, 93.
Lang, D. V. (1980). *J. Phys. Soc. Jpn., Suppl A* **49**, 215.
Lang, D. V., and Henry C. H. (1975). *Phys. Rev. Lett.* **35**, 1525.
Lang, D. V., and Kimerling, L. C. (1976). *Appl. Phys. Lett.* **28**, 248.
Lang, D. V., and Logan, R. A. (1975). *J. Electron. Mater.* **4**, 1053.
Lang, D. V., and Logan, R. A. (1977). *Phys. Rev. Lett.* **39**, 635.
Lang, D. V., Cho, A. Y., Gossard, A. C., Ilegems, M., and Wiegmann, W. (1976). *J. Appl. Phys.* **47**, 2558.
Langer, J. M. (1980). *Lect. Notes Phys.* **122**, 123.
Lannoo, M., and Bourgoin, J. (1981). "Point Defects in Semiconductors I, Solid State Science," Vol. 22. Springer-Verlag, Berlin and New York.
Lax, M. (1959). *J. Phys. Chem. Solids* **8**, 66.
Lax, M., and Shugard, W. J. (1977). *In* "Statistical Mechanics and Statistical Methods in Theory and Application" (U. Landman, ed.), p. 429. Plenum, New York.
Leach, M. F., and Ridley, B. K. (1978). *J. Phys. C* **11**, 2249.
Le Bloa, A., Favennec, P. N., and Colin, Y. (1981). *Phys. Status Solidi A* **64**, 85.
Lefévre, H., and Schulz, M. (1977). *Appl. Phys.* **12**, 45.
Leyral, P., Litty, F., Loualiche, S., Nouailhat, A., and Guillot, G. (1981). *Solid State Commun.* **38**, 333.
Löffler, A. (1969). *Z. Naturforsch.* **24A**, 516, 530.
Lorenz, M. R., Pettit, G. D., and Blum, S. E. (1972). *Solid State Commun.* **10**, 705.
Lucovsky, G. (1965). *Solid State Commun.* **3**, 299.
Majerfeld, A., and Bhattacharya, P. K. (1978). *Appl. Phys. Lett.* **33**, 259.
Makram-Ebeid, S. (1980). *Appl. Phys. Lett.* **37**, 464.
Makram-Ebeid, S. (1981). *In* "Defects in Semiconductors" (J. Narayan and T. Y. Tan, eds.), Materials Research Society Proc., Vol. 2, p. 495. North-Holland, New York.
Makram-Ebeid, S., and Lannoo, M. (1982a). *Phys. Rev. Lett.* **48**, 1281.
Makram-Ebeid, S., and Lannoo, M. (1982b). *Phys. Rev. B* **25**, 6406.
Makram-Ebeid, S., Martin, G. M., and Woodard, D. W. (1980). *J. Phys. Soc. Jpn., Suppl. A* **49**, 287.
Markham, J. J. (1956). *Phys. Rev.* **103**, 588.
Markham, J. J. (1959). *Rev. Mod. Phys.* **31**, 956.
Markvart, T. (1981). *J. Phys. C* **14**, L435, L895.

Martin, G. M. (1980). In "Semi-Insulating III–V Materials, Nottingham 1980" (G. J. Rees, ed.), p. 13. Shiva Publ., Orpington, UK.
Martin, G. M., Mitonneau, A., and Mircea, A. (1977). Electron. Lett. **13**, 192.
Martin, G. M., Mitonneau, A., Pons, D., Mircea, A., and Woodard, D. W. (1980). J. Phys. C **13**, 3855.
Martin, P. A., Streetman, B. G., and Hess, K. (1981). J. Appl. Phys. **52**, 7409.
Martinez, G., Hennel, A. M., Szuszkiewicz, W., Balkanski, M., and Clerjaud, B. (1981). Phys. Rev. B **23**, 3920.
Masterov, V. F., and Samorukov, B. E. (1978). Fiz. Tekh. Poluprovodn. (Leningrad) **12**, 625 [Sov. Phys.—Semicond. (Engl. Transl.) **12**, 363].
Masterov, V. F., Mal'tsev, Yu. V., and Rusanov, I. B. (1980). Fiz. Tekh. Poluprovodn. (Leningrad) **14**, 559 [Sov. Phys.—Semicond. (Engl. Transl.) **14**, 330].
Miller, M. D., and Patterson, D. R. (1977). Rev. Sci Instrum. **48**, 237.
Miller, G. L., Ramirez, J. V., and Robinson, D. A. H. (1975). J. Appl. Phys. **46**, 2638.
Miller, G. L., Lang, D. V., and Kimerling, L. C. (1977). Annu. Rev. Mater. Sci. **7**, 377.
Milnes, A. G. (1973). "Deep Impurities in Semiconductors." Wiley (Interscience), New York.
Mircea, A., and Bois, D. (1979). Conf. Ser. Inst. Phys. No. 46, p. 82.
Mircea, A., and Mitonneau, A. (1975). Appl. Phys. **8**, 15.
Mircea, A., and Mitonneau, A. (1977). J. Phys. (Orsay Fr.) **38**, L41.
Mircea, A., Mitonneau, A., Hallais, J., and Jaros. M. (1977). Phys. Rev. B **16**, 3665.
Mircea, A., Pons, D., and Makram-Ebeid, S. (1980). Lect. Notes Phys. **122**, 69.
Mircea-Roussel, A., and Makram-Ebeid, S. (1981). Appl. Phys. Lett. **38**, 1007.
Mitonneau, A., Martin, G. M., and Mircea, A. (1977a). Conf. Ser. Inst. Phys. No. 33a, p. 73.
Mitonneau, A., Martin, G. M., and Mircea, A. (1977b). Electron. Lett. **13**, 667.
Monemar, B. (1972). J. Lumin. **5**, 472.
Monemar, B. (1981). In "Recent Developments in Condensed Matter Physics" (J. T. Devreese, ed.), Vol. 1, p. 441. Plenum, New York.
Monemar, B., and Samuelson, L. (1976). J. Lumin. **12/13**, 507.
Monemar, B., and Samuelson, L. (1978). Phys. Rev. B **18**, 809.
Monemar, B., Gislason, H. P., Dean, P. J., and Herbert, D. C. (1982). Phys. Rev. B **25**, 7719.
Morante, J. R., Carceller, J. E., Barbolla, J., Cartujo, P. (1982). J. Phys. C **15**, L175.
Morgan, T. N. (1975). J. Electron. Mater. **4**, 1029.
Morgan, T. N. (1978). Phys. Rev. Lett. **40**, 190.
Morgan, T. N., Welber, B., and Bhargava, R. N. (1968). Phys. Rev. **166**, 751.
Mott, N. F. (1977). Philos. Mag. **36**, 979.
Mott, N. F. (1978). Solid State Electron. **21**, 1275.
Mullin, J. B., Ashen, D. J., Roberts, G. G., and Ashby, A. (1977). Conf. Ser. Inst. Phys. No. 33a, p. 91.
Nazareno, H. N., and Amato, M. A. (1982). J. Phys. C **15**, 2165.
Neumark, G. F. (1973). Phys. Rev. B **7**, 3802.
Neumark, G. F. (1974). Phys. Rev. B **10**, 1574.
Neumark, G. F. (1982). Phys. Rev. B **26**, 2250.
Neumark, G. F., DeBitetto, D. J., Bhargava, R. N., and Harnack, P. M. (1977). Phys. Rev. B **15**, 3147; 3156.
Nishizawa, J., Koike, M., Miura, K., and Okuno, Y. (1979). Jpn. J. Appl. Phys. **19**, 25.
Noras, J. M. (1980). J. Phys. C **13**, 4779.
Noras, J. M. (1981a). J. Phys. C **14**, 2341.
Noras, J. M. (1981b). J. Phys. C **14**, L713.
Noras, J. M. (1981c). Solid State Commun. **39**, 1225.
Noras, J. M., and Szawelska, H. R. (1982). J. Phys. C **15**, 2001.

Okamoto, H., Sakata, S., and Sakai, K. (1973). *J. Appl. Phys.* **44**, 1316.
Okumura, T., Takikawa, M., and Ikoma, T. (1976). *Appl. Phys.* **11**, 187.
Okuno, Y., Suto, K., and Nishizawa, J. (1973). *J. Appl. Phys.* **44**, 832.
Okushi, H., and Tokumaru, Y. (1980). *Jpn. J. Appl. Phys.* **19**, L335.
Okushi, H., and Tokumaru, Y. (1981). *Jpn. J. Appl. Phys.* **20**, L45.
Ozeki, M., Komeno, J., Shibatomi, A., and Ohkawa, S. (1979a). *J. Appl. Phys.* **50**, 4808.
Ozeki, M, Shibatomi, A., and Ohkawa, S. (1979b). *J. Appl. Phys.* **50**, 4823.
Pantelides, S. T. (1978). *Rev. Mod. Phys.* **50**, 797.
Pantelides, S. T., ed. (1983). To be published, Gordon and Breach, New York.
Pantelides, S. T., and Grimmeiss, H. G. (1980). *Solid State Commun.* **35**, 653.
Parsey, J. M., Jr., Nanishi, Y., Lagowski, J., and Gatos, H. C. (1981). *J. Electrochem. Soc.* **128**, 936.
Partin, D. L., Chen, J. W., Milnes, A. G., and Vassamillet, L. F. (1979a). *Solid State Electron.* **22**, 455.
Partin, D. L., Chen, J. W., Milnes, A. G., and Vassamillet, L. F. (1979b). *J. Appl. Phys.* **50**, 6845.
Pässler, R. (1974a). *Czech. J. Phys. Sect. B* **24**, 322.
Pässler, R. (1974b). *Phys. Status Solidi B* **65**, 561.
Pässler, R. (1975a). *Czech. J. Phys. Sect. B* **25**, 219.
Pässler, R. (1975b). *Phys. Status Solidi B* **68**, 69.
Pässler, R. (1976a). *Phys. Status Solidi B* **76**, 647.
Pässler, R. (1976b). *Phys. Status Solidi B* **78**, 625.
Pässler, R. (1977a). *Phys. Status Solidi B* **83**, K55.
Pässler, R. (1977b). *Phys. Status Solidi B* **83**, K111.
Pässler, R. (1978). *Phys. Status Solidi B* **85**, 203.
Pässler, R. (1980a). *Phys. Status Solidi B* **101**, K69.
Pässler, R. (1980b). *J. Phys. C* **13**, L901.
Pässler, R. (1981). *Phys. Status Solidi B* **103**, 673.
Pautrat, J. L., Katircioglu, B., Magnea, N., Bensahel, D., Pfister, J. C., and Revoil, L. (1980). *Solid State Electron.* **23**, 1159.
Peka, G. P., Brodovoi, V. A., Zernov, S. L., and Mirets, L. Z. (1975). *Phys. Status Solidi A* **29**, 47.
Peña R. E., and Mattis, D. C. (1981). *J. Phys. C* **14**, 647.
Perlin, Yu. E. (1963). *Usp. Fiz. Nauk.* **80**, 553 [*Sov. Phys.—Usp.* **6**, 542 (1964)].
Petroff, P. M., and Lang, D. V. (1977). *Appl. Phys. Lett.* **31**, 60.
Peuker, K., Enderlein, R., Schenk, A., and Gutsche, E. (1982). *Phys. Status Solidi B* **109**, 599.
Pickin, W. (1978). *Solid-State Electron.* **21**, 1299.
Pickin, W. (1979). *Phys. Status Solidi B* **96**, 617.
Pickin, W. (1980). *Phys. Status Solidi B* **97**, 431.
Pons, D. (1980). *Appl. Phys. Lett.* **37**, 413.
Pons, D., and Makram-Ebeid, S. (1979). *J. Phys. (Orsay, Fr.)* **40**, 1161.
Pons, D., Mooney, P. M., and Bourgoin, J. C. (1980). *J. Appl. Phys.* **51**, 2038.
Pons, D, Mircea, A., Mitonneau, A., and Martin, G. M. (1979). *Conf. Ser. Inst. Phys.* No. 46, p. 352.
Pukhov, K. K., and Sakun, V. P. (1979). *Phys. Status Solidi B* **95**, 391.
Queisser, H. J. (1978). *Solid State Electron.* **21**, 1495.
Queisser, H. J., and Heim, U. (1974). *Annu. Rev. Mater. Sci.* **4**, 125.
Rampacher, H. (1968). *Z. Naturforsch.* **23**a, 401.
Rebsch, J.-T. (1979). *Solid State Commun.* **31**, 377.
Rees, G. J., ed. (1980). "Semi-Insulating III–V Materials, Nottingham, 1980." Shiva Publ., Orpington, UK.

Rees, G. J., Grimmeiss, H. G., Janzén, E., and Skarstam, B. (1980). *J. Phys. C* **13**, 6157.
Ren, S. Y., Dow, J. D., and Wolford, D. J. (1982). *Phy. Rev. B* **25**, 7661.
Riddoch, F. A., and Jaros, M. (1980). *J. Phys. C* **13**, 6181.
Ridley, B. K. (1978a). *J. Phys. C* **11**, 2323.
Ridley, B. K. (1978b). *Solid State Electron.* **21**, 1319.
Ridley, B. K. (1980). *J. Phys. C* **13**, 2015.
Ridley, B. K. (1981). *J. Phys. C* **14**, L717
Ridley, B. K., and Amato, M. A. (1981). *J. Phys. C* **14**, 1255.
Ridley, B. K., and Leach, M. F. (1977). *J. Phys. C* **10**, 2425.
Robbins, D. J. (1980). *J. Phys. C* **13**, L1073.
Robbins, D. J., and Dean, P. J. (1978). *Adv. Phys.* **27**, 499.
Robbins, D. J., and Landsberg, P. T. (1980). *J. Phys. C* **13**, 2425.
Robertson, N., and Friedman, L. (1976). *Philos. Mag.* **33**, 753.
Robertson, N., and Friedman, L. (1977). *Philos. Mag.* **36**, 1013.
Rose, A. (1955). *Phys. Rev.* **97**, 322.
Rose, A. (1957). *Prog. Semicond.* **2**, 109.
Rynne, E. F., Cox, J. R., McGuire, J. B., and Blakemore, J. S. (1976). *Phys. Rev. Lett.* **36**, 155.
Sah, C. T. (1976). *Solid State Electron.* **19**, 975.
Sah, C. T., Forbes, L., Rosier, L. L., and Tasch, A. F., Jr. (1970). *Solid State Electron.* **13**, 759.
Sakai, K, and Ikoma, T. (1974). *Appl. Phys.* **5**, 165.
Samuelson, L., and Monemar, B. (1978). *Phys. Rev. B* **18**, 830.
Schairer, W. (1979). *J. Electron. Mater.* **8**, 139.
Schairer, W., and Grobe, E. (1970). *Solid State Commun.* **8**, 2017.
Schairer, W., and Schmidt, M. (1974). *Phys. Rev. B* **10**, 2501.
Schneider, J., and Kaufmann, U. (1981). *Conf. Ser. Inst. Phys.* No. 59, p. 55.
Schwerdtfeger, G. F. (1972). *Solid State Commun.* **11**, 779.
Seitz, F. (1940). "The Modern Theory of Solids." McGraw-Hill, New York.
Shockley, W., and Read, W. T., Jr. (1952). *Phys. Rev.* **87**, 835.
Sinha, K. P., and DiDomenico, M., Jr. (1970). *Phys. Rev. B* **1**, 2623.
Sinyavskii, É. P. (1966). *Fiz. Tverd. Tela* **8**, 1528 [*Sov. Phys.—Solid State* (*Engl. Transl.*) **8**, 1215].
Sinyavskii, É. P., and Kovarskii, V. A. (1967). *Fiz. Tverd. Tela* **9**, 1464 [*Sov. Phys.—Solid State* (*Engl. Transl.*) **9**, 1142].
Skolnick, M. S., Jain, A. K., Stradling, R. A., Leotin, J., Ousset, J. C., and Askenazy, S. (1976). *J. Phys. C* **9**, 2809.
Smith, B. L., and Carter, M. A. (1975). *J. Phys. D* **8**, 254.
Smith, B. L., Hayes, T. J., Peaker, A. R., and Wight, D. R. (1975). *Appl. Phys. Lett.* **26**, 122.
Stillman, G. E., Wolfe, C. M., and Dimmock, J. O. (1977). In "Semiconductors and Semimetals" (R. K. Willardson and A. C. Beer, eds.), Vol. 12, p. 169. Academic Press, New York.
Stoneham, A. M. (1975). "Theory of Defects in Solids" Oxford Univ. Press, London and New York.
Stoneham, A. M. (1977). *Philos. Mag.* **36**, 983.
Stoneham, A. M. (1979). *J. Phys. C* **12**, 891.
Stoneham, A. M. (1981). *Rep. Prog. Phys.* **44**, 1251.
Strehlow, W. H., and Cook, E. L. (1973). *J. Phys. Chem. Ref. Data* **2**, 163.
Struck, C. W., and Fonger, W. H. (1975). *J. Lumin.* **10**, 1.
Stumpf, H. (1969). *Z. Phys.* **229**, 488.

Stumpf, H. (1971). *Phys. Konden. Mater.* **13**, 101.
Suchkova, N. I., Andrianov, D. G., Omel'yanovskii, E. M., Rashevskaya, E. P., and Solov'ev, N. N. (1975). *Fiz. Tekh. Poluprovodn. (Leningrad)* **9**, 718 [*Sov. Phys.—Semicond. (Engl. Transl.)* **9**, 469].
Sumi, H. (1980a). *J. Phys. Soc. Jpn.* **49**, 1701.
Sumi, H. (1980b). *J. Phys. Soc. Jpn., Suppl. A* **49**, 227.
Sumi, H. (1981). *Phys. Rev. Lett.* **47**, 1333.
Suzuki, K., and Miura, N. (1976). *Solid State Commun.* **18**, 233.
Szawelska, H. R., and Allen, J. W. (1979). *J. Phys. C* **12**, 3359.
Szawelska, H. R., Noras, J. M., and Allen, J. W. (1981). *J. Phys. C* **14**, 4141.
Sze, S. M. (1969). "Physics of Semiconductor Devices," Chap. 3. Wiley (Interscience), New York.
Takikawa, M., and Ikoma, T. (1980). *Jpn. J. Appl. Phys.* **19**, L436.
Tanaka, S., and Toyozawa, Y., eds. (1980). *J. Phys. Soc. Jpn., Suppl. A* **49**.
Tell, B. J., and Kuijpers, F. P. J. (1978). *J. Appl. Phys.* **49**, 5938.
Toyozawa, Y. (1978). *Solid State Electron.* **21**, 1313.
Troxell, J. R., and Watkins, G. D. (1980). *Rev. Sci. Instrum.* **51**, 143.
Tyler, E. H., Jaros, M., and Penchina, C. M. (1977). *Appl. Phys. Lett.* **31**, 208.
Uji, T., and Nishida, K. (1976). *Jpn. J. Appl. Phys.* **15**, 2247.
van der Does de Bye, J. A. W. (1976). *J. Electrochem. Soc.* **123**, 544.
Van Vechten, J. A., and Thurmond, C. D. (1976). *Phys. Rev. B* **14**, 3539.
Vincent, G. (1980). *Appl. Phys.* **23**, 215.
Vincent, G., and Bois, D. (1978). *Solid State Commun.* **27**, 431.
Vincent, G., Chantre, A., and Bois, D. (1979). *J. Appl. Phys.* **50**, 5484.
Vogl, P. (1981). *In* "Feskörperprobleme" (J. Treusch, ed.), Vol. 21, p. 191. Vieweg, Braunschweig.
Wagner, E. E., Hiller, D., and Mars, D. E. (1980). *Rev. Sci. Instrum.* **51**, 1205.
Watts, R. K. (1977). "Point Defects in Crystals" Wiley, New York.
Weissman, Y., and Jortner, J. (1978). *Philos. Mag. B* **37**, 21.
Wessels, B. W. (1975). *J. Appl. Phys.* **46**, 2143.
Wessels, B. W. (1976). *J. Appl. Phys.* **47**, 1131.
Wessels, B. W. (1977). *J. Appl. Phys.* **48**, 1656.
White, A. M. (1980). *In* "Semi-Insulating III–V Materials, Nottingham, 1980" (G. J. Rees, ed.), p. 3. Shiva Publ., Orpington, UK.
White, A. M., Day, B., and Grant, A. J. (1979). *J. Phys. C* **12**, 4833.
White, A. M., Dean, P. J., and Porteous, P. (1976). *J. Appl. Phys.* **47**, 3230.
White, A. M., Krebs, J. J., and Stauss, G. H. (1980). *J. Appl. Phys.* **51**, 419.
Willardson, R. K., and Beer, A. C., eds. (1966–). "Semiconductors and Semimetals," Vols. 1–. Academic Press, New York.
Willmann, F., Bimberg, D., and Blätte, M. (1973). *Phys. Rev. B* **7**, 2473.
Wilson, B. L. H., ed. (1979). *Conf. Ser. Inst. Phys.* No. 43.
Zener, C. (1932). *Proc. R. Soc. London, Ser. A* **137**, 696.
Zylbersztejn, A. (1978). *Appl. Phys. Lett.* **33**, 200.

CHAPTER 2

The Electrical and Photoelectronic Properties of Semi-Insulating GaAs

David C. Look

UNIVERSITY RESEARCH CENTER
WRIGHT STATE UNIVERSITY
DAYTON, OHIO

	List of Symbols	76
I.	Introduction	77
II.	Hall Effect and Magnetoresistance	77
	1. *Apparatus*	78
	2. *Measurement and Analysis Techniques*	80
	3. *Typical Data*	84
III.	Thermal Equilibrium Processes	86
	4. *Donors and Acceptor Statistics*	86
	5. *Temperature-Dependent Hall Effect*	87
	6. *Temperature-Dependent Mobility*	94
IV.	Nonequilibrium Processes	96
	7. *Excitation and Recombination*	96
	8. *Photoconductivity and Photo-Hall Measurements*	100
	9. *Thermally Stimulated Current Spectroscopy*	106
	10. *Photoinduced Transient Current Spectroscopy*	109
V.	Comparison of Techniques	121
	11. *Crystal Quality Determination*	121
	12. *Activation Energy Measurements*	122
	13. *Impurity Concentration Measurements*	123
	14. *Lifetime Measurements*	125
	15. *Impurity and Defect Identifications*	127
	Appendix A. Charge Transport Theory	127
	16. *General Current Equations*	127
	17. *Single-Carrier Hall Effect, Conductivity*	130
	18. *Two-Carrier Hall Effect, Conductivity*	136
	19. *Units*	147
	Appendix B. Semiconductor Statistics	148
	20. *Derivation of Occupation Factors*	148
	21. *Applications*	154
	Appendix C. Derivation of TSC and PITS Equations	162
	22. *TSC Equation: Electrons*	163
	23. *PITS Equation: Electrons*	164
	24. *PITS Equation: Electrons and Holes*	165
	References	167

List of Symbols

a	heating rate	η	reduced energy (ε/kT), thermopower
a_1	symmetry assignment (T_d group)	μ	mobility
b	electron mobility to hole mobility ratio	ν	frequency
B/c	magnetic field (esu)	ξ	magnetoresistance coefficient
c	speed of light, electron concentration to hole concentration ratio	ρ	resistivity
		σ	conductivity, electron or hole capture cross section
d	sample thickness	σ_ν	photon capture cross section
e	*magnitude* of electronic charge, emission rate,	ϕ	volume density of band states per unit energy interval, modified effective density of states
E	electric field, *ionization* energy (positive quantity)		
E_σ	cross-section activation energy	τ	carrier relaxation time, carrier life time
f	Fermi function, lock-in amplifier reference frequency	ω_c	cyclotron resonance frequency
		Subscript	
F	force	0	value of quantity at $t = 0$, value for $B/c = 0$, value at thermal equilibrium, value at beginning of temperature scan
g	degeneracy		
h	Planck's constant		
I	current		
I_0	photon intensity (cm^{-2} s^{-1})		
j	current density	0, 1, 2, 3, ...	number of ionizable electrons (l value)
k	Boltzmann's constant		
m^*	effective mass	av	average
m_0	rest mass of electron	A	acceptor
n	free-electron concentration, bound-electron concentration (occupied centers)	C	conduction band
		Cr	chromium
		D	donor, drift (velocity)
N	total number (or volume density) of centers, effective density of states	F	Fermi energy
		G	band gap
		H	uncorrected Hall-effect measurement
p	free-hole concentration		
q	charge	I	impurity center, ionized-impurity contribution
r	Hall factor		
R	Hall coefficient, reflection coefficient		
S	signal strength, slope	i	intrinsic
t	time	i or k	particular impurity or defect center, particular cell in conduction or valence band
T	absolute temperature, transmission coefficient		
v	velocity	l	value under light irradiation, charge state (number of ionizable electrons)
V	voltage		
W	number of distinguishable permutations		
α	absorption coefficient, reduced Fermi energy ($-\varepsilon_F/kT$), ratio of electron and hole Hall factors	L	lattice contribution
		m	excited state designation, value of temperature at signal maximum, mean value
β	$1/kT$, magneto-Hall coefficient, filling rate, ratio of slopes (S_ρ/S_R)		
		n	electron contribution
ε	energy	O	oxygen

p	hole contribution, pulse time	x	direction parallel to applied electric field
R	Hall coefficient		
S	"shallow" (with respect to E_F), lock-in reference signal	y	direction perpendicular to applied electric field
t	total	z	direction parallel to applied magnetic field
th	thermal		
V	valence band	ρ	resistivity

I. Introduction

The current high interest in semi-insulating (SI) GaAs is apparent to anyone even remotely connected with this field. It arises because the future of the GaAs digital IC and monolithic microwave IC technologies rests on the development of reliable substrate materials (Zucca, 1980; Lee, 1982). To understand SI GaAs we must be able to characterize the electrical properties, and that is the subject of this chapter. Our goal here is not only to discuss the electrical properties themselves, but also the techniques used to measure these properties. The reason is clear: without a proper understanding of the techniques, it is impossible to critically evaluate the derived results. Since the theoretical bases of the various characterization methods are often not discussed at length in the relevant research papers, we have attempted to do that here for several cases of interest. The more lengthy derivations are confined to the Appendixes.

A few precautions are in order. First, because several different topics are discussed, there is some overlap in the meanings of symbols. For example, "σ" is used for both conductivity (Part I) and cross section (Part IV), and "τ" for both relaxation time (Part I) and lifetime (Part IV). The overlap was allowed here because it is also rather universal in the literature. Furthermore, the different meanings are, for the most part, quite widely separated in the text.

A second warning is that no attempt was made to quote every possible reference relevant to a particular subject. Only enough references to illustrate the main points are included. The third warning is rather obvious: certain conjectures, assertions, and opinions may not be universally accepted, but may simply reflect the bias of the author. There are definitely differences of opinion on some aspects of SI GaAs, and that is one thing that makes the field so interesting.

II. Hall Effect and Magnetoresistance

The two most basic electrical parameters of interest, normally, are the carrier concentration and mobility. These quantities may be obtained from the proper measurements of current, electrical field, and applied magnetic field.

How this process is accomplished is of course the subject of transport theory (Beer, 1963; Bube, 1974). Semi-insulating (SI) GaAs poses some special problems in both theory and measurement. The first problem is the high resistivities involved (Cronin and Haisty, 1964), about 10^8-10^9 Ω cm, that lead to typical sample resistances of about 10^{10} Ω at room temperature. The second problem is concerned with mixed conductivity, i.e, conduction by both holes and electrons, which complicates the analysis (Inoue and Ohyama, 1970; Philadelpheus and Euthymiou, 1974; Look, 1975). Because of the importance of this latter subject in dealing with SI GaAs, and because some aspects of it have either not been discussed in the literature, or discussed only very briefly, we have undertaken a rather detailed development of two-carrier transport theory. This discussion is presented in Appendix A, and includes the behavior of the Hall effect and magnetoresistance for both single- and two-carrier cases, with energy-independent and energy-dependent relaxation times.

In Section 1 we will use the theory developed in Appendix A to discuss the various measurement and analysis techniques that have been applied to SI GaAs. We will also discuss the precautions that must be observed in apparatus design, and present an automated Hall-effect and photo-electronic system capable of measuring high resistivity samples.

1. APPARATUS

In Fig. 1 we present the block diagram of a computer-controlled apparatus that is designed to carry out Hall-effect and photoelectronic measurements on high resistivity samples. This system is described in more detail elsewhere (Look and Farmer, 1981). As shown in the figure, there are two standard sample configurations, the Hall bar (Putley, 1968, p. 40) and the van der Pauw (van der Pauw, 1958). The van der Pauw configuration and measurement scheme has the advantage that any sample shape is permissible, although a symmetrically cut sample is nearly always used, in practice. A distinct disadvantage, however, especially for high impedance circuitry, is that switching between current and voltage contacts is required (Hemenger, 1973). The Hall bar has an advantage in that the electric field lines are uniform and parallel to the sample length, if the sample is homogeneous. Any deviation from homogeneity is quickly noted by looking at the respective voltage drops at the three contact points. A caution to be observed with Hall-bar measurements is the possible shorting of the Hall voltage by the large-area current contacts. To avoid this problem the two voltage contacts on the one side should be placed close to the middle, and the length-to-width ratio of the sample should be at least three. For smaller ratios a correction factor can be applied (Bube, 1974, p. 363)

The first concern in performing electrical measurements on SI GaAs is, of course, the high resistances involved, say 10^{10} Ω at room temperature for a

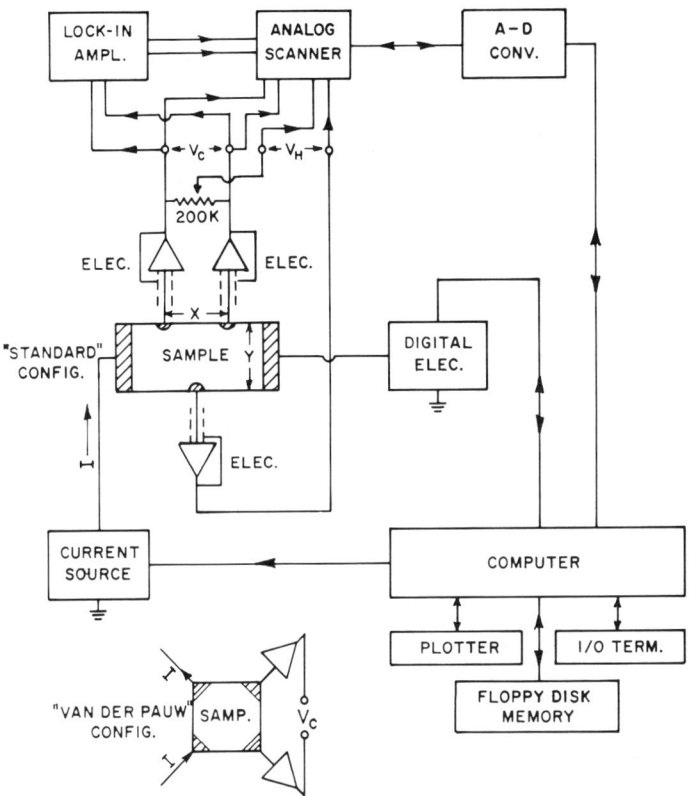

FIG. 1. The relationship of the sample and the various pieces of equipment necessary to automatically measure the current, and the parallel and perpendicular voltages. A "standard" contact configuration is shown in the main drawing, whereas a "van der Pauw" configuration is shown in the inset. The lock-in amplifier is used in ac photoconductivity measurements. [From Look and Farmer (1981), copyright by The Institute of Physics.]

typical size sample. A good way to reduce the effective impedance levels is to interface the contacts with electrometers operating as unity gain amplifiers, as shown in Fig. 1. A separate electrometer for each contact, although not necessary, is desirable because a bad contact may then be immediately spotted. Present-day electrometers, such as the Keithley 602, have very accurate unity-gain output voltages, equal to the input voltage within 10 ppm.

The second concern is the long response times that can result from typical cable and electrometer-input capacitances of 100 pF or more ($RC = 10^{10} \times 10^{-10} \approx 1$ s). These capacitances can be effectively reduced by operating in a guarded mode, i.e., with the inner shield (say, in a triaxial cable system) driven at the same potential as that of the center conductor. (The Keithley

602 does this automatically in its "fast-feedback" mode.) In this way we have been able to reduce the effective capacitance of our system to less than 1 pF. Such a scheme is especially important when carrying out ac measurements, such as ac photoconductivity.

A third concern, with almost any experiment, in fact, is how to reduce data-handling times and inaccuracies. With the advent of low-cost computers and microprocessors this problem is now being solved in many laboratories. The system shown in Fig. 1 and described in more detail in Look et al. (1980) and Look and Farmer (1981) can be used for automatic measurements of electrical properties, including dependences upon temperature, magnetic field, and monochromatic light irradiation. One advantage of this system is that the I/O bus is compatible with the IEEE-488 standard, making possible immediate interfacing with much of the new instrumentation available today. Another advantage is that the components shown in Fig. 1 are all commercially available.

Other Hall-effect systems, both manual and automatic, are described in the recent monograph by Wieder (1979).

2. Measurement and Analysis Techniques

For the purpose of this discussion we will define "semi-insulating" GaAs as material with resistivity in excess of 10^7 Ω cm. Such samples are normally grown by doping with either Cr or O. Sometimes undoped samples exhibit SI characteristics, but nearly any type of characterization (e.g., temperature-dependent Hall measurements) shows that precisely the same impurity or defect center is involved as for the O-doped case. [In fact, oxygen itself may not be a constituent of the center (Huber et al., 1979); this phenomenon will be discussed later.] In our laboratory we have found that *all* such SI GaAs samples, after proper cleaning, have a *negative* Hall coefficient, and thus would naively be interpreted as typical n-type samples. However, such an interpretation is invalid for many Cr-doped crystals, as has long been recognized (Inoue and Ohyama, 1970; Philadelpheus and Euthymiou, 1974), and a simple Hall measurement may therefore give misleading results. Below we will discuss three techniques used for measuring and analyzing such samples and show how to recognize whether or not a simple Hall-effect measurement will suffice. Our discussion will center around results for a typical Cr-doped GaAs crystal and a typical O-doped crystal.

a. Simple Hall-Effect Measurements

If the sample conductivity is dominated by only one type of carrier, then a simple Hall-effect analysis is sufficient. The appropriate equations for a Hall-bar configuration are well known and follow from Section 17 in Appendix A:

$$\rho_n = \sigma_n^{-1} = \left(\frac{E_x}{j_{nx}}\right)_{gs} = \left(\frac{V_x/x}{I_x/yz}\right)_{gs} = \frac{V_x}{I_x}\frac{yz}{x} \quad \Omega\text{ cm,} \tag{1}$$

$$\mu_{Hn} \equiv |R_n\sigma_n| = \left(\frac{E_y}{j_{nx}B/c}\frac{j_{nx}}{E_x}\right)_{gs} = \frac{10^8}{B}\frac{V_y}{V_x}\frac{x}{y}\quad\frac{\text{cm}^2}{\text{V sec}}, \tag{2}$$

$$n_H \equiv 1/R_n e = \left(\frac{j_{nx}B/c}{eE_y}\right)_{gs} = 6.25 \times 10^{10}\frac{IB}{V_y z}\quad\text{cm}^{-3}, \tag{3}$$

where x and y are defined in Fig. 1, z is the sample thickness, "gs" stands for "gaussian," and the *final* units are practical, i.e., V (volts), I (amperes), B (gauss), and x, y, z (centimeters). Other symbols have their usual meanings but are further defined in Appendix A. Note that, contrary to some definitions of R_n, we maintain the "c" (speed of light) with the "B." An example of converting from gaussian to practical units is presented in Section 19 of Appendix A. In the examples given below we will assume that $n_H = n$ and $\mu_{Hn} = \mu_n$, i.e., $r_n = 1$. Equations for a van der Pauw type analysis may be found elsewhere (van der Pauw, 1958; Hemenger, 1973; Wieder, 1979).

Before discussing more complicated situations it is good to consider typical magnitudes of the currents and voltages involved. For SI GaAs, $\rho \simeq 10^8$–$10^9\ \Omega$ cm and a reasonable value of E_x is about 1 V/cm. (Values higher than about 10–50 V/cm will produce nonohmic currents.) Then, if $yz/x \simeq 0.1$, $I \simeq 10^{-10}$–10^{-9} A. The need for the electrometers, discussed earlier, is thus apparent. If $\mu_n \simeq 4000$ cm^2/V s, then a magnetic field of 5000 G will give a Hall voltage, V_y, of about 0.1 V, if $x/y \simeq 2$. Such voltages are easily measured, especially since the dominant noise is often not greater than that due to the electrometers themselves ($\sim 10\ \mu$V). Temperature fluctuations, or microphonics, are sometimes a nuisance, but rarely lead to uncertainties of more than 1 mV over a typical measurement time period.

b. Mixed Conductivity

(1) *Negligible Single-Carrier B Dependence.* For Cr-doped GaAs, having $\rho_0 \gtrsim 5 \times 10^8\ \Omega$ cm, the effects of hole conduction cannot be ignored. For example, the hole current will give a positive contribution to the Hall coefficient, which will subtract from the usual negative value. Thus, a simple Hall measurement will lead to too high a value for n, and too low a value for μ_n [see Eqs. (2) and (3)]. However, by measuring the magnetic field dependences of the resistivity and Hall coefficient, it is possible to obtain the correct values of μ_n, μ_p, n, and p, *if* the single-carrier magnetic field effects are negligible (Look, 1975). This case is discussed in Section 18 of Appendix A, where it is

shown that the magnetic field dependence can be written

$$\frac{c^2}{B^2} + \mu_n^2 Y = S_{\rho m}\frac{\rho_0}{\Delta\rho} = -S_{Rm}\frac{R_0}{\Delta R}. \qquad (4)$$

Here $\Delta\rho$ and ΔR are the changes in resistivity and Hall coefficient, respectively, as B is varied. Plots of c^2/B^2 versus $\rho_0/\Delta\rho$ and c^2/B^2 versus $-R_0/\Delta R$ give slopes $S_{\rho m}$ and S_{Rm}, respectively, where $S_{\rho m}$ and S_{Rm} are defined in Appendix A [Eqs. (A46) and (A48), respectively]. The intercept $\mu_n^2 Y$ is defined by Eq. (A64). The subscript "m" denotes the "mixed-carrier" contribution to the magnetic field dependences. Closed-form solutions of n, p, μ_n, and μ_p can be written in terms of ρ_0, R_0, S_ρ, and S_R, as shown in Eqs. (A65)–(A72).

(2) *Nonnegligible Single-Carrier B Dependences.* When their relaxation times are energy dependent, the single-carrier conductivities and Hall coefficients become dependent on magnetic field, as discussed in Section 17, Appendix A. When the single-carrier magnetic field dependences are comparable in magnitude to those due to the mixed-carrier effects, the analysis in the preceding section breaks down. It is still possible to analyze the B dependence, but only to order $\mu^2 B^2$, without excessive complication. The results are

$$\frac{c^2}{B^2} + A_\rho = (S_{\rho m} + S_{\rho s})\frac{\rho_0}{\Delta\rho}, \qquad (5)$$

$$\frac{c^2}{B^2} + A_R = -(S_{Rm} + S_{Rs})\frac{R_0}{\Delta R}, \qquad (6)$$

where the $S_{\rho m}$ and S_{Rm} terms, defined earlier, are due to mixed-carrier effects, and the $S_{\rho s}$ and S_{Rs}, to single-carrier effects (Look, 1982b). Strictly speaking, A_ρ and A_R should vanish in our approximation (to order $\mu^2 B^2$), but experimentally they sometimes do not. That is, Eqs. (5) and (6) sometimes give more linear curves than the corresponding plots of $\Delta\rho/\rho_0$ versus B^2/c^2 and $\Delta R/R_0$ versus B^2/c^2. As shown in Appendix A, the same magnetic field dependences hold at very high B ($\mu^2 B^2 \gg 1$), but not at intermediate B.

The problem with Eqs. (5) and (6) is that the terms $S_{\rho s}$ and S_{Rs} introduce four new parameters: two magnetoresistance coefficients (ξ_n and ξ_p), and two magneto–Hall coefficients (β_n and β_p). These coefficients are defined by Eqs. (A29) and (A32), respectively, in Appendix A, but unless they are known independently, we cannot uniquely determine n, p, μ_n, and μ_p from this analysis.

Another approach (Look, 1980) is to avoid magnetic field effects altogether by measuring only ρ_0 and R_0. (We are, of course, not considering the magnetic field dependences due to more complicated effects, such as band warping.) These quantities can be related to the electrical parameters according to Eqs. (A39) and (A40), respectively, at $B = 0$:

$$\rho_0^{-1} = \sigma_0 = (\sigma_n + \sigma_p) = e(n\mu_n + p\mu_p), \tag{7}$$

$$R_0 = \frac{\sigma_n^2 R_n + \sigma_p^2 R_p}{(\sigma_n + \sigma_p)^2} = \frac{-n\mu_n^2 + p\mu_p^2}{e(n\mu_n + p\mu_p)}, \tag{8}$$

where, again, we have assumed unity Hall factors ($r_n = r_p = 1$). We then need two more equations to solve for n, p, μ_n, and μ_p. They are

$$np = n_i^2 \simeq (2.6 \times 10^6)^2, \tag{9}$$

$$\mu_p^{-1} \simeq 9 \times 10^{-4} + 13\mu_n^{-1}. \tag{10}$$

Equation (10) is an empirical relationship, which undoubtedly can be improved upon as more data become available. [A theoretical relationship between μ_n and μ_p has also recently been derived (Walukiewicz et al., 1982)]. The solution for the true electron mobility μ_n as a function of the "apparent" electron mobility R_0/ρ_0 is shown in Fig. 2. It is seen that some of the solutions are double valued. Usually, the correct choice can be decided by

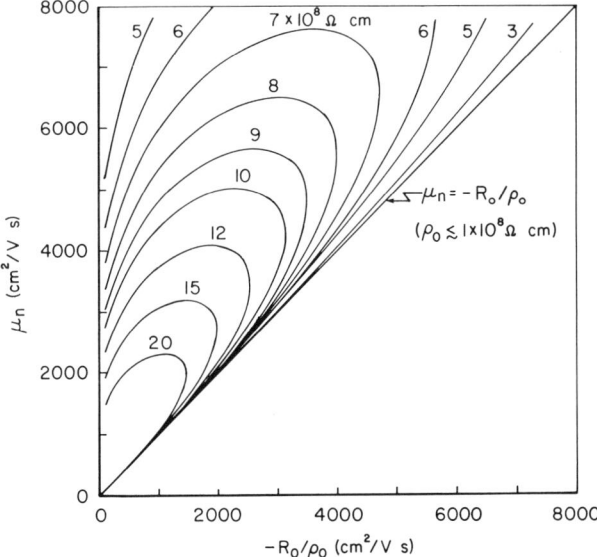

FIG. 2. The relationship of the true mobility μ_n to the measured mobility $-R_0\sigma_0$ at various values of resistivity ρ_0. These curves hold only for samples with negative Hall coefficients, the usual case. [From Look (1980).]

the sign of a thermopower measurement. Also, it is apparent from Fig. 2 that $\mu_n \simeq |R_0/\rho_0|$ for $\rho_0 \lesssim 4 \times 10^8$ Ω cm. Thus, a simple Hall measurement will give correct results for resistivities below this value. This fact is quite important for the analysis of SI GaAs, because nearly *all* O-doped or undoped samples have $\rho_0 \lesssim 4 \times 10^8$ Ω cm.

Equations (7)–(10) have also been fully analyzed by graphical techniques, which are convenient for rapid determination of the electrical parameters (Winter *et al.*, 1982).

3. Typical Data

Results are presented in Table I for a typical Cr-doped and a typical O-doped sample. Each sample is analyzed by three different techniques. *Technique 1* is a simple Hall-effect measurement (Section 2a). Here we assume that only one carrier is important, and since R_0 is always negative for these samples, in our experience, the relevant parameters are μ_n and n, calculated from Eqs. (2) and (3), respectively. (Whether or not both electrons and holes are important can often be determined from the temperature dependence.) In *technique 2* we assume that the sample probably has mixed conductivity, but, in any case, all of the magnetic field dependences of ρ and R are due to mixed-carrier effects. Then we can get $S_{\rho m}$ and S_{Rm} from plots of c^2/B^2 versus $\rho_0/\Delta\rho$

TABLE I

Electrical Parameters at 300°K Calculated by Three Different Techniques for a Typical O-Doped and a Typical Cr-Doped SI GaAs Crystal

	GaAs–Cr Technique			GaAs–O Technique		
	(1)	(2)	(3)	(1)	(2)	(3)
$\rho_0 (10^8$ Ω cm)	9.6	(Same)	(Same)	2.3	(Same)	(Same)
$R_0 (10^{11}$ cm^3/C)	−11.5	(Same)	(Same)	−11.8	(Same)	(Same)
$R_0/\rho_0 (10^3$ cm^2/Vs)	−1.19	(Same)	(Same)	−5.1	(Same)	(Same)
β (unitless)	N/A	0.43	N/A	N/A	−4.9	N/A
$S_{\rho m} (10^6$ cm^4/V^{2}s^2)	N/A	5.2	N/A	N/A	4.2	N/A
$S_{Rm} (10^6$ cm^4/V^{2}s^2)	N/A	11.8	N/A	N/A	−0.45	N/A
$\mu_n (10^3$ cm^2/Vs)	1.19	4.4	(1) 1.21 (2) 4.7	5.1	5.7	5.1
$\mu_p (10^2$ cm^2/Vs)	N/A	4.2	(1) 0.87 (2) 2.8	N/A	17.7	2.9
$n (10^6$ cm^{-3})	5.4	0.49	(1) 5.3 (2) 0.41	5.3	4.3	5.1
$p (10^6$ cm^{-3})	N/A	10.3	(1) 1.28 (2) 16.6	N/A	1.4	1.3
$n_i (10^6$ cm^{-3})	N/A	2.2	(1) 2.6 (2) 2.6	N/A	2.4	2.6

and c^2/B^2 versus $-R_0/\Delta R$, respectively. Finally, the parameters n, p, μ_n, and μ_p are calculated according to Eqs. (A65)–(A68). (Incidentally, the parameter β shown in Table I is the x-axis intercept of an R versus $\Delta\rho/\rho_0$ plot. Since $\beta = S_{\rho m}/S_{Rm}$, we can use this quantity in place of $S_{\rho m}$ or S_{Rm}, if we wish.) Finally, in *technique 3* we measure ρ_0 and R_0, and calculate n, p, μ_n, and μ_p from Eqs. (7)–(10). No magnetic field dependences need be measured when this technique is employed.

Consider first the GaAs–O data. It is immediately apparent that the data in column 1 are almost the same as those in column 3, i.e., mixed-carrier effects are small enough that a simple Hall measurement suffices. This fact could have been predicted from Fig. 2. In fact, any GaAs crystal for which $\rho_0 \lesssim 4 \times 10^8$ Ω cm should exhibit negligible mixed conductivity effects at room temperature (Look, 1980), and this includes all O-doped or undoped samples that we have measured to date. That single-carrier magnetic field dependences must be important here is clear from a glance at the value of μ_p given by technique 2, which is much too high. In general, if the magnetic field dependences are largely due to single-carrier effects, then a mixed-carrier analysis, namely technique 2, will give excessively large values for μ_p, or n_i, or both.

Now consider the GaAs–Cr data. For samples with $\rho > 5 \times 10^8$ Ω cm there are often two solutions obtained from technique 3, as grahically seen in Fig. 2. One of these solutions (no. 1 in column 3) is nearly the same as that in column 1, i.e., it is valid if mixed conductivity is negligible. The other solution (no. 2 in column 3) is correct if mixed-conductivity effects are strong. This solution, in fact, is not too different from the solution given by technique 2, suggesting that indeed single-carrier effects are not very important for this sample. A thermopower measurement confirms that solution 2, under technique 3, is the correct one for this particular GaAs–Cr crystal. In fact, usually the sign of the thermopower alone is sufficient to distinguish between solutions, being positive for solution 2 and negative for solution 1.

Thus, for O-doped, or undoped SI GaAs, techniques 1 and 3 give similar results, because single-carrier effects dominate. In this case we should be able to study the single-carrier magnetoresistance coefficient ξ_n [Eq. (A29)], and the magneto-Hall coefficient β_n [Eq. (A32)], and compare with theory. To our knowledge, no detailed studies of this sort have been carried out. Theoretical work is somewhat hampered by the inapplicability of the Brooks–Herring ionized impurity scattering theory in SI GaAs (see also Section 6), and, less so, by the failure of the relaxation-time approximation for optical-mode lattice scattering. Even if a relaxation time is assumed for the lattice scattering, the various contributions due to the different scattering mechanisms must be properly averaged over energy [Eq. (A23)]. It is safe to say that much experimental and theoretical work in this area remains to be done.

III. Thermal Equilibrium Processes

In Part II we discussed how to measure the electrical parameters n and μ_n (and/or p and μ_p), namely, by means of the conductivity and Hall coefficient. Now we must ask how these parameters relate to the more fundamental quantities of interest, such as impurity concentrations and impurity activation energies. Much can be learned from a consideration of thermal excitation processes only, i.e., processes in which the only variable parameter is temperature. Thus, we are specifically excluding cases involving electron or hole injection by high electric fields or by light. We are also excluding systems that have been perturbed from their thermal equilibrium state and have not yet had sufficient time to return. Some of these "nonequilibrium" situations will be considered in Part IV.

4. Donors and Acceptor Statistics

We begin with a simple-minded discussion of donors and acceptors. In a III–V compound such as GaAs the three valence electrons on the cation (Ga) join with the five valence electrons of the anion (As) to form bonds, either ionic (eight electrons on anion, none on cation), or covalent (four *pairs* of electrons shared by both), or a mixture of these two types. A Group VI impurity element, such as Te, substituting for As, would have one more than enough electrons necessary for bonding, and this additional electron might be bound at some energy E_D below the conduction band. We call such an impurity a "donor" since an absorption of energy E_D will cause the additional electron to be "donated" to the conduction band. Similarly, a Group II impurity element, such as Cd, substituting for Ga, would have one too few electrons to form closed-shell bonding. If an electron from the valence band could absorb energy $E_A < E_G$ and become bound to this impurity, then we would call this impurity an "acceptor." It is easy to see that more complicated situations could arise, e.g., a Group I element substituting for Ga could be a "double" acceptor. In fact, it is thought that the transition-metal impurity Cr can range from a single donor to a double acceptor in GaAs, i.e., four possible charge states (Blakemore, 1980). [The states Cr^{4+}, Cr^{3+}, and Cr^{2+} evidently have bound states within the band gap (Stauss *et al.*, 1980; Blakemore *et al.*, 1982), and Cr^{1+}, slightly above the conduction band (Hennel *et al.*, 1981).] This particular center will be discussed in more detail later. Further information on donors and acceptors can be found in many sources (Bube, 1974, p. 310).

Given some knowledge about the valence and conduction bands, we would like to determine the distribution of available electrons among the various energy states of these bands. *Free* electrons and holes are effectively noninteracting, and it is a common textbook problem to show

2. PROPERTIES OF SEMI-INSULATING GaAs

that the proper distribution function is $f(\varepsilon) = 1/[1 + \exp(\varepsilon - \varepsilon_F)/kT]$, the so-called Fermi function. The inclusion of donor and acceptor levels, however, raises some complications because the electrons on a particular localized level may interact strongly. For example, a second electron attempting to bind at an s-like orbital already containing one electron would experience coulomb repulsion, and it may not have a bound state within the band gap. Such effects can often be accounted for by simply including a "degeneracy factor" K, in the Fermi function for donors or acceptors: $f_{D,A} = 1/[1 + K_{D,A} \exp(\varepsilon_{D,A} - \varepsilon_F)/kT]$. Unfortunately the treatment of K in most references is very brief, often consisting of only a hand-waving derivation for the simplest possible case, a hydrogenic donor. Furthermore, this form of the Fermi function does not even hold when an impurity or defect has more than one charge state within the gap, or excited states within a few kT of the ground state. Because of the mystery and confusion surrounding the degeneracy factor, as well as the possible complications discussed above, it seems worthwhile to carry out a detailed treatment of the electron distribution in a semiconductor. This derivation, which includes both gap levels and the host-lattice energy bands, is given in Appendix B and only the results will be referred to in this section.

5. TEMPERATURE-DEPENDENT HALL EFFECT

We depart briefly from our discussion of SI GaAs to consider an example that better illustrates some of the features of temperature-dependent Hall measurements. This example (Look et al., 1982a) involves bulk GaAs samples that have $\varepsilon_C - \varepsilon_F \simeq 0.15$ eV. We suppose, initially, that the impurity or defect controlling the Fermi level is a donor. Then any acceptors or donors above this energy (by a few kT or more) are unoccupied and any below are occupied. Also, $p \ll n$ for $kT \ll |\varepsilon_G|$. From Eq. (B34), Appendix B, we get

$$n + N_{AS} = N_{DS} + N_D \bigg/ \left(1 + \frac{g_{D1}}{g_{D0}} e^{(\varepsilon_F - \varepsilon_D)/kT}\right), \tag{11}$$

where N_D is the concentration of the donor of interest, ε_D is the energy of this donor when occupied, $g_{D1}(g_{D0})$ is the degeneracy of the occupied (unoccupied) donor state, N_{AS} includes all acceptors lying below ε_F, and N_{DS} includes *all* donors lying above ε_F. Only single-charge-state centers are considered in this section. We assume that there are no significant donor or acceptor concentrations, other than N_D, within a few kT of ε_F. (If there are, it will be apparent in that we will not be able to fit the experimental data properly). Equation (11) can be written, as shown in Appendix B, as

$$(n + \phi_{DC})[n + (N_{AS} - N_{DS})] = \phi_{DC} N_D, \tag{12}$$

where

$$\phi_{DC} = \frac{g_{D0}}{g_{D1}} N'_C T^{3/2} e^{\alpha_D/k} e^{-E_{D0}/kT}. \tag{13}$$

Here $N'_C = 2(2\pi m_n^* k)^{3/2}/h^3$, and α_D is defined by $E_D = E_{D0} - \alpha_D T$ (Van Vechten and Thurmond, 1976). Note that E_D, the "ionization energy," is an inherently positive quantity, defined here with respect to the conduction band. We consider two limiting cases:

Case 1. Low temperature, $n \ll (N_{AS} - N_{DS})$:

$$n = \left(\frac{N_D}{N_{AS} - N_{DS}} - 1\right) \phi_{DC}$$

$$= \left(\frac{N_D}{N_{AS} - N_{DS}} - 1\right) \frac{g_{D0}}{g_{D1}} N'_C T^{3/2} e^{\alpha_D/k} e^{-E_{D0}/kT} \tag{14}$$

or

$$\ln(n/T^{3/2}) = \ln\left[\left(\frac{N_D}{N_{AS} - N_{DS}} - 1\right) \frac{g_{D0}}{g_{D1}} N'_C e^{\alpha_D/k}\right] - \frac{E_{D0}}{kT}. \tag{15}$$

Case 2. High temperature, $\phi_{DC} \gg n$:

$$n = N_D - (N_{AS} - N_{DS}). \tag{16}$$

Thus, we see that a plot of $\ln n/T^{3/2}$ versus $1/T$ should be a straight line of slope $-E_{D0}/k$ at low temperatures. At high temperatures, n approaches a constant, since all available (uncompensated) electrons from the donor of interest have entered the conduction band. The four parameters that may be determined from a fit of Eq. (12) are N_D, $N_{AS} - N_{DS}$, E_{D0}, and $(g_{D0}/g_{D1})\exp(\alpha_D/k)$. We assume that m_n^* is known.

The variation with temperature of n and μ_n for an undoped GaAs crystal grown by the horizontal Bridgman method is shown in Fig. 3. The parameters that give the best fit to Eq. (12) are also shown. The power of this method is illustrated by the small probable errors in the fitted parameters, i.e., less than 15% for N_D and less than 1% for E_{D0}. Very few techniques can lay claim to such accuracy. Note that for maximum reliability it is necessary to know the Hall r factor [Eq. (A17)], since $n = r/eR$. A variational calculation, with N_{AS} as the only undetermined parameter, was used to fit the μ_n versus T data, and also obtain r versus T (Meyer and Bartoli, 1981; Look et al., 1982a).

The data in Fig. 3 can also be analyzed by assuming that the level near 0.15 eV is an acceptor. Then the appropriate charge-balance expression is given by Eq. (B44)–(B46):

$$n + \frac{N_A}{1 + (g_{A0}/g_{A1})e^{(\varepsilon_A - \varepsilon_F)/kT}} + N_{AS} = N_{DS} \tag{17}$$

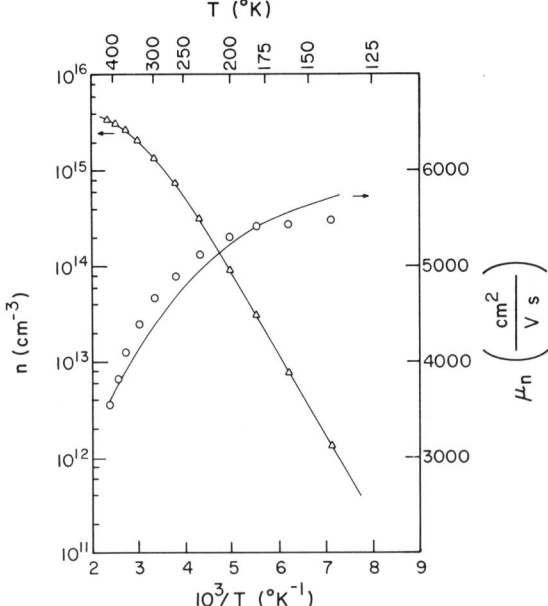

FIG. 3. The temperature dependence of the electron concentration and electron mobility for a GaAs crystal grown by the horizontal Bridgman method. The solid curves are theoretical fits, as described in the text. [From Look et al. (1982a).]

or

$$nN_A = [(N_{DS} - N_{AS}) - n](n + \phi_{AC}), \tag{18}$$

where

$$\phi_{AC} = \frac{g_{A0}}{g_{A1}} N_C' T^{3/2} e^{\alpha_A/k} e^{-E_{A0}/kT}. \tag{19}$$

Here E_{A0} is a positive energy, measured from the *conduction* band. (In most situations, E_A is measured with respect to the valence band, but that is inappropriate here since the relevant transitions are to the conduction band.)

Unfortunately, the fit of Eq. (18) to the data is nearly as good as that of Eq. (12) so that in this sample we cannot clearly distinguish between donor and acceptor behavior by such curve fitting. This point should be taken as a reminder that a level near the conduction band cannot easily be shown to be a donor, nor a level near the valence band, an acceptor. Further discussion on this point is given in Appendix B. Various models for the 0.15-eV level can be compared by determining their predicted values of $(g_0/g_1)\exp(\alpha/k)$, which turns out to be about 1 in this case (Look et al., 1982a). However, further detailed analysis of this particular center is unwarranted here since it was presented only as an example of temperature-dependent Hall-effect analysis.

In SI GaAs, less information can generally be obtained from temperature-dependent Hall-effect measurements. There are three reasons for this situation. (1) Very high temperatures ($>1000°C$) would be required to leave the low temperature regime [Eq. (14)] and make possible the calculations of N_D and $(N_{AS} - N_{DS})$ [or N_A and $(N_{DS} - N_{AS})$]. Thus, only E_{D0} may be determined without ambiguity. (2) Both p and n must often be kept in the charge-balance expression [Eq. (B17)], especially for Cr-doped material, i.e., neither is obviously negligible with respect to the other. (However, sometimes we can drop both in comparison with other terms). (3) In Cr-doped GaAs the deep "O" donor (sometimes called EL2) may also be present in sufficient quantities to demand inclusion in Eq. (B17), and the energy levels of O (EL2) and Cr are close enough (within a few kT, at $300°K$) that each term must include a Fermi function (Lindquist, 1977; Zucca, 1977). That is, the charge-balance equation must now be written, as shown in Appendix B, as

$$n + N_{AS} + \frac{N_A}{1 + \phi_{AC}/n} = \frac{n_i^2}{n} + N_{DS} + \frac{N_D}{1 + n/\phi_{DC}}, \qquad (20)$$

where $\phi_{DC}(O)$ and $\phi_{AC}(Cr)$ are defined as in Eqs. (13) and (19), respectively, and p is written as n_i^2/n. Even though Cr is known to have more than two charge states within the GaAs band gap (Stauss et al., 1980), we will assume that only $Cr^{3+}(g_0)$ and $Cr^{2+}(g_1)$ are important in SI material. Equation (20) is now fourth order in n, rather than only second order, as in the last example [Eq. (12)]. But at all reasonable laboratory temperatures we can ignore n and n_i^2/n with respect to N_{AS} and N_{DS} to get [cf. Eq. (B49)]

$$n = \frac{1}{2B}\{C\phi_{DC} + A\phi_{AC}$$
$$\pm [C^2\phi_{DC}^2 + A^2\phi_{AC}^2 + 2(2N_DN_A - AC)\phi_{AC}\phi_{DC}]^{1/2}\}, \qquad (21)$$

where $A = N_{AS} - N_{DS}$, $B = N_{DS} - N_{AS} - N_A$, and $C = N_A - N_D + N_{AS} - N_{DS}$. Thus, it is clear that when both a donor and acceptor are present in comparable concentrations, with reasonably close energy levels (i.e., less than several kT difference), we can expect a "mixed" temperature dependence. For example, suppose we have a nearly complete compensation case (i.e., $N_{AS} \simeq N_{DS}$, $N_A \simeq N_D$). Then

$$n = \left(\frac{g_{D0}g_{A0}}{g_{D1}g_{A1}}\right)^{1/2} e^{(\alpha_A + \alpha_D)/2k} N_C' T^{3/2} e^{-(E_{D0} + E_{A0})/2kT}, \qquad (22)$$

where again E_{D0} and E_{A0} are *ionization* energies (inherently positive) both with respect to the *conduction* band. Although this situation is, of course, unlikely, it

2. PROPERTIES OF SEMI-INSULATING GaAs

does illustrate that if both a deep donor and a deep acceptor are present, the temperature dependence of n can involve an effective activation energy somewhere between the two individual activation energies. Equation (22) is analogous to that for the intrinsic concentration, Eq. (B16).

We now turn to an analysis of the experimental results for Cr- and O-doped SI GaAs, shown in Table II. First consider the samples doped either with O or nothing. It is obvious that there is excellent consistency in the measured activation energies: $E_{D0} \simeq 0.75 \pm 0.02$ eV. (Only sample MA 287/80 falls outside this range, and it was an inhomogeneous crystal.) Thus, we can safely say that, in general, no other deep level (specifically Cr) of significant concentration influences such samples, a fact that has been confirmed by spark-source mass spectrographic measurements. Therefore, letting N_A, $\phi_{AC} \to 0$ in Eq. (21)

$$n = \left(\frac{N_D}{N_{AS} - N_{DS}} - 1\right) \frac{g_{D0}}{g_{D1}} e^{\alpha_D/k} N'_C T^{3/2} e^{-E_{D0}/kT}$$
$$\equiv Q_D N'_C T^{3/2} e^{-E_{D0}/kT}, \quad (23)$$

where $N'_C = 8.1 \times 10^{13}$ cm^{-3} °K$^{-3/2}$. The values of Q_D obtained are also

TABLE II
TEMPERATURE-DEPENDENT HALL-EFFECT PARAMETERS IN SEMI-INSULATING GaAs[a]

Sample	Doping	E_{D0}(eV)	Q_D	$\mu_n(296°K)$ (cm²/V s)	$\mu_n(420°K)$ (cm²/V s)
MONS O	0	0.76 ± 0.01	630	4500	4100
MA 158/80	0	0.75 ± 0.01	220	4200	3200
W 43/80	None	0.76 ± 0.01	290	5300	3500
MA 285/80	None	0.76 ± 0.01	400	3300	2700
MA 274/80	None	0.76 ± 0.01	150	4500	3000
MA 279/80	None	0.77 ± 0.01	220	5100	3600
MA 288/80	None	0.74 ± 0.01	90	3300	3200
MA 287/80	None	0.68 – 0.74[b]	10–40	1400	1600
		E_{A0}(eV)	Q_A		
LD 6-10	Cr	0.78 ± 0.02	50	3700	2600
SUM 11/76	Cr	0.80 ± 0.02	40	4200	2800 (est.)
VAR	Cr	0.77 ± 0.03	20	4700	3600 (est.)
LD 2	Cr	0.78 ± 0.02	15	2400	2600
MOR 56/76	Cr	0.67 ± 0.05[b]	40	2000	2500
MOR 57/76	Cr	0.75 ± 0.01	130	2500	2300
VAR 66-1	Cr	0.78 ± 0.02	20	4600	3500

[a] Note that energies E_{D0} and E_{A0} are both referenced to the *conduction* band.
[b] Significant curvature in Arrhenius plot; probably inhomogeneous sample.

presented in Table II. Unfortunately, we cannot separate the various terms in Q_D, since only the "low temperature" region is accessible to us. However, some reasonable guesses can be made, and because of their importance will be discussed in some detail below.

First consider the factor $N_D/(N_{AS} - N_{DS})$. It is clear from Eq. (11) (remembering that $n \ll N_{AS} - N_{DS}$) that this factor is simply the inverse of the fractional hole occupation (f) of the deep donor level. In a given group of samples we might expect that the most probable value of f would be about 0.5 since crystals with values very close to either 0 or 1 would no longer be semi-insulating. It is thus reasonable to assume that the mean of the Q_D values for O-doped (or undoped) samples in Table II corresponds to $N_D/(N_{AS} - N_{DS}) \simeq 2$. From the mean value $Q_{Dm} = 290$ (ignoring sample MA 287/80) we then get $(g_{D0}/g_{D1})\exp(\alpha_D/k) \simeq 290$. Recent theoretical studies of O in GaAs indicate a deep donor state with a singly occupied a_1 level in the neutral configuration (Fazzio et al., 1979). Since states with a_1 symmetry are orbitally nondegenerate, only the spin degeneracy is important, giving $g_0 = 1$, $g_1 = 2$. Thus, $\exp(\alpha_D/k) \simeq 580$, or $\alpha_D = 5.5 \times 10^{-4}$ eV/°K, or $E_0 \simeq 0.75 - 5.5 \times 10^{-4} T$. At room temperature, these data predict E (300°K) $\simeq$ 0.59 eV. A somewhat different model, based on the distribution of room-temperature Fermi levels among over 20 samples, gave exactly the same result (Look, 1980). An interesting observation is that the calculated α_D is near the value given for the band-gap temperature variation, although the band-gap variation is not really linear, especially at low temperatures. This observation suggests that the O level is tied to the valence band much more closely than to the conduction band, a conclusion also reached by Yu and Walters from photoluminescence measurements (Yu and Walters, 1982). As discussed earlier, the O (EL2) level may not involve oxygen itself (Huber et al., 1979); however, the preceeding discussion would be unaffected as long as the symmetry were still a_1. Other symmetries, with their associated degeneracies, could easily be incorporated into the model.

We now turn to the Cr-doped GaAs samples. If we ignore sample MOR 56/76, which showed curvature in the Arrhenius plot, the following mean values are obtained: $E_{A0} \simeq 0.78$ eV and $Q_A \simeq 45$. There is more variation in the E_0 for Cr-doped samples than for O-doped samples, possibly because of O contamination in some of the Cr-doped samples. For the Cr acceptor, Q_A is defined somewhat differently than for the donor case (Eq. 23). Thus, if N_D, $\phi_{DC} = 0$ in Eq. (21), we get

$$n = \left(\frac{N_A}{N_{DS} - N_{AS}} - 1\right)^{-1} \frac{g_{A0}}{g_{A1}} e^{\alpha_A/k} N'_C T^{3/2} e^{-E_{A0}/kT}. \quad (24)$$

Now from Eq. (17) it is clear that $N_A/(N_{DS} - N_{AS})$ is simply the inverse of the fractional electron occupation of the acceptor center. Again, we would expect the mean fractional occupation for the group of samples to be about 0.5, so that $[N_A/(N_{DS} - N_{AS}) - 1]^{-1} \simeq 1$. Then $(g_{A0}/g_{A1})\exp(\alpha_A/k) \simeq 45$. A detailed discussion of the factor g_{A0}/g_{A1} is not warranted here, but some crystal-field models (Martinez et al., 1981) of the Cr^{3+} and Cr^{2+} ions, respectively, suggest that $g_{A0} \simeq 4$ and $g_{A1} \simeq 5$. Then $\exp(\alpha_A/k) \simeq 56$, or $\alpha_A = 3.5 \times 10^{-4}$, giving $E_{Cr} \simeq 0.78 - 3.5 \times 10^{-4}T$. At room temperature, the prediction is E_{Cr} (300°K) $\simeq 0.68$ eV. (Remember that these energies are measured with respect to the conduction band.) It is interesting to note that this value is also exactly the same as that obtained from a room temperature Fermi level analysis of some 55 Cr-doped crystals (Look, 1980). However, there is a caveat here. In the room temperature analysis, discussed in Look (1980), it was assumed that $g_{A0}/g_{A1} = 4$, not $\frac{4}{5}$. The assumption that $g_{A0}/g_{A1} = \frac{4}{5}$ would have given E_{Cr} (300°K) $\simeq 0.64$ eV. Another problem is that the determinations of n (or E_F) for the Cr-doped samples are often not as accurate as those for the O-doped samples because of mixed conductivity. Therefore, we must quote a higher uncertainty for E_{Cr}. The final results are

$$E_{Cr} = 0.78 - 3.5 \times 10^{-4}T \pm 0.04 \quad \text{eV}, \tag{25a}$$

$$E_O = 0.75 - 5.5 \times 10^{-4}T \pm 0.02 \quad \text{eV}. \tag{25b}$$

Again it should be noted that we have assumed linear temperature variations because the data do not require a more complicated variation. Recent results that are in basic agreement with Eq. (25a) include the absorption measurements of Martinez et al. (1981) ($E_{Cr} = 0.78$ eV at $T = 4°K$). The fact that the thermal and absorption results are similar suggests that the Franck–Condon shift for the $Cr^{2+} \to Cr^{+3}$ transition is small.

From Eqs. (25a) and (25b) we may conclude that the deep levels due to Cr and O doping are very close at $T = 0$, with the Cr perhaps lying a little below, but that the Cr level is definitely *below* the O (EL2) level at room temperature. These conclusions differ from those of Martin et al. (1980b), who quote

$$E_{Cr}^M \simeq 0.709 - 2.4 \times 10^{-4} \frac{T^2}{T + 204} \quad \text{eV}, \tag{26a}$$

$$E_O^M \simeq 0.759 - 2.37 \times 10^{-4}T \quad \text{eV}, \tag{26b}$$

giving $E_{Cr}^M(300°K) \simeq 0.67$ eV and $E_O^M(300°K) \simeq 0.69$ eV. (The superscript "M" denotes "Martin et al.") Equations (26a) and (26b) were obtained from emission and capture cross-section data. (The authors do not consider these temperature dependences to necessarily be unique.) One problem here is that there is some disagreement in the literature about the temperature dependence of the capture cross sections, at least for Cr. (See Table IV in Section 12.) These problems can have a large impact on relationships such as those given above, and thus should be resolved.

Finally, we quote some recent results of Pons (1980), who also used transient capacitance techniques to determine the temperature dependences of the O (EL2) and Cr energy levels. In the room-temperature range, his results, designated by a superscript "P" are

$$E_{Cr}^{P} \simeq 0.71 \quad \text{eV}, \tag{26c}$$

$$E_{O}^{P} \simeq 0.76 - 2.5 \times 10^{-4}T \quad \text{eV}. \tag{26d}$$

These data may be somewhat inaccurate since they were taken (by us) from a graph (Pons, 1980, Fig. 3), and furthermore are evidently preliminary. The predicted room temperature energies are E_{Cr}^{P} (300°K) $\simeq$ 0.71 eV, and E_{O}^{P} (300°K) $\simeq$ 0.68 eV, again, with respect to the conduction band. Thus, Pons's results are in good agreement with those of Martin et al. (1980b), except for the temperature dependence of the Cr level. He finds E_{Cr} below E_{O} at room temperature, as we do, but the difference is smaller than that predicted by our electrical measurement data described here [Eqs. (25a) and (25b)]. It seems, therefore, that further investigations of the temperature dependences of the O (EL2) and Cr levels will be necessary.

Before closing the discussion on temperature-dependent Hall measurements in GaAs–Cr it should be noted that another interpretation concerning the observed low temperature slope has been forwarded (Ashby et al., 1976). By combining temperature-dependent conductivity data with space-charge-limited current data it has been shown that the 0.7-eV slope could really be due to two different levels, at 0.40 and 0.98 eV. While this conjecture may be true, under certain conditions the weight of other evidence, including photoconductivity (Allen, 1968), photocapacitance (Kocot et al., 1979; Szawelska and Allen, 1979), DLTS (Lang and Logan, 1975), and PITS (Fairman et al., 1979) results, seems to leave no doubt that indeed there is a Cr-related level near 0.7 eV. The data of this section strongly suggest that this level controls the electrical properties of SI GaAs–Cr.

As a final remark, we remind the reader that Eqs. (25a), (26a), and (26c) refer to the $Cr^{2+} \rightarrow Cr^{3+}$ transition. Very recently, the $Cr^{4+} \rightarrow Cr^{3+}$ thermal transition has also been studied (Look et al., 1982), and shown to obey $E(4+ \rightarrow 3+) - E_V = (0.324 - 1.4 \times 10^{-4}T)$ eV.

6. Temperature-Dependent Mobility

In Section 17, Appendix A, it is shown that the Hall mobility μ_H and conductivity mobility μ are related by

$$\mu_H = |R\sigma| = r\mu = (e/m^*)\langle \tau^2 \rangle / \langle \tau \rangle, \tag{27}$$

where τ is the relaxation time, and the averages over energy are defined in Eqs. (A23) and (A24). For several independent electron-scattering mechanisms, a, b,

2. PROPERTIES OF SEMI-INSULATING GaAs

$c, \ldots$, we set $\tau^{-1} = \tau_a^{-1} + \tau_b^{-1} + \cdots$ and then take the averages required in Eq. (27). However, in GaAs, polar optical-mode scattering is strong near room temperature, and for this inelastic scattering mechanism a relaxation time cannot even be defined (Putley, 1968, p. 142). This problem has been overcome by different methods (Ehrenreich, 1960; Rode, 1970; Nag, 1980; Meyer and Bartoli, 1981), but a further problem, the treatment of ionized-impurity scattering, still prohibits a full quantitative understanding of SI GaAs. Fortunately, a qualitative picture of temperature-dependent mobility is easily formulated in terms of Matthiessen's approximation (Bube, 1974, pp. 232 and 238), i.e., $\mu^{-1} = \mu_a^{-1} + \mu_b^{-1} + \cdots$. This approximation is generally quite good except in regions where two or more scattering mechanisms are of comparable strength. At any rate, in reasonably pure GaAs the lattice scattering dominates at room temperature, and $\mu_L \propto T^{-n}$, where $n \simeq 2$. Below liquid-nitrogen temperature, ionized-impurity scattering usually dominates and $\mu_I \propto T^{3/2}$. Thus, Matthiessen's rule would give $\mu^{-1} \simeq AT^n + BT^{-3/2}$, and the mobility should go through a maximum at some temperature, lower for the purer samples. Indeed, this phenomenon is nearly always observed in *conductive* samples. (In Fig. 3, the maximum is just being reached at the lowest temperatures of measurement.) Thus, mobility is a rather sensitive determinant of sample purity, or at least sample *quality*, since defects and inhomogeneous regions can also limit the mobility.

The electron mobilities at 296 and 420°K are given for several Cr-doped and O-doped samples in Table II. The data for the Cr-doped crystals should be considered less accurate since a mixed-conductivity analysis was necessary in most cases (Look, 1980). However, the temperature dependences are not unlike those of conductive GaAs samples with similar impurity concentrations (10^{16}–10^{18} cm^{-3}). At least two of the crystals (MA 287/80 and MOR 56/76) appeared to be inhomogeneous, as evidenced by nonlinear Arrhenius plots. However, it is doubtful that the bulk of the data require a percolation-type conduction mechanism to be operative, as has been suggested (Robert et al., 1979). That is, it appears that normal band conduction, with lattice-phonon, ionized-impurity, and possibly space-change (or localized-potential) (Podor, 1983) scattering, is sufficient to explain most of the mobility results.

Even if the qualitative aspects of the electron scattering seem to be well understood, the quantitative aspects are not, as mentioned above. One problem is that the treatment of electron screening employed in the Brooks–Herring formulation of ionized-impurity scattering breaks down at the low electron concentrations in SI GaAs. And the Conwell–Weisskopf formulation, which treats the screening more correctly, requires a somewhat arbitrary "screening-cutoff" parameter. Some progress has been made in reconciling the two formulations (Ridley, 1977), but a detailed theoretical treatment of electron scattering in SI GaAs has yet to be performed. From an experimental point of view, low temperature mobility measurements would be useful. However, the usual technique of Hall-effect measurements in the dark is not feasible because of the high resistivities involved (estimated at 10^{45} Ω cm, 77°K). Hall-effect measurements in the light are possible, but care must be taken that the equilibrium Fermi level is not significantly changed (Haloulos et al., 1980).

Several recent works also explore temperature-dependent mobility measurements in SI GaAs, with mixed conductivity (Hrivnak et al., 1982; Kang et al., 1982; Walukiewicz et al., 1982).

IV. Nonequilibrium Processes

In thermal equilibrium, the fractional occupations of all states, including those in the valence band, the conduction band, and the band gap, can be deduced from a single parameter, the Fermi energy [cf. Eq. (B11)]. States with energy more than a few kT above ε_F are empty, whereas those below ε_F are full. Since it is difficult to experimentally learn anything about completely empty or completely full states, we are effectively limited to states near ε_F when we do equilibrium measurements. This situation may be overcome by various forms of excitation, with light being the most important for SI materials. We will restrict our theoretical discussion to a simple phenomenological approach that illustrates the main points. More complicated results will simply be quoted where necessary and referenced. The major experimental techniques will be discussed in some detail, especially with regard to their application to SI GaAs.

7. Excitation and Recombination

Consider a semiconductor in thermal equilibrium in the dark. When we give values for the fractional occupations of the various energy levels, they are valid only in an average sense because thermal excitation and recombination processes are constantly causing small fluctuations in the equilibrium occupations. These fluctuations are, of course, the basis of generation–recombination (G–R) noise. The excitation process may be thought of as arising from blackbody photons, in equilibrium with the lattice phonons. Recombination is simply the capture and annihilation of free electrons or holes by free or bound holes or electrons, respectively. At equilibrium the averaged emission and capture rates must of course be equal so that the averaged time derivatives of all of the various state occupations will vanish. In particular, this is true for conduction-band states, giving

$$\frac{dn}{dt} = 0 = \sum_i e_{ni} n_i - n \sum_i \sigma_{ni} v_n (N_i - n_i) \qquad (28)$$

where n_i is the electron density at center i (or cell i in the valence band), e_{ni} is the thermal emission rate from center i to the conduction band, σ_{ni} is the capture cross section of center i for electrons, v_n is the mean thermal velocity of the electrons, and $(N_i - n_i)$ is the unoccupied density of center i. Recombination of free electrons with free holes is neglected. We can perform a gedanken experiment in which only one center i is present in the sample. Such a sample could, in principle, be prepared. Equation (28) then gives

$$n_i = \frac{N_i}{1 + e_{ni}/(n\sigma_{ni}v_n)} = \frac{N_i}{1 + \dfrac{e_{ni}}{N_C\sigma_{ni}v_n} e^{-\varepsilon_F/kT}} \qquad (29)$$

since at equilibrium $n = N_C \exp(\varepsilon_F/kT)$ in the Boltzmann approximation (for ε_F more than a few kT from the conduction band). Now comparison with Eq. (B11c) yields a very important result:

$$e_{ni} = (g_{i0}/g_{i1})N_C\sigma_{ni}v_n e^{-E_i/kT}, \qquad (30)$$

where $E_i \equiv \varepsilon_C - \varepsilon_i$ is the ionization energy with respect to the conduction band. If we take N_C from Eq. (B14), use $v_n = \sqrt{8kT/\pi m_n^*}$, and allow σ_{ni} to be "thermally activated," i.e., $\sigma_{ni} = \sigma_{ni}(\infty)\exp(-E_{\sigma i}/kT)$, then

$$e_{ni} = \frac{16\pi m_n^* k^2}{h^3} \frac{g_{i0}}{g_{i1}} \sigma_{ni}(\infty) T^2 e^{\alpha_i/k} e^{-(E_{i0} + E_{\sigma i})/kT}, \qquad (31)$$

where we have assumed $E_i = E_{i0} - \alpha_i T$ (cf. Section 5). For GaAs, $16\pi m_n^* k^2/h^3 \simeq 2.0 \times 10^{20}$ s^{-1} cm^{-2} °K^{-2}. The great utility of Eq. (31) lies in the fact that an Arrhenius plot of e_{ni}/T^2 will give $E_{i0} + E_{\sigma i}$, and if $E_{\sigma i} = 0$ or if it can at least be independently measured, then E_{i0} can be determined. Furthermore, if g_{i0}/g_{i1} and α_i are known, then $\sigma_{ni}(\infty)$ can also be determined. However, in carrying out emission-related experiments, such as deep level transient spectroscopy (DLTS) (Lang, 1974), it must always be remembered that the slope of an Arrhenius plot gives $E_{i0} + E_{\sigma i}$, not E_{i0} alone. The $E_{\sigma i}$ can be quite appreciable; e.g., the deep electron trap related to Cr in GaAs has $E_\sigma \simeq 0.25$ eV, as measured by one group (Lang and Logan, 1975), or somewhat less, as measured by other groups (Mitonneau et al., 1979; Jesper et al., 1980). Any uncertainty in $E_{\sigma i}$, of course, leads to the same uncertainty in E_{i0}. The measurement of e_{ni} will be discussed in detail later, but we should note here that a similar relationship for hole emission can easily be derived [cf. Eq. (C26)]. It should also be noted that Eq. (28), and what will follow below, is valid only if band-to-band recombination is not important. It could be included but the appropriate capture term would have to take account of the hole velocity as well as the electron velocity.

We now add the effects of light excitation. Suppose a flux of I_0 photons/cm^2 s is incident upon the sample, and the cross section for photon capture by the occupied center i is σ_{vni}. Then Eq. (28) is modified as follows:

$$\sum_i (e_{ni} + I_0\sigma_{vni})n_i = n \sum_i \sigma_{ni}v_n(N_i - n_i). \tag{32}$$

In the presence of the light let $n_i \to n_{i0} + \Delta n_i$ and $n \to n_0 + \Delta n$. Since Eq. (28) holds in equilibrium, we can eliminate several terms in Eq. (32) and get

$$\sum_i \left[\frac{\sigma_{ni}v_n n_0 N_i}{n_{i0}} + I_0\sigma_{vni} \right] \Delta n_i + I_0\sigma_{vni}n_{i0}$$

$$= \Delta n \sum_i \sigma_{ni}v_n[(N_i - n_{i0}) - \Delta n_i]. \tag{33}$$

For SI GaAs it is almost always true that $\Delta n_i \ll n_{i0}$ (unless center i is a long-lifetime trap). If this inequality is sufficiently strong, then the first term on the left-hand side of Eq. (33) can be ignored, and

$$\Delta n = I_0 \sum_i \sigma_{vni}n_{i0} \Big/ \sum_i \sigma_{ni}v_n(N_i - n_{i0}). \tag{34a}$$

A similar relationship for holes, to be used later, can easily be derived:

$$\Delta p = I_0 \sum_i \sigma_{vpi}(N_i - n_{i0}) \Big/ \sum_i \sigma_{pi}v_p n_{i0}. \tag{34b}$$

Each σ_{vni} is a strong function of monochromatic light energy hv as a threshold $hv \simeq E_i$ is approached. This phenomenon will be discussed in the next section. However, for completeness we should relate Eq. (34a) to a common expression involving the absorption coefficient α_n and electron lifetime τ_n:

$$\Delta n = I_0 \alpha_n \tau_n. \tag{35}$$

Comparing Eqs. (34a) and (35) it is clear that

$$\alpha_n = \sum_i \sigma_{vni}n_{i0} \tag{36a}$$

$$\tau_n^{-1} = \sum_i \sigma_{ni}v_n(N_i - n_{i0}). \tag{36b}$$

A more detailed analysis of extrinsic (below band gap) light excitation (Look, 1977b) gives

$$\Delta\sigma = \frac{eI_0}{\alpha d}(1 - e^{-\alpha d})(\alpha_n\mu_n\tau_n + \alpha_p\mu_p\tau_p), \tag{37}$$

where $\Delta\sigma$ is the change in conductivity (the photoconductivity) due to both hole and electron excitation, e the electronic charge (*not* emission rate), d is the sample thickness, and α_n and α_p the absorption coefficients for electron and hole excitation, respectively. (Note that it is possible for α to be greater than

$\alpha_n + \alpha_p$ if light is absorbed by other processes besides electron and hole excitation.) The assumptions used in deriving Eq. (37) can be found in Look (1977b). If $\alpha d \ll 1$, and if the absorption of light is due entirely to electronic transitions (i.e., $\alpha_n = \alpha$, $\alpha_p = 0$), then $\Delta\sigma = e\mu_n I_0 \alpha_n \tau_n = e\mu_n \Delta n$, in agreement with Eq. (35).

Before discussing photoconductivity in more detail it should be noted that absorption measurements themselves can be quite useful. From Eq. (36a) it is seen that if the σ_{vni} are known and if α_n can be measured, then the n_{i0} can be determined. Further, if the E_i are known (or can be measured) and the equilibrium free-electron concentration n_0 is measured (giving E_F), then the N_i can also be determined from Eq. (B11c). Usually α is calculated from a transmission measurement; i.e.,

$$T = (1 - R)^2 e^{-\alpha d}/(1 - R^2 e^{-2\alpha d}), \tag{38}$$

where T is the transmission coefficient, R the reflection coefficient, and d the sample thickness (Pankove, 1971). Often $R^2 \exp(-2\alpha d) \ll 1$ so that

$$\alpha \simeq (1/d)\ln[(1 - R)^2/T]. \tag{39}$$

For GaAs, $R \simeq 0.3$ over a wide range of wavelength, so that $\alpha \simeq d^{-1}\ln(0.5/T)$. Note that $R^2 \exp(-2\alpha d) \lesssim R^2 \simeq 0.09$; therefore, our above approximation is fairly good, even for $\alpha = 0$. Martin and co-workers (1979) have used absorption measurements at 1.34 μm to determine the Cr concentration in GaAs. Typical transmission spectra are shown in Fig. 4 for a sample doped with

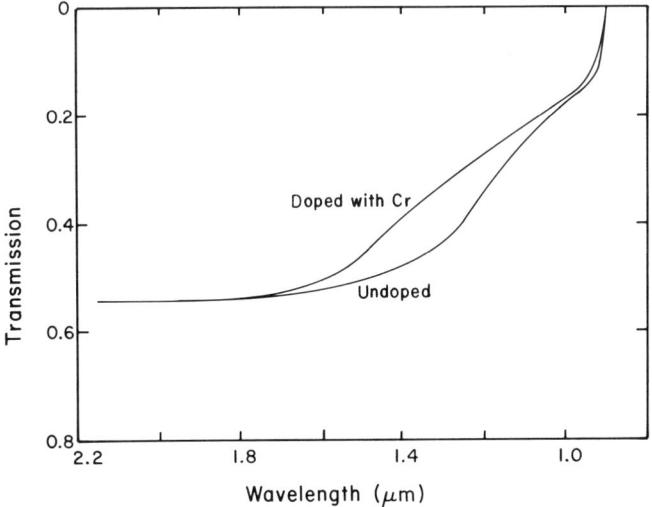

FIG. 4. Room temperature optical transmission of two SI GaAs crystals, undoped and Cr-doped, respectively. Sample thickness, 0.473 cm. [From Martin et al. (1979).]

about 3×10^{16} Cr/cm^3. Martin (1981) has also calibrated the concentration of the electron trap EL2 by similar methods.

8. Photoconductivity and Photo-Hall Measurements

The conductivity at low magnetic field is given in Eq. (A39) as $\sigma = e(n\mu_n + p\mu_p)$. Then the photoconductivity $\Delta\sigma$ is

$$\Delta\sigma = e(\mu_{n0}\Delta n + n_0 \Delta\mu_n + \mu_{p0}\Delta p + p_0 \Delta\mu_p). \tag{40}$$

Here it is assumed that $\Delta\mu_n \ll \mu_{n0}$ and $\Delta\mu_p \ll \mu_{p0}$ so that the terms $\Delta n \Delta\mu_n$ and $\Delta p \Delta\mu_p$ are negligible. These conditions generally hold, as will be seen later. However, equivalent inequalities involving Δn and n_0, or Δp and p_0, are not always true, since it is often possible to make $\Delta n \gg n_0$, or $\Delta p \gg p_0$, in SI material (Look, 1977a). To estimate the relative importance of the various terms in Eq. (40), it is instructive to consider how μ_n and μ_p will vary with the light excitation. If the sample is quite pure, then we would expect $\Delta\mu_n, \Delta\mu_p \simeq 0$ since the light will not affect the lattice-phonon modes much. Suppose, however, a high concentration of ionized impurities is present, such that they completely dominate the carrier scattering. Then to a first approximation, μ_n, $\mu_p \propto N_I^{-1}$, where, for simplicity, we will assume that only one type of impurity center is present, with N_I being the *ionized* concentration of this impurity. Then, $\Delta\mu_{n0} = -\mu_n \Delta N_I/N_I$, $\Delta\mu_p = -\mu_{p0}\Delta N_I/N_I$, and Eq. (40) becomes

$$\Delta\sigma = e\left[\mu_{n0}\left(\Delta n - \frac{n_0}{N_I}\Delta N_I\right) + \mu_{p0}\left(\Delta p - \frac{p_0}{N_I}\Delta N_I\right)\right]. \tag{41}$$

But for SI GaAs, $n_0/N_I, p_0/N_I \ll 1$, and ΔN_I cannot be much larger than n or p, except for very long lifetime traps. For example, typically $n_0 \simeq 10^6$–10^7 cm^{-3}, and $N_I \simeq 10^{16}$ cm^{-3}. Thus, to a good approximation,

$$\Delta\sigma = e(\mu_{n0}\Delta n + \mu_{p0}\Delta p). \tag{42}$$

Equation (42) should hold as long as there is not a *significant* charge redistribution, i.e., as long as the occupations of trapping states are not significantly altered. For the case of a single carrier, Eq. (42), in conjunction with Eq. (37), gives

$$\Delta n = I_0 \alpha_n \tau_n (1 - e^{-\alpha d})/\alpha d \tag{43}$$

in the extrinsic excitation regime, with a similar equation for Δp. Generally, $\alpha \simeq 1$–10 cm^{-1} for SI GaAs (Look, 1977b), so that if $d \ll 0.1$ cm, then $\alpha d \ll 1$ and Eq. (42) becomes

$$\Delta\sigma = eI_0\left(\frac{\mu_{n0}\sum_i \sigma_{vni}n_{i0}}{\sum_i \sigma_{ni}v_n(N_i - n_{i0})} + \frac{\mu_{p0}\sum_i \sigma_{vpi}(N_i - n_{i0})}{\sum_i \sigma_{pi}v_p n_{i0}}\right), \tag{44}$$

where we have used Eqs. (34a) and (34b) for Δn and Δp, respectively.

It is often the goal of photoconductivity studies to measure the photoexcitation cross sections, i.e., the σ_{vni} and σ_{vpi}. Although it is hard to determine the *magnitudes* of these cross sections, because of the other terms in Eq. (44) that are unknown, it is often easy to determine their *energy dependences*. The reason is that none of the other terms are usually very energy dependent. In 1965, Lucovsky published an oft-quoted equation for σ_{vi}, applicable to deep centers:

$$\sigma_{vi} = CE_i^{1/2}(hv - E_i)^{3/2}/(hv)^3, \qquad (45)$$

where E_i is a "threshold" energy, i.e., the energy at which photoconduction processes should begin. Equation (45) predicts a sharp onset at $hv = E_i$ and a relatively broad maximum at $hv = 2E_i$. If there are several centers in the band gap capable of emitting holes or electrons, then, according to Eq. (44), several onsets should occur. (Note that centers well above the Fermi level will have $n_{i0} \simeq 0$ and therefore cannot emit electrons, and centers well below E_F cannot emit holes because $N_i - n_{i0} \simeq 0$.) Photoconductivity data for three Cr-doped GaAs samples are shown in Fig. 5. The predicted sharp threshold occurs near 0.5 eV, but the onset is not quite as sharp as expected from the Lucovsky equation, also shown. This difference is not too surprising for two reasons: (1) Lucovsky employed a δ-function potential, which may be too simple, even for very deep centers; and (2) he did not allow for phonon interactions, which makes the formula applicable only at $T = 0$. In recent years, both of these conditions have been relaxed, and progress is being made toward the fitting of photoionization cross sections at elevated temperatures (Noras, 1980; Ridley, 1980; Chantre *et al.*, 1981; Martinez *et al.*, 1981). It is often an advantage to be able to work at higher temperatures (especially room temperature!), not only because of the obvious experimental simplicities, but also because long-lifetime trapping is not as significant. In any case, it is usually possible to identify at least one E_i and sometimes several, especially at lower temperatures, as seen in Fig. 6. If phonons are properly taken into account in the theory, it should be possible to determine E_i versus T, although few studies of this type have been carried out as yet. A more complete discussion of photoionization cross sections is given by Neumark and Kosai, in this volume.

The strong peak shown at 0.88 eV for sample A in Fig. 5 is not consistent with the Lucovsky theory, nor, indeed, with most of the other theories that assume bound-state-to-band excitations. [Some recent calculations show the possibility of a sharp peak in the cross section if certain conditions are met, but it is not clear that these conditions are reasonable (Blow and Inkson, 1980).] One suggestion is that free-hole excitation above 0.88 eV causes a drastically reduced electron lifetime (Masut and Penchina, 1981) [i.e., larger denominator in the first term of Eq. (44)], but this interpretation is questionable because the

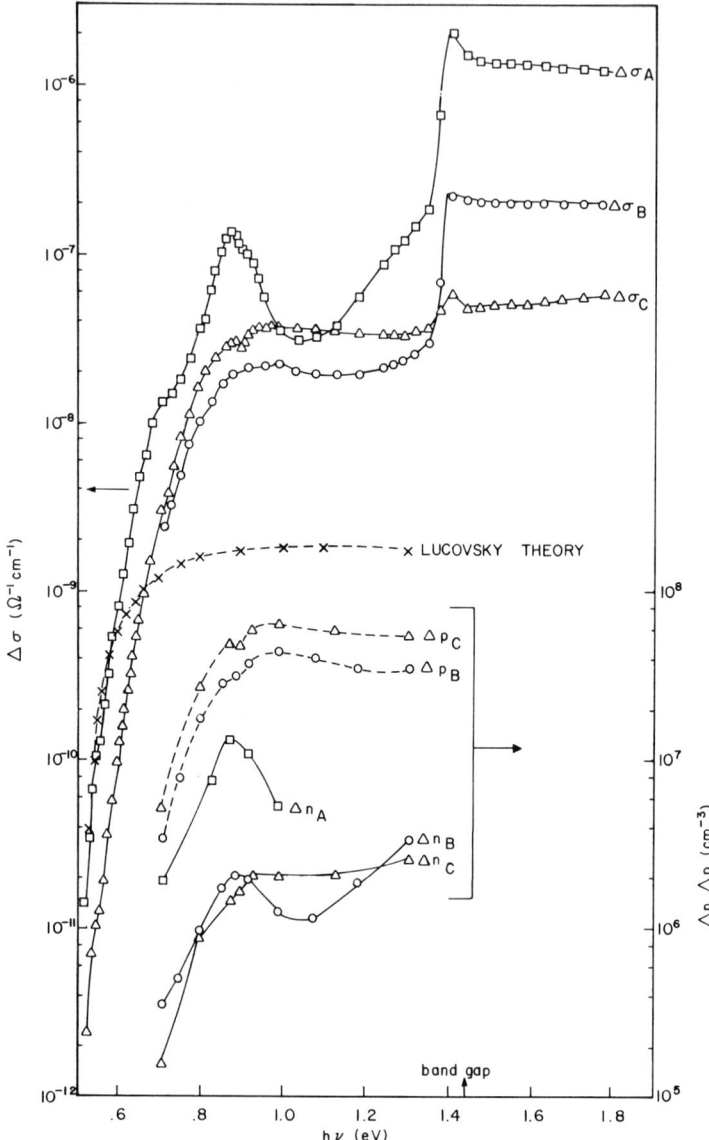

FIG. 5. Photoconductivity, $\Delta\sigma$, and photoexcited-carrier-concentration (Δn and Δp) spectral data for three GaAs–Cr samples, A, B, and C. Also shown is the Lucovsky photoionization cross section [Eq. (45)] for excitation from a level at 0.52 eV. The theory is fitted to the data at 0.53 eV. [From Look (1977a).]

2. PROPERTIES OF SEMI-INSULATING GaAs   103

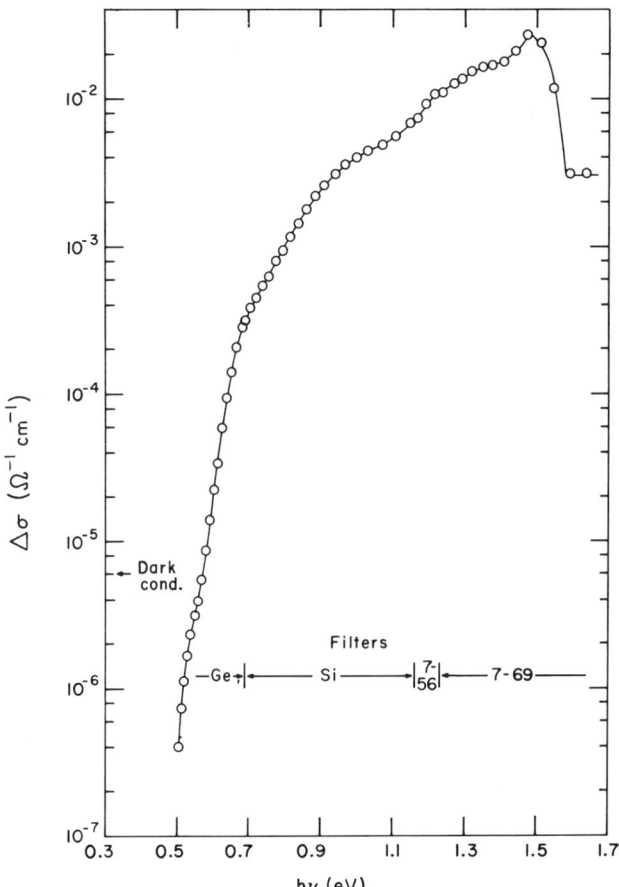

FIG. 6. Photoconductivity of a Bridgman O-doped GaAs crystal at 79°K. ($I_0 = 2 \times 10^{14}$ photons/cm s.)

peak is also seen in absorption spectra (Lin and Bube, 1976). Another suggestion of long standing is that the peak represents an intracenter excitation in the Cr^{2+} ion itself, with photoconductivity then caused by an autoionization of the Cr excited state, resonant with the conduction band (Ippolitova et al., 1976). This idea is supported by the data of Fig. 5, which show that the peak is stronger in the more n-type SI samples, which would be expected to have a higher fraction of Cr^{2+} (over Cr^{3+}). However, a criticism of the intracenter-excitation model is that 0.88 eV is too large a crystal-field splitting, compared to the more established Cr^{2+} data in some of the II–VI compounds (Kaufmann and Schneider, 1976). Some more detailed models lead to smaller crystal-field splittings. It is safe to say that this controversy is certainly not over yet, and may not be for some time. Several recent works discuss the GaAs–Cr situation more fully (Blakemore, 1980; White, 1980).

Although we argued earlier that μ_n and μ_p probably do not vary much under light excitation, it is certainly not true that their "combined" mobility, i.e., the total $R\sigma$, will be constant, especially in SI material. The reason is immediately obvious from Eqs. (A39) and (A40):

$$R_0\sigma_0 = (-n\mu_n^2 + p\mu_p^2)/(n\mu_n + p\mu_p). \tag{46}$$

This quantity can actually change sign even while μ_n and μ_p are themselves remaining nearly constant (Putley, 1968, p. 116). An example of the strong dependence of $R\sigma$ on photoexcitation energy is shown in Fig. 7. It is clear that the samples represented here, to a varying degree, become more "p-type" above $h\nu \simeq 0.7$ eV, and then more "n-type" again above 1.0 eV. Although μ_n is changing somewhat, as shown by mixed-conductivity measurements (Section 18, Appendix A), the dominant contribution that makes $R\sigma$ more positive is a hole excitation near 0.7 eV, i.e., a valence-band-to-bound-state transition in which the bound state is about 0.7 eV above the valence band. The relevant band edge would not have been known without the photo-Hall measurements. (In this case the problem is academic, since 0.7 eV is about in the middle of the band gap.) The sharp rise in $R\sigma$ at 1.0 eV is probably due to an electron transition from a level 1.0 eV below the conduction band. In this case the photo-Hall data are of more than academic interest, since photoconductivity results alone could not have distinguished between a valence-band transition or a conduction-band transition. The 0.7-eV level is likely due to the transition $Cr^{3+} \rightarrow Cr^{2+} + e^+$, since it occurs only in Cr-doped GaAs, and since it is stronger in the more "p-type" (lower E_F) samples, which would be expected to have a higher fraction of Cr^{3+}. The excitation at 1.0 eV is not necessarily due to Cr and, in fact, its origin is unknown. Also, the involvement of higher conduction bands, L and X, cannot be ruled out here (Chantre et al., 1981).

Although most of the measurements described above are dc, a brief discussion of ac photoconductivity measurements is in order. The apparatus shown in Fig. 1 incorporates a lock-in amplifier for this purpose, and in some cases (e.g., the Fe^{2+} photoconductivity resonance in InP–Fe) it is a convenient means of increasing the signal-to-noise ratio. However, with SI GaAs, extreme caution must be used, because the limiting lifetimes in the system may not be sample emission and recombination times, but the RC time constant due to the high sample resistance ($\sim 10^{10}$ Ω at 300°K), and cable and meter capacitances (perhaps > 100 pF, without guarding). In a high excitation regime ($\Delta n \gg n_0$), such as is true for much of the data in Figs. 5 and 6, R itself is a function of light energy, and a sharp feature may look entirely different than its real shape. Such problems can be isolated by varying the light intensity, for example, but their possible presence must be kept in mind.

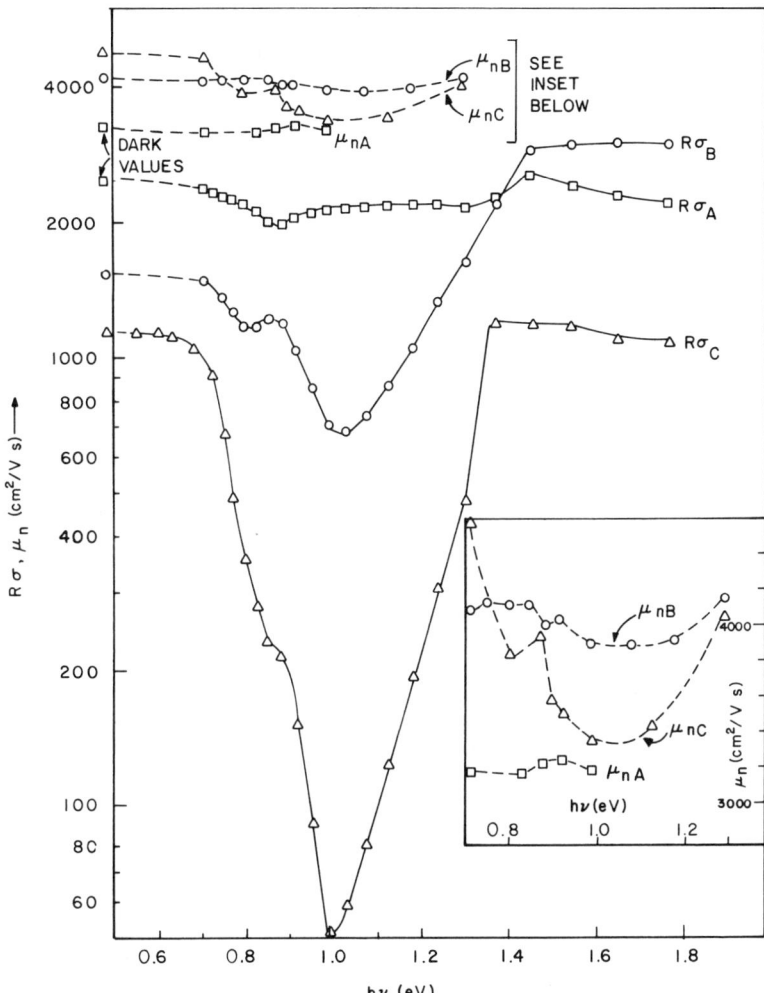

FIG. 7. Photo-Hall $R\sigma$ and electron mobility μ_n spectral data for three GaAs–Cr samples, A, B, and C (same samples as in Fig. 5). An expanded version of the data is shown in the inset. [From Look (1977a).]

We must again emphasize that in spite of a vast amount of electrical and optical data on Cr-doped GaAs, there is still not general agreement on many matters (White, 1980). Of recent help is an increasing amount of theoretical activity that may help clarify some of these issues (Ll'in and Masterov, 1977; Hemstreet and Dimmock, 1979; Clark, 1980; Hemstreet, 1980).

9. Thermally Stimulated Current Spectroscopy

Up until now we have ignored the effects of trapping, i.e., the capture and slow release of electrons by centers lying above E_F, or of holes by centers below E_F. It is easy to include these effects formally; if hole emission and capture are ignored, at first, then

$$\frac{dn_i}{dt} = -e_{ni}n_i + n(N_i - n_i)\sigma_{ni}v_n, \qquad (47)$$

where n is found according to Eq. (28):

$$\frac{dn}{dt} = \sum_i e_{ni}n_i - n \sum_i \sigma_{ni}v_n(N_i - n_i). \qquad (48)$$

Here the summation index i covers both traps and recombination centers. We suppose initially that there is only one trap and that the rest of the impurities or defects act as recombination centers. A trap, of course, is by definition more likely to release its charge to the appropriate band than to hold it until annihilation by subsequent capture of the opposite charge. The reverse is true for recombination centers. We can often lump the sum effect of all the recombination centers into a single lifetime τ_n:

$$\tau_n^{-1} = \sum_{j,\,\text{recomb. centers}} \sigma_{nj}v_n(N_j - n_j). \qquad (49)$$

Evidently, the constancy of τ_n is a good approximation in SI GaAs (Castagne et al., 1980). Then Eq. (48) becomes

$$\frac{dn}{dt} = e_{ni}n_i - n\sigma_{ni}v_n(N_i - n_i) - \frac{n}{\tau_n}. \qquad (50)$$

From this point on the index i will refer to the particular trap of interest. Equations (47) and (50) form a pair of coupled differential equations, but unfortunately they are not linear because of the terms in $(n)(n_i)$; thus, analytic solutions are not easily found. However, it is possible to make reasonable approximations, and these will be described below in the context of a common experiment.

We first suppose that *band-gap* light is shined on a sample at a temperature low enough that e_{ni} is negligible in Eq. (47) or Eq. (50). The light will add a driving term $I_0\alpha_n$ to Eq. (50). Then, if $\tau_n^{-1} \gg \sigma_{ni}v_n(N_i - n_i)$, Eq. (50) is easily solved to give $n = I_0\alpha_n\tau_n[1 - \exp(-t/\tau_n)]$. Since τ_n is very small (10^{-9}–10^{-8} s) in GaAs, much smaller, for example, than trap filling or emission rates that can be commonly measured, n may be considered constant while the light is on, yielding, for Eq. (47),

$$n_i = N_i(1 - e^{-n\sigma_{ni}v_n t}) = N_i(1 - e^{-I_0\alpha_n\tau_n\sigma_{ni}v_n t}). \qquad (51)$$

2. PROPERTIES OF SEMI-INSULATING GaAs

For Cr-doped GaAs it is usually easy to make $n \simeq 10^7$ cm^{-3}, with light from a typical monochromator (Look, 1977a), and if $\sigma_i \simeq 10^{-14}$ cm^2, and $v_n \simeq 10^7$ cm s^{-1}, the trap will be completely filled in a few seconds. Upon shutting off the light, n will fall quickly to near zero, and if e_{ni} is also quite small at the filling temperature, Eq. (47) shows that $dn_i/dt \simeq 0$, i.e., the trap remains filled ($n_i = N_i$).

To release the electrons from the trap we simply raise the temperature, usually at a linear rate with time, i.e.,

$$dT = a\, dt. \tag{51a}$$

Initially, we will assume that as the electrons are released from the trap, they quickly are captured at recombination centers and are not retrapped, i.e., $\tau_n^{-1} \gg (N_i - n_i)\sigma_{ni}v_n$. Then the average n in the conduction band will remain small and Eq. (47) becomes

$$\frac{dn_i}{dt} = a\frac{dn_i}{dT} = -e_{ni}(T)n_i(T), \tag{52}$$

which is easily solved:

$$n_i(T) = n_i(T_0)\exp\left(\int_{T_0}^{T} -\frac{e_{ni}}{a}\, dT\right). \tag{53}$$

Under the same approximation as above Eq. (50) can now be written

$$\frac{dn}{dt} = a\frac{dn}{dT} = e_{ni}n_i - \frac{n}{\tau_n}, \tag{54}$$

where a is the heating rate [Eq. (51a)], and e_{ni} is given by Eq. (31). Equation (54) is solved in Appendix C, with the result

$$n = n_{i0}\tau_n e_{ni}\exp\left(-\int_{T_0}^{T}\frac{e_{ni}}{a}\, dT'\right). \tag{55}$$

Since e_{ni} is an increasing function of temperature, n will go through a maximum, at which $dn/dT = 0$:

$$\frac{dn}{dT} = n_{i0}\tau_n\left(e^{-\int(e_{ni}/a)\, dT'}\frac{de_{ni}}{dT} - e_{ni}\frac{e_{ni}}{a}e^{-\int(e_{ni}/a)\, dT'}\right) = 0 \tag{56}$$

or

$$de_{ni}/dT = e_{ni}^2/a \quad \text{at} \quad T = T_m.$$

By writing $e_{ni} = D_i T^2 \exp(-E_{i0}/kT)$ as in Eq. (31), we get

$$\left(2T_m + \frac{E_{i0}}{k}\right) = \frac{D_i}{a}T_m^4 e^{-E_{i0}/kT_m}, \tag{57}$$

where $D_i = (16\pi m_n^* k^2/h^3)(g_{i0}/g_{i1})\sigma_{ni}(\infty)\exp(\alpha_i/k)$. Here we have assumed that

σ_{ni} is temperature independent. Often $E_{i0}/k \gg 2T_m$ so that

$$E_{i0} = kT_m \ln \frac{16\pi m_n^* k^3 (g_{i0}/g_{i1}) \sigma_{ni} e^{\alpha_i/k} T_m^4}{h^3 a E_{i0}}. \tag{58}$$

This equation can be written

$$\frac{E_{i0}}{kT_m} = \ln \frac{T_m^4}{a} + \ln \frac{16\pi m_n^* k^3 \sigma_{ni}(g_{i0}/g_{i1}) e^{\alpha_i/k}}{h^3 E_{i0}} \tag{59}$$

so that a plot of $\ln[T_m^4/a]$ versus T_m^{-1} has a slope of E_{i0}/k and an intercept of $\ln[16\pi m_n^* k^3 \sigma_{ni}(g_{i0}/g_{i1}) \exp(\alpha_i/k)/h^3 E_{i0}]$. For GaAs, this latter term becomes $\ln[1.7 \times 10^{16}(g_{i0}/g_{i1}) \exp(\alpha_i/k) \sigma_{ni} E_{i0}^{-1}]$, or for $(g_{i0}/g_{i1}) \simeq 1$, $\alpha_i \simeq 0$, and $\sigma_{ni} \simeq 10^{-15}$ cm^2, the term becomes $\ln(17/E_{i0})$, where E_{i0} is measured in electron volts. If $T_m \gtrsim 100°$K, then it is clear that for most reasonable values of σ_{ni} the first term in Eq. (59) will dominate the second and, to a first approximation, $E_{i0} \simeq kT_m \ln(T_m^4/a)$. However, it is better to vary the heating rate (a) in order to carry out the Arrhenius plot suggested by Eq. (59). Then a rough value for σ_{ni} may also be obtained from the intercept. Fillard et al. (1978) have given a critical evaluation of TSC measurements in general.

Some recent thermally stimulated current (TSC) data obtained by Martin and Bois (1978) are shown in Fig. 8. Here the maximum occurs at 264°K, giving $E_{i0} \simeq 0.61$ eV, for $\sigma_{ni} \simeq 10^{-15}$ cm^2. Evidently, "a" was not varied in this experiment. Note that the dark current begins to increase rapidly near room temperature in these data. Here our analysis must break down since capture

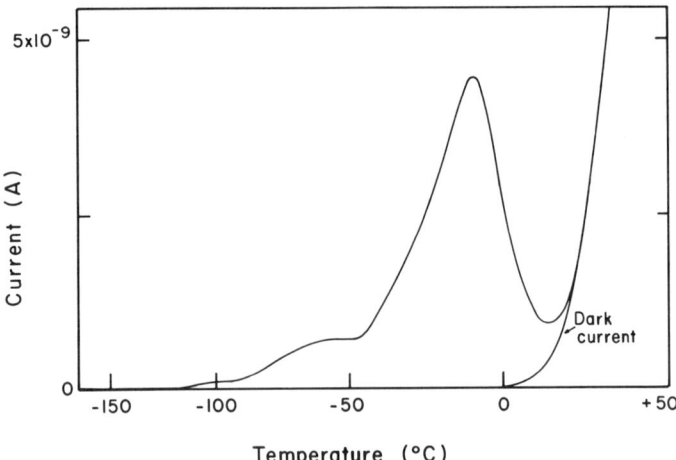

FIG. 8. TSC spectrum for SI GaAs crystal. Thickness, 28 μm; heating rate, 0.045°K/s. [From Martin and Bois (1978). This figure was originally presented at the Spring 1978 Meeting of the Electrochemical Society, Inc. held in Seattle, Washington.]

processes will no longer be negligible. However, the peaks from traps emitting in this temperature region will be obscured by the dark current anyway, so that the problems in analysis are somewhat academic. Other TSC results in SI GaAs have been given by Castagne et al. (1980).

Some mention should be given of another very simple method of determining E_i. If it is assumed that the trap is constantly in equilibrium with the conduction band, then the maximum carrier concentration will occur approximately when E_F crosses E_i, i.e.,

$$E_i \simeq E_F(T_m) \simeq kT_m \ln \frac{N_C(T_m)}{n(T_m)}, \tag{60}$$

where N_C is given by Eq. (B14). This simple formulation must inherently include retrapping (Bube, 1960), since quasiequilibrium is assumed, but nevertheless it should be used with caution. Discussion is given in the books by Bube (1960) and Milnes (1973).

Finally, some other points should be brought out. (1) We generally measure the conductivity σ not n, the carrier concentration. Since $\sigma = en\mu_n$, the temperature dependence of the mobility should be taken into account. However, this has only a minor effect on Eq. (58) since the temperature dependence of the mobility is generally not strong. (2) Although we have dealt only with electron traps in our analysis, the case for hole traps would follow immediately with obvious notation changes. (3) We cannot tell the difference between electron and hole traps simply by measuring the TSC, especially for SI material with E_F near midgap. This problem will be dealt with further in the next section.

10. Photoinduced Transient Current Spectroscopy

As discussed in the previous section, the TSC experiment consists of filling the traps at low temperature, via light excitation, and then sweeping the temperature upward to empty them. Conversely, one could fill the traps and then watch them empty at the *same* temperature, if the emission rate were high enough at that temperature. But if the emission rate were too high, then of course the traps could never be filled at the chosen temperature. To quantify these matters we again return to Eq. (47):

$$\frac{dn_i}{dt} = -e_{ni}n_i + n(N_i - n_i)\sigma_{ni}v_n. \tag{47}$$

As before, we assume that due to independent recombination processes the electrons have a lifetime τ_n with $\tau_n^{-1} \gg (N_i - n_i)\sigma_{ni}v_n$. Then, while the light is on, $n = I_0\alpha_n\tau_n$, as shown in the discussion preceeding Eq. (51), and

$$\frac{dn_i}{dt} = -e_{ni}n_i + I_0\alpha_n\tau_n(N_i - n_i)\sigma_{ni}v_n, \tag{61}$$

where α_n is the absorption coefficient for all excitations that put electrons into the conduction band. An assumption here, of course, is that $n \simeq I_0 \alpha_n \tau_n \gg n_0$, the thermal equilibrium value. The solution to Eq. (61) is

$$n_i = \frac{N_i}{1 + e_{ni}/\beta_{ni}} (1 - e^{-(e_{ni} + \beta_{ni})t}), \qquad (62)$$

where $\beta_{ni} \equiv I_0 \alpha_n \tau_n \sigma_{ni} v_n$, and the boundary condition is $n_i = 0$ at $t = 0$. Here β_{ni} is an effective "filling" rate. Equation (62) shows that the trap can be completely saturated with electrons only if $e_{ni} \ll \beta_{ni}$. Although it appears as if this condition would always be fulfilled at a high enough light intensity (I_0), we must remember that there is a limit to the density of electrons that can be excited. That is, as I_0 is increased, both α_n and τ_n must eventually become dependent upon I_0. Also, we cannot always fulfill $e_{ni} \ll \beta_{ni}$ simply by going to a low enough temperature, because some traps are too shallow for this condition to be satisfied at any reasonable temperature, even 4°K. In general, traps shallower than 0.1 eV are difficult to investigate by emission-spectroscopic methods.

To continue this discussion we will assume that a light pulse of time t_p has been incident upon the sample. When the light is shut off, Eq. (61) will become $dn_i/dt = -e_{ni} n_i$ and the solution will be

$$n_i(t) = \frac{N_i}{1 + e_{ni}/\beta_{ni}} (1 - e^{-(e_{ni} + \beta_{ni})t_p}) e^{-e_{ni}(t - t_p)} \qquad (63)$$

for times $t \geq t_p$. (Note that $t = 0$ is at the *beginning* of the light pulse.) If $t_p \gg (e_{ni} + \beta_{ni})^{-1}$, then

$$n_i(t) = \frac{N_i}{1 + e_{ni}/\beta_{ni}} e^{-e_{ni}(t - t_p)}. \qquad (64)$$

The process described by Eqs. (62) and (63) is shown graphically in Fig. 9a. We can now also solve Eq. (50) for n, again assuming that $\tau_n^{-1} \gg \sigma_{ni} v_n (N_i - n_i)$. The result, as shown in Appendix C, is

$$n(t) = \frac{e_{ni}}{\tau_n^{-1} - e_{ni}} \frac{N_i}{1 + e_{ni}/\beta_{ni}} (1 - e^{-(e_{ni} + \beta_{ni})t_p})(e^{-e_{ni}(t - t_p)} - e^{-(t - t_p)/\tau_n})$$
$$+ n(t_p) e^{-(t - t_p)/\tau_n} \qquad (65)$$

for $t \geq t_p$. It is perhaps not obvious that $n(t)$ is a monotonically decreasing function of t. An extremum in n versus t would appear if dn/dt vanished, or if

$$e_{ni} \tau_n e^{-(t_m - t_p)/\tau_n} = 1 - n(t_p)/M, \qquad (66)$$

where M includes the first three (time-independent) factors on the right-hand side of Eq. (65), and we have again assumed that $\tau_n^{-1} \gg e_{ni}$. Now $n(t_p) \simeq I_0 \alpha_n \tau_n$

2. PROPERTIES OF SEMI-INSULATING GaAs

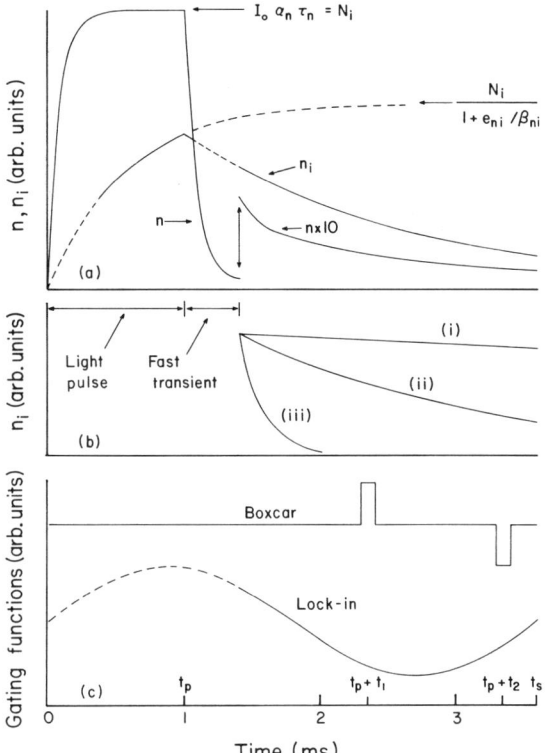

FIG. 9. The behavior of the occupied trap concentration n_i [Eq. (63)] and the free electron concentration n [Eq. (65)] during and after a light pulse of duration t_p. For part (a) the parameters are $e_{ni} = 0.6$ ms^{-1}, $\beta_{ni} = 1.2$ ms^{-1}, and $\tau_n^{-1} = 11.8$ ms^{-1}. For part (b) the parameters are $e_{ni} = 0.06$, 0.6, and 6 ms^{-1}, respectively, for curves (i), (ii), and (iii). The choice of parameters is for illustrative purposes only and may not reflect a realistic situation. The shape of n_i is only approximately correct in the dotted portions. Part (c) shows the gating functions for boxcar and lock-in amplifiers, respectively.

under our assumption, $\tau_n^{-1} \gg \sigma_{ni} v_n (N_i - n_i)$. Therefore, $n(t_p) \simeq \beta_{ni}/\sigma_{ni} v_n$ and, from the definition of M,

$$\frac{n(t_p)}{M} = \frac{\beta_{ni}}{\sigma_{ni} v_n} \frac{1 - e_{ni}\tau_n}{e_{ni}\tau_n} \frac{\beta_{ni} + e_{ni}}{\beta_{ni} N_i} \frac{1}{(1 - e^{-(e_{ni}+\beta_{ni})t_p})}$$

$$\geq \frac{1 - e_{ni}\tau_n}{\tau_n \sigma_{ni} v_n N_i}. \tag{67}$$

The last inequality follows because the factor left out is obviously greater than unity. By our previous assumption, the denominator is much less than unity, and since for practical cases $e_{ni}\tau_n \ll 1$ also, it is clear that $n(t_p)/M \gg 1$. But then there can be no extremum in n versus t because the right-hand side of Eq. (66) is negative and the left-hand side is always positive. The time dependence of $n(t)$, according to Eq. (65), will thus always be monotonically decreasing after the pulse and, in fact, should look something like that plotted schematically in Fig. 9b. Once the terms involving $\exp(-t/\tau_n)$ have fallen to zero, the time constant will be e_{ni}^{-1}, the same as for $n_i(t)$ in Eq. (64).

There are several ways to measure the resulting transient in n. It would seem logical to pick a convenient temperature such that the transient is neither too fast nor too slow to measure [e.g., curve (ii) in Fig. 9b], and then record the exponential curve, by some means, for later analysis. By carrying out this process for several temperatures it would be possible to plot e_{ni}/T^2 versus T^{-1} and determine E_i (if the capture cross section is independent of T). Unfortunately this process is time consuming. In 1974, Lang published a much more convenient way to handle this problem. He suggested using a double-gated boxcar integrator to measure $n(t)$ at two different times after the light pulse, as shown in Fig. 9c. (Actually, Lang was measuring a capacitance transient, not a current transient, but the analysis is basically the same. Differences will be discussed later.) Then from Eq. (65)

$$\Delta n \equiv n_1 - n_2 = Ce_{ni}(e^{-e_{ni}t_1} - e^{-e_{ni}t_2}), \tag{68}$$

where C is a constant. It is easily seen from Eq. (65) that C will be independent of e_{ni} as long as $\tau_n^{-1} \gg e_{ni}$, $e_{ni} \ll \beta_{ni}$, and $(e_{ni} + \beta_{ni})t_p \gg 1$. Equation (68) will apply after the initial fast transient, with time constant τ_n, has died down. By using repetitive light pulses the boxcar method allows the signal-to-noise ratio to be increased, by integration. (Again, Lang used forward-bias voltage pulses, instead of light pulses, but the same considerations hold.) Then, rather than searching for a convenient temperature at which to measure the emission rate, he simply swept temperature. It is apparent from Fig. 9b that Δn should go through a maximum, because both at very low T (case i) and very high T (case iii), $\Delta n \simeq 0$. The emission rate at the maximum is given by

$$0 = \frac{d\Delta n}{dT} = \frac{d\Delta n}{de_{ni}} \frac{de_{ni}}{dT}$$

$$= \frac{de_{ni}}{dT} \left[e_{ni}(-t_1 e^{-e_{ni}t_1} + t_2 e^{-e_{ni}t_2}) + (e^{-e_{ni}t_1} - e^{-e_{ni}t_2}) \right] \tag{69}$$

with the result

$$e^{-e_{ni}^*(t_2-t_1)} = \frac{1 - e_{ni}^* t_1}{1 - e_{ni}^* t_2}, \tag{70}$$

where e_{ni}^* is the value of e_{ni} at the temperature (T_m) of maximum Δn. Note that Eq. (70) is independent of the *magnitude* of n. Thus, we need not know the baseline of the signal. By using different values of t_1 and t_2, different T_m will be produced, and a plot of e_{ni}^*/T_m^2 versus T_m^{-1} will then give E_{i0}. In practice, t_1/t_2 is often kept constant, whereas t_1 and t_2 are both varied. As an example, if $t_1/t_2 = 0.5$, then a maximum will occur at $e_{ni}^* t_1 = 1.44$, according to Eq. (70). A word of caution should be noted here. The condition for a maximum is often given (Lang, 1974) as $e_{ni}^* = [\ln(t_1/t_2)]/(t_1 - t_2)$, instead of Eq. (70). The difference is that for *capacitance*-transient measurements, n_i [Eq. (63)] is important, whereas for current-transient measurements (our case), n [Eq. (65)] is the valid parameter. The most important difference is an extra factor e_{ni} in the equation for n that will shift the temperature maximum upward. For the example above, the maximum in Δn_i will occur at $e_{ni}^* t_1 = 0.693$ instead of 1.44. The various relationships for maxima, such as Eq. (70), would of course become more complex if we relaxed the approximations stated immediately after Eq. (68). Furthermore, τ_n and β_{ni} themselves can vary with temperature, but probably not strongly enough to appreciably shift the temperature maxima.

This experiment may also be carried out by means of lock-in detection, as schematically illustrated in Fig. 9c. If $t_p \ll t_s$, where t_s is the period of the lock-in reference signal, the signal S will be given by

$$S = A \int_0^{t_s} n(t) \sin(2\pi t/t_s) \, dt, \tag{71}$$

where A is a constant depending on amplifier gain, etc. It is usually necessary to blank the first part of each cycle containing the pulse and initial fast transient. Then the terms involving $\exp(-t/\tau_n)$ in Eq. (65) will not contribute, and Eq. (71) yields

$$S_i = G_i \frac{(2\pi f)^2}{e_{ni}^2 + (2\pi f)^2} (1 - e^{-e_{ni}/f}), \tag{72}$$

where $f = t_s^{-1}$ is the lock-in reference frequency and

$$G_i = A \frac{e_{ni}}{\tau_n^{-1} - e_{ni}} \frac{N_i}{1 + e_{ni}/\beta_{ni}} (1 - e^{-(e_{ni} + \beta_{ni})t_p}). \tag{73}$$

By setting $dS/dT = (dS/de_{ni})(de_{ni}/dT) = 0$, we find the condition for maximum S:

$$e^{e_{ni}/f} = 1 + \frac{e_{ni}^*}{f} \frac{(e_{ni}^*/f)^2 + (2\pi)^2}{(e_{ni}^*/f)^2 - (2\pi)^2}. \tag{74}$$

In deriving Eq. (74) we have assumed the previous conditions $\tau_n^{-1} \gg e_{ni}$, $\beta_n \gg e_{ni}$, and $(e_{ni} + \beta_{ni})t_p \gg 1$. An approximate solution is $e_{ni}^* = 6.35 f$. As before, by varying f, we can obtain e_{ni}^* as a function of T_m and thus determine E_{i0}, as shown by Eq. (59). We must again note that the n_i

maximum would occur at a different temperature than the n maximum. In fact, if we had integrated $n_i(t)$, instead of $n(t)$, in Eq. (71), we would have found that $S(n_i)$ would go through a maximum at $e_{ni}^* = 2.36f$, instead of $6.35f$.

Let us now consider an example. Suppose the material is GaAs and there are three traps, of energies $E_{i0} = 0.3, 0.4,$ and 0.5 eV, with the other parameters as follows, for all three traps: $g_{i0}/g_{i1} = 1, \sigma_{ni} = 10^{-15}$ cm^2, $\alpha_i = 0$, and $E_{\sigma i} = 0$ [cf. Eq. (31)]. Let us also set $f = 250$ Hz. According to Eq. (74), these traps should give maxima in S at 222, 287, and 350°K, respectively. An important point here is that the traps should appear well separated in temperature, and that is because the e_{ni} are relatively strong functions of temperature, for deep levels. It is this fact that makes temperature spectroscopy useful. When more than one trap is contributing to $n(t)$, as in this example, it is straightforward to show that their contributions should simply be summed, i.e., $S = \sum S_i$, as long as the electrons are much more likely to recombine than be retrapped. [Equation (54) is solved for the more general case with a summation sign in front of the first term on the right-hand side, and then the solution, Eq. (65), also has the summation.] The curve $S = \sum S_i$ is plotted in Fig. 10 for the three traps discussed above.

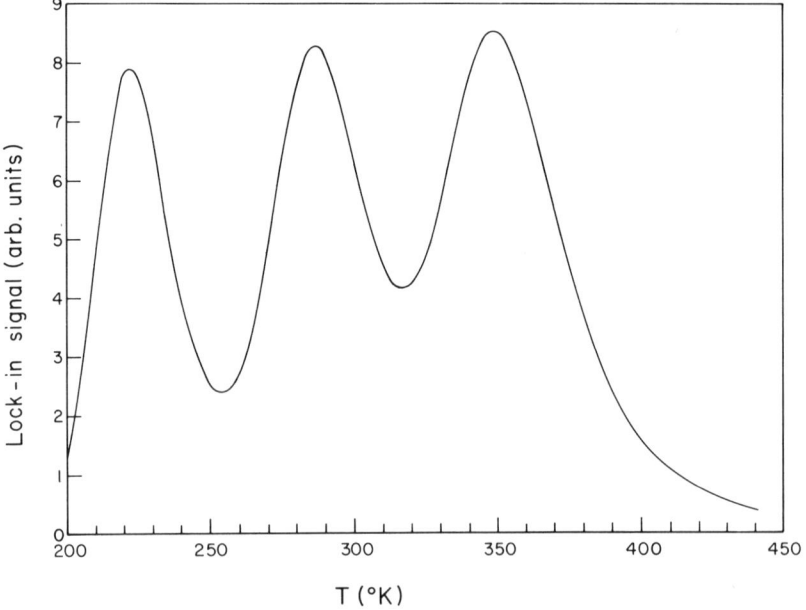

FIG. 10. Theoretical plot of the emission signal for three traps in GaAs with activation energies 0.3, 0.4, and 0.5 eV, respectively. Other parameters are given in the text.

Before discussing the experimental aspects in more detail a word about the measurement of capture cross sections is in order. At a time slightly greater than t_p, just after the initial fast transients [$\exp(-t/\tau_n)$ terms] have died out, the rate of carriers entering the conduction band will be $e_{ni}n_i(t_p)$, so that

$$n(t_p) = e_{ni}\tau_n n_i(t_p) \simeq N_i e_{ni}\tau_n \frac{1 - e^{-\beta_{ni}(1 + e_{ni}/\beta_{ni})t_p}}{1 + e_{ni}/\beta_{ni}} \tag{75}$$

from Eq. (63). Then, from the definitions of e_{ni} [Eq. (30)], and $\beta_{ni} = I_0\alpha_n\tau_n\sigma_{ni}v_n$, we get

$$\frac{e_{ni}}{\beta_{ni}} = \frac{(g_{i0}/g_{i1})N_C e^{\alpha_i/k} e^{-E_{i0}/kT}}{I_0\alpha_n\tau_n}. \tag{76}$$

Equation (75) can be rewritten by relating $n(t_p)$ to $n(\infty)$:

$$n(t_p) = n(\infty)(1 - e^{-\beta_{ni}(1 + e_{ni}/\beta_{ni})t_p}) \tag{77}$$

since $n(\infty) = N_i e_{ni}\tau_n/(1 + e_{ni}/\beta_{ni})$. For a deep enough trap $e_{ni}/\beta_{ni} \ll 1$ and β_{ni} can then be determined from Eq. (77) by varying t_p. In any case, e_{ni}/β_{ni} can, in principle, be determined by measuring $n(\infty)$ as a function of I_0, since $e_{ni}/\beta_{ni} \propto I_0^{-1}$ if α_n and τ_n are independent of I_0. We assume that n_1, the carrier concentration with the light on, is much greater than n_0, the equilibrium value in the dark. In fact, if this were not true, it would make little sense to talk about filling traps with the light pulse. Then $n_1 \simeq I_0\alpha_n\tau_n$ [cf. discussion preceeding Eq. (51)], and n_1 can be determined by a Hall measurement, or estimated from a conductivity measurement. Since $\beta_{ni} = I_0\alpha_n\tau_n\sigma_{ni}v_n \simeq n_1\sigma_{ni}v_n$, our measurements of β_{ni} and n_1 will give σ_{ni}. It is often necessary to measure σ_{ni} versus T in order to accurately determine E_{i0}, because if σ_{ni} is thermally activated, then the measured emission activation energy is actually $E_{i0} + E_{\sigma ni}$. This point was discussed before in connection with Eq. (31).

Another way to determine σ_{ni} is from the intercept of a $\ln(T_m^4/a)$ versus T_m^{-1} plot, as discussed previously in connection with Eq. (59). Besides σ_{ni}, the intercept includes the factor E_{i0}, which is determined from the slope, but it also includes the factor (g_{i0}/g_{i1}) and $\exp(\alpha_i/k)$, which in general are not known. However, the biggest problem, perhaps, is that σ_{ni} varies with the *exponential* of the intercept, so that a small error in the intercept is greatly magnified in the determination of σ_{ni}. Therefore, if possible, it is better to determine σ_{ni} from the charging curve, as described above, rather than the decay curve.

We now discuss some of the experimental aspects of temperature spectroscopy. Lang (1974) called his original method deep level transient spectroscopy (DLTS), and he measured capacitance transients produced by voltage pulses in diodes made from conductive materials. However, in SI materials, this method is not feasible and an alternate method, involving current transients produced by light pulses in bulk material (or Schottky structures), was recently introduced, first by Hurtes et al. (1978a) (HBMB), then by Martin and Bois (1978) (MB), and independently by Fairman et al. (1979) (FMO). Note that a good discussion of capacitance transients is given elsewhere in this volume by Neumark and Kosai.

The FMO method, called PITS (photoinduced transient spectroscopy), involves pulsing the light with a chopper; i.e., the sample is illuminated when a chopper hole is in front of the monochromatic light beam, and the emission transient appears when a blade stops the light. The resulting photocurrent is amplified by a high speed preamplifier, fed into a sample/hold amplifier, and finally to a lock-in detector (Fairman et al., 1979). The analysis of this processs is much like that resulting in Eqs. (72) and (74). It is also possible to obtain useful data by recording the *rise* in photocurrent, while the sample is being illuminated, as well as the decay. The details of the FMO method have evidently not yet been published.

The HBMB-M method, called OTCS (optical transient current spectroscopy), is somewhat different, as illustrated in Fig. 11. Here the sample is illuminated through a semitransparent layer of Cr, and the other side is covered with a thicker Cr layer, producing somewhat of a Schottky–Schottky structure, although with a high series resistance (Hurtes et al., 1978a). The light is pulsed, by means of a shutter, and the emission transient is probed at two times, t_1 and t_2; thus, Eq. (70) applies. Typical spectra, for SI GaAs–Cr, are shown in Fig. 11c. By observing the peak height as the bias polarity is changed, it is often possible to determine whether the detected level is an electron or hole trap. However, it must be noted that this identification is not as clear as it is when a capacitance transient can be measured.

Some recently published PITS and OTCS data are shown in Table III. A naive comparison of these data, and the spectra themselves, would seem to indicate that the PITS method gives sharper line shapes and yields more levels, suggesting greater sensitivity. However, a true comparison would have to involve the same samples and take into account such experimental parameters as integration times. There appears to be almost a continuum of energy levels from about 0.15 to 0.90 eV. Some are identified with DLTS-observed levels, mostly by comparison of respective activation energies. It should be noted here that the activation energies in Table III, designated E_a, are "apparent" activation energies, i.e., simply the slopes of $\ln(e_{ni}^*/T^2)$ versus T_m^{-1} plots. These slopes really give $E_a = E_{i0} + E_{\sigma i}$, not just E_{i0}, as seen by Eq. (31). If σ_{ni} is temperature independent, then $E_{\sigma i} = 0$, but this has to be explicitly shown, perhaps by one of the methods discussed earlier. Unless it is known that $E_a = E_{i0}$, it is not wise to compare these values with energy levels deduced from completely different experiments, e.g., temperature-dependent Hall measurements. Indeed, detailed analyses of $E_{\sigma i}$ have been carried out for

2. PROPERTIES OF SEMI-INSULATING GaAs

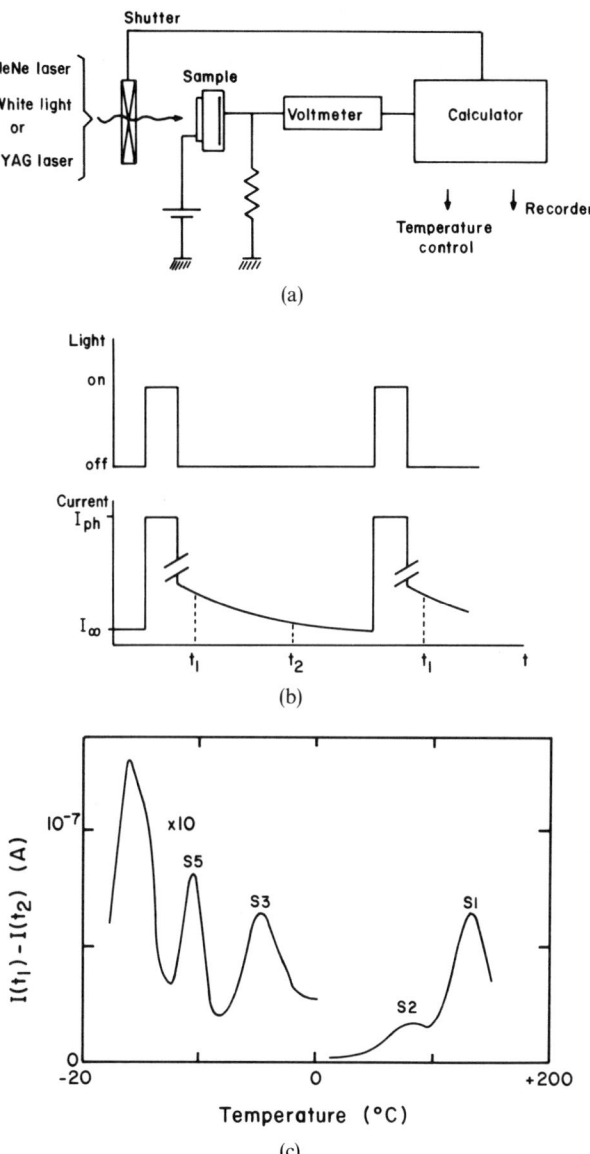

FIG. 11. (a) Block diagram of the OTCS apparatus. (b) Timing diagram. Compare Eq. (68) and Fig. 9. (c) OTCS spectrum for SI GaAs crystal with YAG laser excitation. Sample thickness, 28 μm; bias voltage, 8 V; $t_1 = 34$ ms. [From Martin and Bois (1978). These figures were originally presented at the Spring 1978 Meeting of the Electrochemical Society, Inc. held in Seattle, Washington.]

TABLE III

Levels Observed by OTCS and PITS in SI GaAs

Desig.	E_a (eV)	σ_0 (cm^2)	Hole or electron trap	DLTS ident.	Impurity ident.	Expt. method	Ref.[a]
	0.14	1×10^{-16}				PITS	a
	0.15	8×10^{-14}				PITS	a
	0.22	2×10^{-15}				PITS	b
	0.26	2×10^{-12}				PITS	a
S6	0.27	2×10^{-14}	h	HL12		OTCS	c
	0.30	7×10^{-14}	h	HL12		PITS	a
C6	0.32			EL6		OTCS	d
	0.34	6×10^{-13}				PITS	a
	0.34	4×10^{-14}	e	EL6		PITS	a
S5	0.34	3×10^{-14}	e	EL6		OTCS	c
S4	0.35	5×10^{-15}	e	EL5		OTCS	c
C5	0.41				Cu	OTCS	d
	0.46	2×10^{-13}				PITS	b
	0.48	7×10^{-12}				PITS	b
	0.48	6×10^{-14}				PITS	b
	0.51	1×10^{-12}	e	EL4		PITS	a
C4	0.54			EL3		OTCS	d
	0.55	1×10^{-13}				PITS	b
	0.55	7×10^{-15}				PITS	b
C3	0.56				Fe	OTCS	d
S3	0.57	5×10^{-13}		EL3	Fe	OTCS	c
	0.60	1×10^{-12}	e	EL3		PITS	a
	0.65	1×10^{-13}			Cr	PITS	a
S2	0.74	6×10^{-15}	e	EL2	"O"	OTCS	c
	0.75	1×10^{-11}			"O"	PITS	b
C2	0.81			EL2	"O"	OTCS	d
	0.83	2×10^{-13}	h	HL10		PITS	a
	0.90	2×10^{-14}	h	HL1	Cr	PITS	a
	0.90	2×10^{-14}	h	HL1	Cr	PITS	a
S1	0.90	2×10^{-14}	h	HL1	Cr	OTCS	c
C1	0.90				Cr	OTCS	d

[a](a) Freeman and Oliver (1980); (b) Fairman *et al.* (1979); (c) Martin and Bois (1978); (d) Hurtes *et al.* (1978b).

some of the more common traps, such as S2 (or EL2), a dominant electron trap in undoped or O-doped SI GaAs, and S1 (or C1, or HL1), a dominant hole trap in Cr-doped GaAs (Mitonneau *et al.*, 1979; Martin *et al.*, 1980a).

Two deficiencies of the current-transient spectroscopy, as compared with the capacitance-transient spectroscopy, involve the determinations of the trap concentrations, and the relevant band to which the emission is occurring. Progress is being made on these problems, especially the latter (Martin and Bois, 1978). Thus, several of the centers listed in Table III are designated as "electron" or "hole" traps, although it is unclear how firm these identifications are. As far as trap concentrations are concerned, they can be determined if τ_n is known [cf., for example, Eq. (65), (73), or (75)]. The lifetime τ_n can be estimated from the relationship $n_1 \simeq I_0 \alpha_n \tau_n$, if α_n is known since n_1 can be determined from a Hall-effect measurement or estimated from a conductivity measurement. However, these ambiguities do not exist in capacitance spectroscopy.

One further complication, which applies to all forms of emission spectroscopy, occurs when both electron and hole emission rates are nonnegligible. Then Eq. (47) must be modified as follows:

$$\frac{dn_i}{dt} = -e_{ni}n_i + e_{pi}(N_i - n_i) + n(N_i - n_i)\sigma_{ni}v_n$$
$$- pn_i\sigma_{pi}v_p. \tag{78}$$

Equation (48) remains the same, as long as recombination of free electrons with free holes is negligible:

$$\frac{dn}{dt} = \sum_i e_{ni}n_i - n \sum_i \sigma_{ni}v_n(N_i - n_i) \tag{48}$$

and we can add

$$\frac{dp}{dt} = \sum_i e_{pi}(N_i - n_i) - p \sum_i \sigma_{pi}v_p n_i. \tag{79}$$

These equations can be solved by invoking the same constraints as before, namely: (1) there is only one dominant trap, (2) free carriers are much more likely to recombine than be trapped, and (3) while the light is on, n and p are essentially constant, with $n \simeq I_0 \alpha_n \tau_n$ and $p \simeq I_0 \alpha_p \tau_p$. Then Eq. (78) can be solved for the time the light pulse is on, and Eq. (48) [now the same as Eq. (54)], for the time after the light pulse. Thus for $t \geq t_p$, as shown in Appendix C,

$$n = \frac{e_{ni}\tau_n N_i}{1 + e_{ni}/e_{pi}} + \frac{e_{ni}\tau_n N_i}{1 - (e_{ni} + e_{pi})\tau_n} \left(\frac{1}{1 + Z_i} - \frac{1}{1 + e_{ni}/e_{pi}}\right)$$
$$\times (1 - e^{-e_{npi}t_p})e^{-(e_{ni}+e_{pi})(t-t_p)} + C_n e^{-(t-t_p)/\tau_n}, \tag{80}$$

where $Z_i = (e_{ni} + \beta_{pi})/(e_{pi} + \beta_{ni})$, $e_{npi} = (e_{ni} + e_{pi} + \beta_{ni} + \beta_{pi})$, and C_n is a constant of integration. The most important point to note about this formula is the time constant, $(e_{ni} + e_{pi})^{-1}$. If e_{ni} and e_{pi} are comparable in magnitude, it clearly will be impossible to measure an unambiguous activation energy, at least by simple experiments of the types discussed earlier in this section. This problem should be most prevalent for traps near the middle of the gap, since for shallower traps the exponential dependence on energy [cf. Eq. (30)] should heavily weight one of the emission terms with respect to the other. Actually, for traps near midgap, Eq. (80) is of doubtful validity anyway, as discussed in Appendix C.

For completeness, we also solve Eq. (79) for p under the same restrictions as before. The result is

$$p = \frac{e_{pi}\tau_p N_i}{1 + e_{pi}/e_{ni}} + \frac{e_{pi}\tau_p N_i}{1 - (e_{ni} + e_{pi})\tau_p} \left(\frac{1}{1 + e_{ni}/e_{pi}} - \frac{1}{1 + Z_i} \right)$$
$$\times (1 - e^{-e_{npi}t_p})e^{-(e_{ni}+e_{pi})(t-t_p)} + C_p e^{-(t-t_p)/\tau_p}, \quad (81)$$

where C_p is a constant of integration. (Note that the terms involving τ_n and τ_p in both Eqs. (80) and (81) are expected to decay much faster than the other time-dependent terms in these equations, because $(e_{ni} + e_{pi})\tau_{n,p} \ll 1$.) In the limit e_{ni} and $\beta_{ni} \to 0$, we have

$$p = \frac{e_{pi}\tau_p}{1 + e_{pi}/\beta_{pi}} \frac{N_i}{1 - e_{pi}\tau_p} (1 - e^{-(e_{pi}+\beta_{pi}t_p)})e^{-e_{pi}(t-t_p)} + C_p e^{-(t-t_p)/\tau_p}, \quad (82)$$

which corresponds exactly to Eq. (65) (for n), as it should.

The total current is now given by

$$I(t) = C_1(n\mu_n + p\mu_p)$$
$$= C_1 N_i \left(\frac{\mu_n e_{ni}\tau_n}{1 + e_{ni}/e_{pi}} + \frac{\mu_p e_{pi}\tau_p}{1 + e_{pi}/e_{ni}} \right)$$
$$+ C_1 N_i \left(\frac{\mu_n e_{ni}\tau_n}{1 - (e_{ni} + e_{pi})\tau_n} - \frac{\mu_p e_{pi}\tau_p}{1 - (e_{ni} + e_{pi})\tau_p} \right)$$
$$\times \left(\frac{1}{1 + Z_i} - \frac{1}{1 + e_{ni}/e_{pi}} \right)(1 - e^{-e_{npi}t_p})e^{-(e_{ni}+e_{pi})(t-t_p)}$$
$$+ C_1 \mu_n C_n e^{-(t-t_p)/\tau_n} + C_1 \mu_p C_p e^{-(t-t_p)/\tau_p}, \quad (83)$$

where C_1 is a constant depending on sample dimensions and applied voltage. It can easily be shown that the first term is always greater than the second, so that the total current is always positive, as of course it must be. However, the *differential* current will be negative if the second term is negative. Suppose that $\tau_n = \tau_p$, $(e_{ni} + e_{pi})\tau_n \ll 1$, $\beta_{pi} \gg e_{ni}$, and $\beta_{ni} \gg e_{pi}$. Then the negative differential current will occur for $\mu_p e_{pi} > \mu_n e_{ni}$ and $\sigma_{pi} v_p / \sigma_{ni} v_n < e_{ni}/e_{pi}$, or vice versa. This phenomenon has been observed (Hurtes et al., 1978b; Deveaud and Toulouse, 1980) and the data can be fitted with Eq. (83), although perhaps not uniquely (Hurtes et al., 1978b).

Before closing our discussion of transient phenomena it should be remarked that the emission transients are sometimes nonexponential, due to electric field effects (Makram-Ebeid, 1980), nonuniform doping, and other causes. When this problem exists, the standard boxcar or lock-in techniques (Fig. 9c) will give spurious results (White et al., 1979). Thus, the transients themselves should always be examined before any data are taken. Methods of dealing with nonexponential behavior are discussed elsewhere (Kirchner et al., 1981).

To summarize this section, it appears that the techniques involving emission transients are very powerful, and can yield much useful information. However, in analyzing the data, it is necessary to be aware of the approximations involved in the various theoretical formulations. Fortunately, much progress is being made in the understanding of some of these techniques and formulations, making them even more useful.

V. Comparison of Techniques

The advantages and disadvantages of most of the techniques dealt with in this report have already been discussed in their respective sections. However, it is perhaps useful to compare their relative utilities in determining certain useful quantities, such as, e.g., impurity concentrations, or ionization energies. The discussion in this section will be brief and mostly qualitative, and it must be remembered that some of the conclusions drawn may be more subjective than hard and firm.

11. CRYSTAL QUALITY DETERMINATION

It is useful to have a means of determining overall electrical quality, i.e., the ability of the sample to conduct current in the intended manner. In general, electrical performance is limited by two problems: (1) unwanted impurities or defects, and (2) inhomogeniety. Both of these problems can affect the carrier mobility, which thus becomes an important parameter to measure. In bulk, SI GaAs it is not too uncommon to measure room temperature electron mobilities of 1500 cm^2/V s or less (cf., for example, sample MA 287/80 of Table II). One must first be sure that this low a value is not simply due to mixed conductivity, a phenomenon discussed in Sections 2 and 18. If $\rho \lesssim 4 \times 10^8 \, \Omega$ cm, then mixed conductivity is probably not important (Look, 1980). Furthermore, impurity concentrations are seldom high enough to reduce the mobility this much, so that the most likely explanation is inhomogeniety (Look, 1982c). This problem is not too difficult to understand in samples for which the deep compensating level is either nearly filled or nearly empty, for then slight variations in shallow donor or acceptor concentrations can produce n- or p-type regions, interspersed with the SI regions.

12. Activation Energy Measurements

The activation energy, i.e., the energy necessary to excite a carrier to its respective band, is one of the most important determinants of the effect that a particular impurity or defect will have on electrical properties. Shallow centers make possible high conductivity, and junctions. Deep centers can produce SI material, but can also lead to carrier traps, limiting device speed and increasing noise. We have discussed several ways of measuring activation energies in SI crystals: (1) temperature-dependent Hall-effect (TDH) measurements, (2) thermally stimulated current (TSC) spectroscopy, (3) photoinduced current transient spectroscopy (PITS or OTCS), and (4) photoconductivity (PC) measurements. In comparing energy levels determined by these techniques two precautions must always be kept in mind. (1) The first three (TDH, TSC, and PITS) involve Arrhenius plots, and thus, if the energy varies linearly with temperature, it is effectively a value at $T=0$. Photoconductivity, on the other hand, gives an energy at the temperature of measurement. (2) The TDH method determines the energy with respect to a known band (conduction or valence), whereas the other techniques, measuring only total current, normally do not. Another advantage of TDH is that either shallow or deep levels can be studied while PITS and TSC are usually limited to levels deeper than about 0.1 eV. Photoconductivity is also difficult to measure for energies less than about 0.3 eV because monochromators with sufficient intensity in this range are scarce. (However, sometimes the complementary energy, $E_G - E_i$, can be seen.) A disadvantage of TDH is that often only one dominant level, near E_F, is amenable to treatment, whereas the other methods may pick out several levels.

Another important factor is the speed of measurements. Here TDH usually comes out ahead of TSC and PITS, because the latter methods involve a separate temperature *scan* for each data point in the Arrhenius plot, since e_{ni} is determined at the spectral peak. However, it is also possible, with more sophisticated equipment, to actually measure e_{ni} from the current transient itself, at each temperature (Kirchner *et al.*, 1981). Then only one temperature scan is necessary for the Arrhenius plot of e_{ni}. However, in this case it is necessary to know the precise position of the transient base line.

Finally, the reliability of the various activation energy measurements should be judged. The TDH measurement is unambiguous as long as the Hall factor [Eq. (A17)] is either close to unity or else is not very temperature dependent (or both), and if mixed conductivity effects are either small or can be taken into account. Usually, neither of these problems is very important as far as a major change in the slope of the Arrhenius plot is concerned. The emission experiments, on the other hand, lead to an "apparent" activation energy of $E_{i0} + E_{\sigma i}$, where $E_{\sigma i}$ is given by the relationship $\sigma_{ni} = \sigma_{ni}(\infty)\exp(-E_{\sigma i}/kT)$. Thus, it is necessary to determine $E_{\sigma i}$, from capture experiments, before E_{i0} can be accurately determined from the emission experiments.

In Table IV we present $E_{\sigma i}$ and E_{i0} data on two important deep centers in GaAs, Cr, and "O" (EL2). The results from three different laboratories are included, but no attempt was made to show everything available in the literature. It is clear that neither the $E_{\sigma i}$ results nor the E_{i0} results agree well for Cr, but are not too bad for O. In contrast, the TDH measurements of E_{i0}, shown in Table II, are much more consistent. It should be noted that the TDH samples (Table II) were semi-insulating, whereas the emission-spectroscopy samples (Table IV) were conducting in order that capacitance transient (DLTS) experiments could be performed. The PITS and OTCS techniques applied to these samples would have been unable to clearly distinguish between hole and electron traps.

Thus, it must be concluded that TDH measurements lead to more reliable values of activation energy than DLTS. This statement is also true with respect to PITS and OTCS data, as seen in Table III. However, it can be expected that emission-spectroscopic inconsistencies will be gradually eliminated as more work is carried out.

We should also briefly discuss the reliability of PC threshold measurements for determining E_i. The problem here is that for each temperature the spectroscopic data ($\Delta\sigma$ versus $h\nu$) must in general be fitted to a theoretical expression, such as that found in Eq. (45). Above $T = 0$, when phonons are involved, there is no well-accepted theory of the photon capture cross section to use for data fitting. In fact, even at $T = 0$, the theoretical situation is ambiguous because the exact form of the impurity potential is not known. Thus, unless the PC threshold data are very steep, the energies derived from them may be hard to determine with great accuracy.

13. IMPURITY CONCENTRATION MEASUREMENTS

Because of the very important role of impurities in determining semiconductor properties, it is desirable to know their concentrations, at least of the electrically active ones. Of course, the techniques we have discussed in this chapter never make a positive identification of a particular impurity without confirmation by one of the established analytical techniques, such as spark-source mass spectroscopy (SSMS) or secondary-ion mass spectroscopy (SIMS). Once such confirmation is established, however, then a particular technique can be considered as somewhat of a "secondary standard" for analysis of the impurity that has been confirmed. It must be remembered here that an analytical method such as SSMS will see the *total* amount of the impurity in question, no matter what the form in the lattice, whereas an electrical technique will see only that fraction that is electrically active.

The TDH method, discussed in Section 5, can give an accurate value for the concentration of the dominant donor or acceptor near the Fermi level.

TABLE IV

DLTS Activation Energy and Capture Cross Section Data for Deep Cr- and "O"-Related Centers in GaAs.[a]

Label	Carrier	E_{ni} (eV)	E_{pi} (eV)	σ_{ni} (cm^2)	σ_{pi} (cm^2)	Ref.[b]
Cr-related center						
	e	0.61		$9 \times 10^{-16} \exp(-0.277/kT)$ (340–435°K)		a
HL1	e			$1 \times 10^{-17} \exp(-0.115/kT)$ (230–390°K)		b
EM1	e			$1-5 \times 10^{-20}$ (50–450°K)		c
	h		0.78		$5 \times 10^{-19} \exp(0.125kT)$ (370—410°K)	a
HL1	h		0.89		5×10^{-17} (150–370°K)	b
EM1	h		0.8		3×10^{-17} (temp. indep.—range not given)	c
O-related center						
	e	0.75		$3 \times 10^{-15} \exp(-0.075/kT)$ (150–245°K)		a
EL2	e	0.82		$6 \times 10^{-15} \exp(-0.066/kT)$ (50–275°K)		b
EL2	h				2×10^{-18} (320°K)	b

[a] The ionization energies E_{ni} and E_{pi} are with respect to the conduction band, and valence band, respectively.
[b] (a) Lang and Logan (1975); (b) Mitonneau et al. (1979); (c) Jesper et al. (1980).

An example was shown in Fig. 3. Here, besides the value of the dominant donor concentration, it was also possible to determine $N_{AS} - N_{DS}$, where N_{AS} is the concentration of *all* acceptors below E_F, and N_{DS}, the concentration of all donors above E_F. An analysis of the mobility can help separate N_{AS} and N_{DS}. For SI GaAs, however, the TDH data are essentially always in a "low temperature" regime, at all reasonable laboratory temperatures, and, in this case, we can at best determine $N_D/(N_{AS} - N_{DS})$, according to Eq. (23). Thus, the TDH method, by itself, is not good for determining impurity concentrations in SI material, although it can be somewhat more useful in conjunction with a mobility analysis.

The TSC signal strength will be proportional to the appropriate trap density, as seen in Eq. (55). Unfortunately, the lifetime τ_n (or τ_p for hole emission) also enters the equation, and this quantity will depend on other traps and recombination centers in the sample. If τ_n could be separately determined, then the TSC method could be "calibrated" for trap density. Essentially the same considerations hold for the transient-current methods (PITS and OTCS), as seen in Eq. (65). A further complication enters if e_{ni} and e_{pi} are of comparable magnitude, and e_{ni}/e_{pi} is unknown [cf. Eq. (83)].

The only transient-spectroscopic techniques capable of determining concentration are those involving capacitance, such as the original DLTS. Even for DLTS this measurement is not always possible, and, in any case, it is not feasible to use the capacitive techniques with SI material. Thus, none of the various methods discussed here are, by themselves, very useful for determining impurity or defect concentrations in SI GaAs. That is, they are more qualitative than quantitative.

A nonelectronic method of measuring impurity concentrations is that of absorption spectroscopy. From Eq. (36a) it is seen that $\alpha_{ni} = \sigma_{vni} n_{i0}$, where α_{ni} is the absorption constant due to electronic transitions from level i to the conduction band. The *total* impurity concentration N_i can be related to n_{i0} by a knowledge of E_F. The photon-capture cross section σ_{vni} can be determined for once and for all by doping experiments or by independently measuring N_i in some sample. This process has been carried out for Cr impurity (Martin, 1979) as well as O (EL2) (Martin, 1981) in GaAs. The same considerations hold for photoconductivity measurements, except that τ_n also needs to be known, as seen from Eq. (35).

14. LIFETIME MEASUREMENTS

By "lifetime" we mean the average time that an excess carrier exists before annihilation by a carrier of the opposite sign. This is as opposed to "relaxation" time, the average time between collisions, or "trapping" time, the average time in a band before being trapped. We have used the same symbol, τ, to represent both the lifetime and the relaxation time because this symbol is used for each of these parameters in the bulk of the literature. For the most part, the different meanings do not appear in the same section.

In our discussion of emission experiments we have assumed that $\tau_n(e_{ni} + e_{pi}) \ll 1$ so that we could ignore transients involving τ_n as being over before the emission-related transients had progressed significantly. This is usually a very good assumption in GaAs since $\tau_n \simeq 10^{-8}$ s, typically, and we can set our boxcar or lock-in rate window to pick up e_{ni}^{-1} of anywhere from 10^{-5} to 1 s. If, indeed, we would like to measure this initial fast transient, a very good electronic detection system is necessary. With present-day technology, it is not possible, of course, to use the lock-in method, and the pulse method requires nanosecond or shorter pulses and an equally fast detection system. It is no problem to get nanosecond laser pulses, but commercial transient recorders are limited to about a 2-ns integration time in order to get 8-bit resolution (1 part in 256). Such recorders may be good enough in many cases, but are also somewhat expensive. Thus, direct measurement lifetime transients may often be possible, but is rarely convenient.

There are several indirect methods of measuring τ_n or τ_p. One is the photomagnetoelectric (PME) technique, which is discussed thoroughly elsewhere (Holeman and Hilsum, 1961; Look, 1977b). The PME method, in conjunction with the PC method, gives an unambiguous value for τ_n, *as long as* $\tau_n = \tau_p$ (i.e., if there is no significant trapping). If $\tau_n \neq \tau_p$, then each lifetime can sometimes be measured separately, but the experiment is more complex (Li and Huang, 1972). A big advantage of the PME method is that it is trivial to implement if PC is already being performed and if a magnet is available. Furthermore, short lifetimes (say 10^{-10} s) are no problem.

Other indirect methods for measuring lifetimes often involve device structures such as $p-n$ junctions. The electron-beam-induced current (EBIC) technique, for example, measures the increase in junction current as an impinging electron beam moves close to the junction, i.e., within a few minority-carrier diffusion lengths. If a diffusion constant can be estimated, say by knowledge of the minority-carrier mobility, then the minority-carrier lifetime can be calculated. However, SI GaAs does not form good junctions, so such methods are really not applicable.

There is one other means of determining lifetime, available if both photo-Hall (PH) and absorption experiments can be carried out. This possibility is simply illustrated by Eq. (35). Here the PH measurement gives Δn, the absorption measurement gives α_n, and I_0 can be easily measured with a calibrated light detector. An obvious caveat here, of course, is that we must assume that $\alpha = \alpha_n$, i.e., that all of the light absorption is due to electronic transitions. For above-band-gap light this assumption will almost certainly be true. It was seen before that absorption measurements can be useful in determining impurity concentrations. Thus, a combination of PH and absorption data may yield both N_i and τ_n. If the carrier mobility can be estimated, then PC data may be sufficient in place of the PH data. It goes without saying that all results from these experiments must be analyzed carefully to make sure that any simple interpretations are indeed valid.

15. IMPURITY AND DEFECT IDENTIFICATIONS

The identification of an impurity, defect, or impurity-defect complex by some particular technique must nearly always be accomplished in conjunction with doping experiments. Thus, the well-known, sharp, zero-phonon photoluminescence lines at 0.84 eV in GaAs are almost certainly associated with Cr, as established by Cr-doping experiments (Koschel *et al.*, 1976). However, some care must be taken here. For example, a dominant electron trap (*EL*2) in O-doped GaAs is probably not associated with O, according to recent experiments (Huber *et al.*, 1979). Thus, the doping must be accompanied by a positive identification of the relevant impurity concentration, say by SSMS, or SIMS. These general considerations apply to all the techniques discussed below.

Each of the methods considered in this report has a certain impurity "fingerprinting" value. By this we mean that one or more distinctive parameters determined by the method will have the same value every time a particular impurity or defect is in the sample. For TDH measurements the distinctive parameter is essentially E_{i0}, the activation energy at $T = 0$. For PC and absorption measurement it is E_i, the value at the temperature of measurement, and for the TSC, PITS, and OTCS methods it is basically the temperature of maximum peak height (for a given rate window). The latter three methods also yield the preexponential factor in the emission expression [cf. Eq. (30)], and this factor is often useful in distinguishing between traps of roughly equal activation energies.

In comparing the impurity-identification capabilities of these techniques, one important fact stands out: the TDH, PC, and absorption methods will normally give information on only one, or at best a few, impurities in the same sample, whereas the TSC, PITS, and OTCS methods often see many. Thus, the latter three techniques are more useful for general survey studies. Unfortunately, however, only a few impurities, and no defects, can be positively identified in GaAs by any of these methods at the present time. This problem must continue to be addressed.

Appendix A. Charge Transport Theory

16. GENERAL CURRENT EQUATIONS

Instead of the usual Boltzmann transport equation approach, we will present a more phenomenological method of attaining the desired results (Bube, 1974). The hope here is that the physics will not so easily be obscured in the mathematics. We begin with the Lorentz force on a charged particle moving in electric and magnetic fields:

$$\mathbf{F} = m^* \frac{d\mathbf{v}}{dt} = q\left(\mathbf{E} + \frac{\mathbf{v} \times \mathbf{B}}{c}\right), \tag{A1}$$

where the use of a position and energy-independent m^* implies not only the validity of effective-mass theory, but also isotropic, homogeneous material and spherical equal-energy surfaces. The other symbols in Eq. (A1) have their usual meanings. Gaussian units are employed in this section.

We will first consider only electrons ($q = -e$) and later generalize to include holes. By letting the external magnetic field $\mathbf{B}$ define the z axis of our coordinate system, i.e., $\mathbf{B} = B\hat{z}$, we can write the components of Eq. (A1) as

$$\frac{dv_{nx}}{dt} = -\frac{e}{m_n^*} E_x - \omega_{cn} v_{ny}, \tag{A2a}$$

$$\frac{dv_{ny}}{dt} = -\frac{e}{m_n^*} E_y + \omega_{cn} v_{nx}, \tag{A2b}$$

where $\omega_{cn} \equiv eB/m_n^* c$, the cyclotron resonance frequency. There is no z component of v in Eq. (A2) because there are no forces on the electron in this direction. Note that we have defined e and ω_c as positive quantities.

To solve the above equations we can take the derivative of Eq. (A2a), and then substitute Eq. (A2b), getting

$$\left(\frac{d^2}{dt^2} + \omega_{cn}^2\right) v_{nx} = \frac{e\omega_{cn}}{m_n^*} E_y. \tag{A3}$$

The general solution to Eq. (A3) is (Ford, 1955)

$$v_{nx} = Ae^{i\omega_{cn}t} + Be^{-i\omega_{cn}t} + \frac{e}{m_n^* \omega_{cn}} E_y, \tag{A4}$$

where A and B are complex constants. Now the condition of reality requires that $B = A^*$. By expanding the complex exponentials in terms of sines and cosines, and letting $A = C/2 - iD/2$, we get

$$v_{nx} = C \cos \omega_{cn} t + D \sin \omega_{cn} t + \frac{e}{m_n^* \omega_{cn}} E_y, \tag{A5}$$

where C and D are now real. A similar procedure leads a solution for v_{ny}:

$$v_{ny} = E \cos \omega_{cn} t + F \sin \omega_{cn} t - \frac{e}{m_n^* \omega_{cn}} E_x, \tag{A6}$$

where E and F are real constants. To relate C, D, E, and F, we can substitute

Eqs. (A5) and (A6) into Eq. (A2a):

$$-C\omega_{cn}\sin\omega_{cn}t + D\cos\omega_{cn}t = -\frac{e}{m_n^*}E_x - \omega_{cn}$$
$$\times \left(E\cos\omega_{cn}t + F\sin\omega_{cn}t - \frac{e}{m_n^*\omega_{cn}}E_x\right).$$

By matching coefficients we see that $E = -D$ and $F = C$. Therefore

$$v_{nx} = C\cos\omega_{cn}t + D\sin\omega_{cn}t + \frac{e}{m_n^*\omega_{cn}}E_y, \quad (A7a)$$

$$v_{ny} = -D\cos\omega_{cn}t + C\sin\omega_{cn}t - \frac{e}{m_n^*\omega_{cn}}E_x. \quad (A7b)$$

At $t = 0$ these expressions give

$$v_{nx0} = C + \frac{e}{m_n^*\omega_{cn}}E_y, \quad (A8a)$$

$$v_{ny0} = -D - \frac{e}{m_n^*\omega_{cn}}E_x. \quad (A8b)$$

The final solution can then be written

$$v_{nx} = \left(v_{nx0} - \frac{e}{m_n^*\omega_{cn}}E_y\right)\cos\omega_{cn}t$$
$$- \left(v_{ny0} + \frac{e}{m_n^*\omega_{cn}}E_x\right)\sin\omega_{cn}t + \frac{e}{m_n^*\omega_{cn}}E_y, \quad (A9a)$$

$$v_{ny} = \left(v_{ny0} + \frac{e}{m_n^*\omega_{cn}}E_x\right)\cos\omega_{cn}t$$
$$+ \left(v_{nx0} - \frac{e}{m_n^*\omega_{cn}}E_y\right)\sin\omega_{cn} - \frac{e}{m_n^*\omega_{cn}}E_x. \quad (A9b)$$

Equations (A9) show that the electron exhibits the expected cyclotron motion in the presence of the magnetic field. However, collisions must also be taken into account. Let $N(t)$ be the number of particles that have *not* experienced a collision for time t (after some arbitrary beginning time, $t = 0$). Then it is reasonable to assume that the rate of decrease of $N(t)$ will be given by $dN \propto -N\,dt = -N\,dt/\tau$. The solution of this equation is $N(t) = N_0 \exp(-t/\tau)$, where N_0 is the total number of particles. It can easily be shown that τ is simply the mean time between collisions. The probability of having not experienced a collision in time t is, of course, $N(t)/N_0 = \exp(-t/\tau)$. The average velocity components of an electron can now be written as

$$\bar{v}_{nx} = \frac{\int_0^{N_0} v_{nx} \, dN(v_{nx})}{\int_0^{N_0} dN(v_{nx})} = \frac{\int_\infty^0 v_{nx}\left(-\frac{1}{\tau_n}\right) N_0 e^{-t/\tau_n} dt}{N_0} = \frac{1}{\tau_n}\int_0^\infty v_{nx} e^{-t/\tau_n} dt, \quad \text{(A10)}$$

with a similar equation for v_{ny}. Before solving Eq. (A10) it should be noted that we really ought to average over the initial velocities, v_{nx0} and v_{ny0}, also. However, it is easy to see that this process will cause the terms involving v_{nx0} and v_{ny0} in Eq. (A9) to vanish, while not affecting the other terms. The reason is that positive and negative velocities must be equal in number, and thus cancel, as long as the system has been free from external forces for a period of at least several τ_n immediately preceeding $t = 0$.

We can now solve for the components of current density by using Eqs. (A9) and (A10):

$$j_{nx} \equiv -ne\bar{v}_{nx} = \frac{ne^2}{m_n^*}\left(\frac{\tau_n}{1 + \omega_{cn}^2\tau_n^2} E_x - \frac{\omega_{cn}\tau_n^2}{1 + \omega_{cn}^2\tau_n^2} E_y\right), \quad \text{(A11a)}$$

$$j_{ny} \equiv -ne\bar{v}_{ny} = \frac{ne^2}{m_n^*}\left(\frac{\omega_{cn}\tau_n^2}{1 + \omega_{cn}^2\tau_n^2} E_x + \frac{\tau_n}{1 + \omega_{cn}^2\tau_n^2} E_y\right). \quad \text{(A11b)}$$

Similar expressions can be generated for holes simply by letting $\omega_c \to -\omega_c$. The use of a simple "relaxation" time τ_n needs justification, which will not be attempted here. Suffice it to say that this assumption is not bad for elastic scattering processes, which include most of the important mechanisms. A well-known exception is polar optical-phonon scattering, at temperatures below the Debye temperature (Putley, 1968, p. 138). We have further assumed here that τ_n is independent of energy, although this condition will be relaxed later.

17. Single-Carrier Hall Effect, Conductivity

a. Single-Carrier, Energy-Independent τ

With an electric field applied in the x direction and a magnetic field in the z direction our boundary condition is $j_{ny} = 0$. Then, from Eq. (A11b), $E_y = -\omega_{cn}\tau_n E_x$, and from Eq. (A11a),

$$j_{nx} = \frac{ne^2}{m_n^*}\left(\frac{\tau_n E_x}{1 + \omega_{cn}^2\tau_n^2} + \frac{\omega_{cn}^2\tau_n^3 E_x}{1 + \omega_{cn}^2\tau_n^2}\right)$$

$$= \frac{ne^2\tau_n}{m_n^*} E_x \equiv ne\mu_n E_x \equiv \sigma_n E_x. \quad \text{(A12)}$$

Thus, we have the well-known result that the magnetic field has no effect on the resistance (no magnetoresistance) if $\tau \neq \tau(\varepsilon)$. (Note that our equations were derived for spherical equal-energy surfaces.) This is because the Hall field E_y exactly balances the Lorentz force for *all* of the carriers, thereby keeping their trajectories straight across the sample (except for the helical motion, of course). Note in Eq. (A12) that we have also defined $\mu_n = e\tau_n/m_n^*$ and $\sigma_n = ne\mu_n$, following standard practice.

From Eq. (A12) and the relationship $E_y = -\omega_{cn}\tau_n E_x$ we can derive the Hall coefficient:

$$R_n \equiv \frac{E_y}{j_{nx}(B/c)} = -\frac{1}{ne}, \qquad (A13)$$

where, again, e is a positive quantity. We thus see that even in relatively high magnetic fields, where $\omega_{cn}\tau_n \gg 1$, the Hall coefficient and resistance are unchanged from their low field values ($\omega_{cn}\tau_n \ll 1$).

It is perhaps instructive at this point to consider units and orders of magnitude. Consider the mobility $\mu_n = e\tau_n/m_n^*$ cm^2/statvolt s, where $e = 4.8 \times 10^{-10}$ esu, and m_n^* is expressed in grams. The practical unit for mobility is cm^2/V s, and the conversion is therefore μ_n (cm^2/V s) = $(300)^{-1}\mu_n$(cm^2/statvolt s). For SI GaAs, a typical value of electron mobility is 4000 cm^2/V s at 300°K. Also, $m_n^* \simeq 0.067 m_0$, so the relaxation time is $\tau_n \simeq 1.5 \times 10^{-13}$ s. For a magnetic field strength of 5000G, we can calculate $\omega_{cn} = eB/m_n^*c \simeq 1.3 \times 10^{12}$ s^{-1}. The product $\omega_{cn}\tau_n \simeq 0.2$ in this case. We must begin to worry about the validity of our theory when $\omega_{cn}\tau_n \simeq 1$ [or μ_n (cm^2/V s) $\times B$ (G) $\simeq 10^8$], because then the energy levels of the system may be significantly perturbed, requiring a full quantum-mechanical treatment. However, as long as $kT \gg \hbar\omega_{cn}$ (or $T \gg 10°$K for our example above) the present classical treatment is still valid.

To continue with our example we will assume a value for $n = 10^7$ cm^{-3}, again typical for SI GaAs. Then $R_n = (ne)^{-1} = 6.2 \times 10^{11}$ cm^3/C and the resistivity $\rho_n \equiv \sigma_n^{-1} = (ne\mu_n)^{-1} = 1.6 \times 10^8 \,\Omega$ cm. For an applied electric field of 1 V/cm, the average velocity parallel to the field (drift velocity) is $v_D = \mu_n E_x \simeq 4 \times 10^3$ cm s^{-1}, compared with a mean thermal velocity of $v_{th} = \sqrt{8kT/\pi m_n^*} \simeq 4 \times 10^7$ cm s^{-1}. We can see that for high enough electric fields ($E_x \simeq 10^4$ V/cm, in this case) $v_D \to v_{th}$, and our theory will again break down, because τ_n (and therefore μ_n) must then become dependent upon E_x.

b. *Energy-Dependent τ*

For most scattering processes the relaxation times will depend on particle energy. Thus, we must average Eqs. (A11a) and (A11b) over energy. To simplify the notation we will write

$$j_{nx} = \sigma_{nxx} E_x + \sigma_{nxy} E_y, \qquad (A14a)$$

$$j_{ny} = \sigma_{nyx} E_x + \sigma_{nyy} E_y, \qquad (A14b)$$

where

$$\sigma_{nxx} = \sigma_{nyy} = \frac{ne^2}{m_n^*}\left\langle\frac{\tau_n}{1+\omega_{cn}^2\tau_n^2}\right\rangle, \tag{A15a}$$

$$\sigma_{nxy} = -\sigma_{nyx} = -\frac{ne^2}{m_n^*}\left\langle\frac{\omega_{cn}\tau_n^2}{1+\omega_{cn}^2\tau_n^2}\right\rangle. \tag{A15b}$$

Here the angle brackets denote an average over energy. The reader is again reminded that we are dealing with isotropic solids. The condition $j_{ny} = 0$ gives $E_y = -\sigma_{nyx}E_x/\sigma_{nyy}$, and the insertion of this result into Eq. (A14a) yields, for the Hall coefficient,

$$R_n \equiv \frac{E_y}{j_{nx}(B/c)} = -\frac{1}{(B/c)}\frac{\sigma_{nyx}}{\sigma_{nxx}\sigma_{nyy}-\sigma_{nxy}\sigma_{nyx}} \equiv -\frac{r_n}{ne}, \tag{A16}$$

where

$$r_n = \left\langle\frac{\tau_n^2}{1+\omega_{cn}^2\tau_n^2}\right\rangle\bigg/\left(\left\langle\frac{\tau_n}{1+\omega_{cn}^2\tau_n^2}\right\rangle^2 + \omega_{cn}^2\left\langle\frac{\tau_n^2}{1+\omega_{cn}^2\tau_n^2}\right\rangle^2\right). \tag{A17}$$

This is the well-known "Hall factor" or "r factor," which, in the low magnetic field limit, makes the "Hall mobility" different from the "conductivity mobility." To see this relationship, consider the limit $\omega_{cn} \to 0$ (or $B \to 0$). Then $r_{n0} = \langle\tau_n^2\rangle/\langle\tau_n\rangle^2$. The Hall mobility is defined as $|R_n\sigma_n| = (r_n/ne)(ne\mu_n) = r_n\mu_n$, where $\mu_n = e\langle\tau_n\rangle/m_n^*$ is the conductivity mobility. At high magnetic fields, however, we see that $r_n = 1$ (for $\omega_{cn}\tau_n \gg 1$), and thus $|R_n\sigma_n| = \mu_n$.

The average of some quantity Q over energy will be given by

$$\langle Q\rangle = \int_0^\infty Qf(\varepsilon)\phi(\varepsilon)\,d\varepsilon \bigg/ \int_0^\infty f(\varepsilon)\phi(\varepsilon)\,d\varepsilon, \tag{A18}$$

where $f(\varepsilon)$ is the energy distribution function and $\phi(\varepsilon)$ is the density of states. [The denominator of Eq. (A18) is simply n.] Now the presence of the electric field will shift $f(\varepsilon)$ slightly from the equilibrium Fermi function since the electrons will be moving preferentially in the $-E_x$ direction (with an average velocity $\bar{v}_{nx}$ calculated earlier):

$$f(\varepsilon) = f_0(\varepsilon) + \bar{v}_{nx}\frac{\partial f_0(\varepsilon)}{\partial v_{nx}} + \cdots. \tag{A19}$$

We can write

$$\frac{\partial f_0}{\partial v_{nx}} = \frac{\partial f_0}{\partial\varepsilon}\frac{\partial\varepsilon}{\partial v_{nx}} \simeq 3m_n^*v_{nx}\frac{\partial f_0}{\partial\varepsilon} \tag{A20}$$

since $\varepsilon = \frac{1}{2}m_n^*(v_{nx}^2 + v_{ny}^2 + v_{nz}^2) \simeq \frac{3}{2}m_n^*v_{nx}^2$ (note that the thermal velocity $v_{nx} \gg \bar{v}_{nx}$, the drift velocity). Then

$$\langle j_{nx} \rangle = -e \int_0^\infty v_{nx}\left(f_0 + 3m_n^*\bar{v}_{nx}v_{nx}\frac{\partial f_0}{\partial \varepsilon}\right)\phi_n \, d\varepsilon$$

$$= -e \int_0^\infty \left(v_{nx}f_0\phi_n + 2\varepsilon\bar{v}_{nx}\frac{\partial f_0}{\partial \varepsilon}\phi_n\right) d\varepsilon. \quad (A21)$$

The first term will vanish since negative velocities will exactly cancel positive velocities under the equilibrium distribution, f_0. [This fact would have been more obvious if we had started with a $dv_{nx}dv_{ny}dv_{nz}$ integration in Eq. (A18), instead of a $d\varepsilon$ integration.] For a Boltzmann distribution, $\partial f_0/\partial \varepsilon = -(kT)^{-1}\exp(\varepsilon_F - \varepsilon)/kT$, and we will also assume $\phi_n(\varepsilon) \propto \varepsilon^{1/2}$; then

$$\langle j_{nx} \rangle \propto \int_0^\infty \bar{v}_{nx}\varepsilon^{3/2}e^{-\varepsilon/kT} \, d\varepsilon, \quad (A22)$$

which shows the proper function of energy necessary for averaging the quantities in Eq. (A17).

Finally, in order to calculate r_n we must assume a particular energy dependence for τ_n. Often τ_n can be written (Putley, 1968, p.72) in the form $\tau_n = a\varepsilon^{-s}$, where a and s are constants. Then

$$\langle \tau_n \rangle = \frac{\int_0^\infty \tau_n \varepsilon^{3/2}e^{-\varepsilon/kT} \, d\varepsilon}{\int_0^\infty \varepsilon^{3/2}e^{-\varepsilon/kT} \, d\varepsilon} = a\frac{\int_0^\infty \varepsilon^{(3/2)-s}e^{-\varepsilon/kT} \, d\varepsilon}{\int_0^\infty \varepsilon^{3/2}e^{-\varepsilon/kT} \, d\varepsilon}$$

$$= a(kT)^{-s}\frac{\Gamma(\frac{5}{2} - s)}{\Gamma(\frac{5}{2})}, \quad (A23)$$

where $\Gamma(x)$ is the so-called Γ function. Similarly,

$$\langle \tau^2 \rangle = a^2\frac{\int_0^\infty \varepsilon^{(3/2)-2s}e^{-\varepsilon/kT} \, d\varepsilon}{\int_0^\infty \varepsilon^{3/2}e^{-\varepsilon/kT} \, d\varepsilon} = a^2(kT)^{-2s}\frac{\Gamma(\frac{5}{2} - 2s)}{\Gamma(\frac{5}{2})}. \quad (A24)$$

Then

$$r_{n0} \equiv \frac{\langle \tau^2 \rangle}{\langle \tau \rangle^2} = \frac{\Gamma(\frac{5}{2})\Gamma(\frac{5}{2} - 2s)}{[\Gamma(\frac{5}{2} - s)]^2}. \quad (A25)$$

For lattice acoustic-mode deformation potential scattering, $s = \frac{1}{2}$, giving $r_{n0} = 3\pi/8 = 1.18$. For ionized-impurity scattering, $s = -\frac{3}{2}$, giving $r_{n0} = 315\pi/512 = 1.93$. For a mixture of independent scattering processes we must calculate τ from $\tau^{-1} = \tau_1^{-1} + \tau_2^{-1} + \cdots$. The integrals in Eqs. (A23) and (A24) must now, in general, be performed numerically, although a rough approximation, known as Matthiessen's rule (Bube, 1974, pp. 232 and 238), can sometimes be used: $\mu^{-1} = \mu_1^{-1} + \mu_2^{-1} + \cdots$.

Consider a mixture of acoustic-mode (τ_L) and ionized-impurity (τ_I) scattering. For $\tau_L \ll \tau_I$ we would expect $r_{n0} = 1.18$ and for $\tau_I \ll \tau_L$, $r_{n0} = 1.93$. But for intermediate mixtures, r_{n0} goes through a minimum value, dropping to about 1.05 at 15% ionized-impurity scattering (Nam, 1980). For this special case ($s_L = \frac{1}{2}$, $s_I = -\frac{3}{2}$), the integrals can be evaluated in terms of tabulated functions (Bube, 1974). For optical-mode scattering the relaxation-time approach is not valid, at least below the Debye temperature, but r_n may still be obtained by such theoretical methods as a variational calculation (Ehrenreich, 1960; Nag, 1980) or an iterative solution of the Boltzmann equation (Rode, 1970), and typically varies between 1.0 and 1.4 as a function of temperature (Stillman et al., 1970; Debney and Jay, 1980).

As mentioned earlier, for high magnetic fields ($\omega_{cn}\tau_n \gg 1$), $r_n = 1$. In fact, a measurement of $R(B \to 0)/R(B \to \infty)$, when possible, is the best way (Stillman et al., 1970) to determine r_{n0}. It should be mentioned again that all of the equations developed in the preceeding paragraphs, which involve averages over energy, are predicated upon nondegenerate statistics, certainly valid for SI GaAs. For the other extreme, the case of highly degenerate electron distributions, $\tau_n = \tau_n(\varepsilon_F)$, a constant independent of energy, and thus $r_n = 1$ (for spherical equal-energy surfaces.)

It is apparent from our discussion so far that the Hall factor is sometimes difficult to calculate and even more difficult to measure. However, it rarely varies by more than 20% from a value of 1.2 (Stillman et al., 1970; Debney and Jay, 1980). Such a situation is entirely tolerable, especially when one realizes the range of semiconductor carrier concentrations (10^3–10^{20} cm^{-3} in our laboratory) that can be measured by this simple technique.

We now consider the magnetic field dependence of the Hall coefficient and conductivity in more detail. By setting $j_y = 0$ in Eq. (A14b), we can write Eq. (A14a), with subscript n suppressed, as

$$j_x = \left(\sigma_{xx} + \sigma_{xy}\frac{\sigma_{yx}}{\sigma_{yy}}\right)E_x$$

$$= \frac{ne^2 E_x}{m^*}\left(\left\langle\frac{\tau}{1+\omega_c^2\tau^2}\right\rangle + \left\langle\frac{\omega_c\tau^2}{1+\omega_c^2\tau^2}\right\rangle^2 \bigg/ \left\langle\frac{\tau}{1+\omega_c^2\tau^2}\right\rangle\right). \quad \text{(A26)}$$

To order $\omega_{cn}^2\tau_n^2$ the above equation can be written

$$j_{nx} = \frac{ne^2 E_x}{m_n^*}\left[\langle\tau_n\rangle - \omega_{cn}^2\left(\langle\tau_n^3\rangle - \frac{\langle\tau_n^2\rangle^2}{\langle\tau_n\rangle}\right)\right]. \quad \text{(A27)}$$

2. PROPERTIES OF SEMI-INSULATING GaAs

Then, from the definition $\sigma_n \equiv j_{nx}/E_x$, we get

$$\frac{\sigma_{n0} - \sigma_n}{\sigma_{n0}} = \frac{\Delta\sigma_n}{\sigma_{n0}} = \frac{\Delta\rho_n}{\rho_n} \simeq \frac{\Delta\rho_n}{\rho_{n0}} = \omega_{cn}^2 \left(\frac{\langle\tau_n^3\rangle\langle\tau_n\rangle - \langle\tau_n^2\rangle^2}{\langle\tau_n\rangle^2} \right)$$

$$= \omega_{cn}^2 \frac{\langle\tau_n^2\rangle^2}{\langle\tau_n\rangle^2} \left(\frac{\langle\tau_n^3\rangle\langle\tau_n\rangle}{\langle\tau_n^2\rangle^2} - 1 \right) = \xi_n R_{n0}^2 \sigma_{n0}^2 \frac{B^2}{c^2}, \qquad (A28)$$

where the relationships $\sigma_{n0} = ne^2 \langle\tau_n\rangle/m_n^*$ and $R_{n0} = -\langle\tau_n^2\rangle/\langle\tau_n\rangle^2 ne$ have been used. The magnetoresistance coefficient ξ_n is given by

$$\xi_n = \frac{\langle\tau_n^3\rangle\langle\tau_n\rangle}{\langle\tau_n^2\rangle^2} - 1 = \frac{\Gamma(\tfrac{5}{2} - 3s)\Gamma(\tfrac{5}{2} - s)}{[\Gamma(\tfrac{5}{2} - 2s)]^2} - 1. \qquad (A29)$$

The last relationship here assumes, of course, that $\tau_n = a\varepsilon^{-s}$. For acoustic-mode scattering, $s = \tfrac{1}{2}$ and $\xi = 4/\pi - 1 \simeq 0.273$. For ionized-impurity scattering, $s = -\tfrac{3}{2}$ and $\xi \simeq 0.577$.

It is interesting to calculate the magnetoresistance in high magnetic field, i.e., $\omega_{cn}\tau_n \gg 1$. Following the same procedure as above it is easy to show that $j_{nx} = ne^2 E_x/m_n^* \langle\tau_n^{-1}\rangle$, giving

$$\frac{\sigma_{n0}}{\sigma_{n\infty}} = \frac{\rho_{n\infty}}{\rho_{n0}} = \langle\tau_n\rangle\langle\tau_n^{-1}\rangle \equiv \gamma_n = \Gamma(\tfrac{5}{2} + s)\Gamma(\tfrac{5}{2} - s)/[\Gamma(\tfrac{5}{2})]^2. \qquad (A30)$$

Again, for $s = \tfrac{1}{2}$, $\gamma_n \simeq 1.13$, and for $s = -\tfrac{3}{2}$, $\gamma_n \simeq 3.4$. Equations (A28) and (A30) predict a decrease in σ_n to a saturation value, $\sigma_{n\infty}$, as B is increased. Although the initial decrease is nearly always observed, the saturation phenomenon is not (Putley, 1968, p. 97).

We next return to the Hall coefficient. It is evidently rather commonly believed that, to the approximation considered above ($\omega_{cn}^2 \tau_n^2$), the Hall coefficient is independent of B (Putley, 1968, p. 96). However, this statement is somewhat misleading. Suppose we rewrite Eq. (A16) as $R_n = -f(\sigma)/(B/c)$. Then, to examine B^2 terms in R_n we must expand $f(\sigma)$ to order B^3, because of the factor B in the denominator of R_n. This process can be shown to yield (Look, 1982b)

$$\frac{R_{n0} - R_n}{R_{n0}} = -\frac{\Delta R_n}{R_{n0}} = (\beta_n - 2\xi_n) R_{n0}^2 \sigma_{n0}^2 \frac{B^2}{c^2}, \qquad (A31)$$

where

$$\beta_n = \frac{\langle\tau_n^4\rangle\langle\tau_n\rangle^2}{\langle\tau_n^2\rangle^3} - 1. \qquad (A32)$$

Some typical values of β_n, ξ_n, r_n, and γ_n are given in Table A1. However, it is not known to what extent inhomogeneities influence the experimental results.

18. Two-Carrier Hall-Effect, Conductivity

We now include hole conduction along with the electron conduction. All formulas that we have derived so far for electrons go over immediately to holes simply by letting $e \to -e$ and $n \to p$. However, it is the simultaneous existence of hole and electron conductivities in the same sample that leads to interesting results. For example, it will be seen that the conductivity and Hall coefficient become dependent on B even when τ_n and τ_p are independent of energy.

We first write current equations for holes analogous to those for electrons [Eqs. (A11) and (A14)]:

$$j_{px} = \sigma_{pxx} E_x + \sigma_{pxy} E_y, \tag{A33a}$$

$$j_{py} = \sigma_{pyx} E_x + \sigma_{pyy} E_y, \tag{A33b}$$

where

$$\sigma_{pxx} = \sigma_{pyy} = \frac{pe^2}{m_p^*} \left\langle \frac{\tau_p}{1 + \omega_{cp}^2 \tau_p^2} \right\rangle, \tag{A34a}$$

$$\sigma_{pxy} = -\sigma_{pyx} = \frac{pe^2}{m_p^*} \left\langle \frac{\omega_{cp} \tau_p^2}{1 + \omega_{cp}^2 \tau_p^2} \right\rangle. \tag{A34b}$$

Here $\omega_{cp} \equiv eB/m_p^* c$, m_p^* is the hole effective mass, and τ_p, the hole relaxation time. The total current is simply the sum of the individual currents:

$$j_x = j_{nx} + j_{px} = (\sigma_{nxx} + \sigma_{pxx}) E_x + (\sigma_{nxy} + \sigma_{pxy}) E_y, \tag{A35a}$$

$$\begin{aligned} j_y &= j_{ny} + j_{py} = (\sigma_{nyx} + \sigma_{pyx}) E_x + (\sigma_{nyy} + \sigma_{pyy}) E_y \\ &= -(\sigma_{nxy} + \sigma_{pxy}) E_x + (\sigma_{nxx} + \sigma_{pxx}) E_y, \end{aligned} \tag{A35b}$$

where, in Eq. (A35b), we have explicitly made use of the symmetry relationships among the various σ_{ij}. Again, the boundary condition for our experiment is $j_y = 0$, giving

$$E_y = \frac{\sigma_{nxy} + \sigma_{pxy}}{\sigma_{nxx} + \sigma_{pxx}} E_x \tag{A36}$$

so that, from Eq. (A35a), and the definition of σ and R,

$$\sigma \equiv \frac{j_x}{E_x} = \frac{(\sigma_{nxx} + \sigma_{pxx})^2 + (\sigma_{nxy} + \sigma_{pxy})^2}{(\sigma_{nxx} + \sigma_{pxx})}. \tag{A37}$$

$$R \equiv \frac{E_y}{j_x(B/c)} = \frac{1}{(B/c)} \frac{(\sigma_{nxy} + \sigma_{pxy})}{(\sigma_{nxx} + \sigma_{pxx})^2 + (\sigma_{nxy} + \sigma_{pxy})^2}. \tag{A38}$$

a. Energy-Dependent τ, Low B

We first consider the case $B = 0$. Then Eqs. (A15) and (A34) yield $\sigma_{pxx} = pe^2 \langle \tau_p \rangle / m_p^* = ep\mu_p$, $\sigma_{pxy} = (pe^2/m_p^*)(eB/m_p^* c)\langle \tau_p^2 \rangle = ep\mu_p^2 r_p (B/c)$, $\sigma_{nxx} = en\mu_n$, and $\sigma_{nxy} = -en\mu_n^2 r_n(B/c)$, where $r_i \equiv \langle\tau_i^2\rangle/\langle\tau_i\rangle^2$, and where we have retained the (B/c) term in σ_{pxy} and σ_{nxy} because of the cancelling (B/c) term in the denominator of Eq. (A38). Equations (A37) and (A38) now become

$$\sigma_0 = e(n\mu_n + p\mu_p) = \sigma_n + \sigma_p, \tag{A39}$$

$$R_0 = \frac{r_p p\mu_p^2 - r_n n\mu_n^2}{e(n\mu_n + p\mu_p)^2} = \frac{\sigma_n^2 R_n + \sigma_p^2 R_p}{(\sigma_n + \sigma_p)^2}, \tag{A40}$$

where $\sigma_n = en\mu_n$, $\sigma_p = ep\mu_p$, $R_n = -r_n/ne$, and $R_p = r_p/pe$. From this point on, we will assume that all quantities with a subscript "n" or "p" are measured at $B = 0$; i.e., $r_n \to r_{n0}$, $\sigma_n \to \sigma_{n0}$, etc. However, the *total* quantities, σ and R, will explicitly be denoted by a subscript "0," if they are to be measured at $B = 0$.

Equation (A40) demonstrates the well-known vanishing of the Hall coefficient when $r_p p\mu_p^2 = r_n n\mu_n^2$. This effect has often been observed as temperature is raised, usually in a sample which is p-type at low temperature (Look, 1975). At this point we should mention the various definitions of p- and n-type.

(1) From a *particle density* point of view a sample is *n-type if $n > p$*.
(2) From a conductivity [Eq. (A39)] or thermopower [Eq. (A75)] point of view, n-type means $n\mu_n > p\mu_p$.
(3) Finally, from a *Hall coefficient* [Eq. (A40)] perspective, n-type means $n\mu_n^2 > p\mu_p^2$.

One must always be cognizant of these various definitions when reading the literature.

We next investigate the behavior of σ and R at finite, but low B. Terms to order $\mu^2 B^2$ may be obtained by expanding the σ_{nij} and σ_{pij} to order $\omega_c^2 \tau^2$ in Eq. (A37), and $\omega_c^3 \tau^3$ in Eq. (A38). [The higher order expansion in Eq. (A38) is necessary because of the B/c term in the denominator of that equation. This point has already been explained.] For example, σ_{pxx} becomes

$$\sigma_{pxx} = \frac{pe^2}{m_p^*}\left\langle\frac{\tau_p}{1+\omega_{cp}^2\tau_p^2}\right\rangle \simeq \frac{pe^2}{m_p^*}(\langle\tau_p\rangle - \omega_{cp}^2\langle\tau_p^3\rangle)$$

$$= \frac{pe^2\langle\tau_p\rangle}{m_p^*}\left[1 - \left(\frac{eB}{m_p^* c}\right)^2 \frac{\langle\tau_p^3\rangle}{\langle\tau_p\rangle}\right]$$

$$= \frac{pe^2\langle\tau_p\rangle}{m_p^*}\left[1 - \left(\frac{e}{m_p^*}\right)^2 \left(\frac{\langle\tau_p^2\rangle^2\langle\tau_p\rangle}{\langle\tau_p\rangle\langle\tau_p^2\rangle^2}\right)\frac{\langle\tau_p^3\rangle}{\langle\tau_p\rangle}\left(\frac{B}{c}\right)^2\right]$$

$$= \sigma_p[1 - R_p^2 \sigma_p^2 (\xi_p + 1)(B/c)^2], \tag{A41}$$

where, as previously defined, $\xi_p = \langle\tau_p^3\rangle\langle\tau_p\rangle/\langle\tau_p^2\rangle^2 - 1$. The other σ_{ij} can be similarly approximated, and, after much algebra (Look, 1982b), Eqs. (A37) and

(A38) yield

$$-\frac{\Delta\sigma}{\sigma_0} \simeq \frac{\Delta\rho}{\rho_0} = \left[\frac{\sigma_n\sigma_p(R_n\sigma_n - R_p\sigma_p)^2}{(\sigma_n + \sigma_p)^2} + \frac{\xi_n R_n^2\sigma_n^3 + \xi_p R_p^2\sigma_p^3}{(\sigma_n + \sigma_p)}\right]_0 \frac{B^2}{c^2}$$

$$= \left[\frac{c(\alpha b + 1)^2}{\alpha^2 b(1 + bc)^2} + \frac{\xi_n bc}{1 + bc} + \frac{\xi_p}{\alpha^2 b^2(1 + bc)}\right]_0 r_{n0}^2\mu_n^2 \frac{B^2}{c^2}, \quad (A42)$$

$$-\frac{\Delta R}{R_0} = \left[\frac{\sigma_n^2\sigma_p^2(R_n + R_p)(R_n\sigma_n - R_p\sigma_p)^2}{(\sigma_n + \sigma_p)^2(R_n\sigma_n^2 + R_p\sigma_p^2)} + \frac{\beta_n R_n^3\sigma_n^4 + \beta_p R_p^3\sigma_p^4}{R_n\sigma_n^2 + R_p\sigma_p^2}\right.$$

$$\left. - \frac{(2\xi_n R_n^2\sigma_n^3 + 2\xi_p R_p^2\sigma_p^3)}{(\sigma_n + \sigma_p)}\right]_0 \frac{B^2}{c^2}$$

$$= \left[\frac{c(1 - \alpha c)(1 + \alpha b)^2}{\alpha^2(1 + bc)^2(\alpha b^2 c - 1)} + \frac{\beta_n \alpha b^2 c}{(\alpha b^2 c - 1)} - \frac{\beta_p}{\alpha^2 b^2(\alpha b^2 c - 1)}\right.$$

$$\left. - \frac{2\xi_n bc}{(1 + bc)} - \frac{2\xi_p}{\alpha^2 b^2(1 + bc)}\right]_0 r_{n0}^2\mu_n^2 \frac{B}{c}, \quad (A43)$$

where $c \equiv n/p$ (except when appearing as B/c), $b \equiv \mu_n/\mu_p$, $\alpha \equiv r_n/r_p$, and the r_i's, ξ_i's, and β_i's were defined earlier. The subscript "0" on the large brackets denotes that all quantities inside are to be evaluated at $B = 0$. Equations (A42) and (A43) were recently derived by Betko and Merinsky (1979), and by Look (1980), except that in these works the β_i factors were included only phenomenologically. In the present work, the β_i's are given a precise meaning [Eq. (A32)], and thus can be calculated and compared with experiment.

Equations (A42) and (A43) are clearly split into mixed-carrier terms (first term in each expression), which vanish as $c \to 0$ or $c \to \infty$, and single-carrier terms (those involving the ξ_i's and β_i's), which vanish if τ_n and τ_p are independent of energy. The single-carrier terms of course approach their proper limits as $c \to 0$ or $c \to \infty$ [cf. Eqs. (A28) and (A31)]. Suppose we write Eqs. (A42) and (A43) in the form

$$\frac{c^2}{B^2} + A_\rho = (S_{\rho m} + S_{\rho s})\frac{\rho_0}{\Delta\rho}, \quad (A44)$$

$$\frac{c^2}{B^2} + A_R = -(S_{Rm} + S_{Rs})\frac{R_0}{\Delta R}, \quad (A45)$$

where

$$S_{\rho m} = \frac{\sigma_n\sigma_p(R_n\sigma_n - R_p\sigma_p)^2}{(\sigma_n + \sigma_p)^2}, \quad (A46)$$

$$S_{\rho s} = \frac{\xi_n R_n^2\sigma_n^3 + \xi_p R_p^2\sigma_p^3}{(\sigma_n + \sigma_p)}, \quad (A47)$$

and

$$S_{Rm} = \frac{\sigma_n^2 \sigma_p^2 (R_n + R_p)(R_n \sigma_n - R_p \sigma_p)^2}{(\sigma_n + \sigma_p)^2 (R_n \sigma_n^2 + R_p \sigma_p^2)} \tag{A48}$$

$$S_{Rs} = \frac{\beta_n R_n^3 \sigma_n^4 + \beta_p R_p^3 \sigma_p^4}{R_n \sigma_n^2 + R_p \sigma_p^2} - \frac{(2\xi_n R_n^2 \sigma_n^3 + 2\xi_p R_p^2 \sigma_p^3)}{(\sigma_n + \sigma_p)}. \tag{A49}$$

Here the subscripts "m" and "s" denote "mixed" and "single," respectively. The intercepts A_ρ and A_R should really be zero here, of course, but we have inserted them for a purpose, which will become clearer in the next section. In fact, we will find that the *form* of Eqs. (A44) and (A45) is quite general, holding also for *high* magnetic fields ($\mu^2 B^2 \gg 1$) in the energy-*dependent*-τ case, and for *all* magnetic fields in the energy-*independent*-τ case. Thus, it is perhaps no surprise that experimental plots of $\rho_0/\Delta\rho$ and $R_0/\Delta R$ versus c^2/B^2 are so often straight lines. However, we must emphasize that the present form is not *completely* general, in that it does not hold theoretically for *medium*-strength magnetic fields in the energy-*dependent*-τ case.

It is interesting to study the dependence of $\Delta\rho/\rho_0$ and $-\Delta R/R_0$ on $c(=n/p)$, in order to see the relative contributions of the single- and mixed-carrier terms. In Fig. 12 we have plotted $S_{\rho m}$, $S_{\rho s}$, and S_ρ, along with S_{Rm}, S_{Rs}, and S_R, normalized by the factor $r_{n0}^2 \mu_n^2$. The assumptions are $b = 10$, $r_{n0} = r_{p0} = 1.18$, $\beta_n = \beta_p = 1.546$, and $\xi_n = \xi_p = 0.273$; these values of r, β, and ξ hold for acoustic deformation potential scattering, as shown in Table A1. It is seen

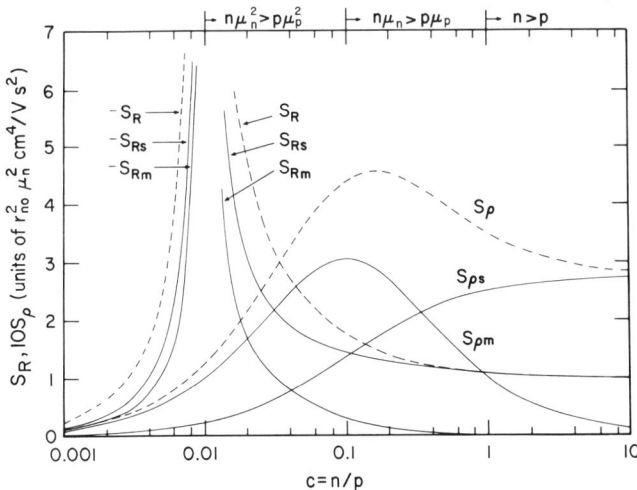

FIG. 12. Plots of S_ρ, $S_{\rho s}$, $S_{\rho m}$, S_R, S_{Rs}, and S_{Rm} for acoustic deformation potential scattering under the conditions $\mu_n/\mu_p = 10$; $r_{n0} = r_{p0} = 1.18$; $\xi_n = \xi_p = 0.273$; $\beta_n = \beta_p = 1.546$.

TABLE A1

Theoretical Values of r_n, ξ_n, β_n, and γ_n for Several Scattering Mechanisms ($\tau_n = a_n \varepsilon^{-s}$), and Experimental Values for an O-Doped, 10^7-Ω cm GaAs Crystal, and an S-Doped, 10^{-1}-Ω cm GaAs Layer, 3 μm Thick

Type of scattering	s	r_n	ξ_n	β_n	$\beta_n - 2\xi_n$	γ_n
Acoustic deformation potential	$\frac{1}{2}$	1.18	0.273	1.546	1.00	1.13
Acoustic piezoelectric potential	$-\frac{1}{2}$	1.10	0.0865	0.268	0.0950	1.13
Ionized impurity	$-\frac{3}{2}$	1.93	0.577	2.52	1.37	3.39
Experimental: 10^7 Ω cm			0.79	1.94	0.36	
10^{-1} Ω cm			0.31	0.33	-0.29	

that $S_{Rs} > S_{Rm}$ for all values of c in this case, but that the relative magnitudes of S_{ps} and S_{pm} cross over at about $c = 0.35$. Note also that S_R becomes very large for $\alpha b^2 c \simeq 1$, and changes sign. In fact, R itself changes sign here.

In a later section (18d) we will find that the mixed-conductivity terms in fact allow σ and R to be magnetic field dependent *even* when τ is *independent* of energy. This is because the Hall field E_y can never exactly balance the magnetic field forces for both holes and electrons at the same time.

b. Energy-Dependent τ, High B

At high fields ($\omega_c^2 \tau^2 \gg 1$) we can use Eqs. (A15) and (A34) to write $\sigma_{pxx} = pm_p^* \langle \tau_p^{-1} \rangle / (B/c)^2$, $\sigma_{nxx} = nm_n^* \langle \tau_n^{-1} \rangle / (B/c)^2$, $\sigma_{pxy} = ep/(B/c)$, and $\sigma_{nxy} = -en/(B/c)$. Then Eq. (A37) becomes

$$\sigma_{B \to \infty} = \frac{(nm_n^* \langle \tau_n^{-1} \rangle + pm_p^* \langle \tau_p^{-1} \rangle)^2 / (B/c)^4 + e^2(-n + p)^2 / (B/c)^2}{(nm_n^* \langle \tau_n^{-1} \rangle + pm_p^* \langle \tau_p^{-1} \rangle)/(B/c)^2}. \quad \text{(A50)}$$

At $B = \infty$ we get

$$\sigma_\infty = \frac{e^2(p - n)^2}{nm_n^* \langle \tau_n^{-1} \rangle + pm_p^* \langle \tau_p^{-1} \rangle}$$

$$= \frac{e\mu_n \mu_p (p - n)^2}{n\mu_p \langle \tau_n \rangle \langle \tau_n^{-1} \rangle + p\mu_n \langle \tau_p \rangle \langle \tau_p^{-1} \rangle}$$

$$= \frac{e\mu_n \mu_p (p - n)^2}{n\mu_p \gamma_n + p\mu_n \gamma_p}, \quad \text{(A51)}$$

2. PROPERTIES OF SEMI-INSULATING GaAs

where we have used the relationships $\mu_n = e\langle\tau_n\rangle/m_n^*$ and $\mu_p = e\langle\tau_p\rangle/m_p^*$, and have defined $\gamma_i \equiv \langle\tau_i\rangle\langle\tau_i^{-1}\rangle$. Representative values of γ_i are found in Table A1. Similarly, from Eq. (A38),

$$R_{B\to\infty} = \frac{1}{(B/c)} \frac{e(p-n)/(B/c)}{(nm_n^*\langle\tau_n^{-1}\rangle + pm_p^*\langle\tau_p^{-1}\rangle)^2/(B/c)^4 + e^2(p-n)^2/(B/c)^2} \quad (A52)$$

and the value at $B = \infty$ is

$$R_\infty = 1/e(p-n). \quad (A53)$$

Before proceeding further we note that both σ and R should approach constant values at high magnetic field, at least within our approximation in this paper of spherical equal-energy surfaces. For n-type samples, the saturation of R has been reported in several works (Stillman et al., 1970), but a similar saturation in σ is seldom observed. For p-type samples, or mixed n and p conductors, the magnetic fields necessary for saturation are very high, since both $\mu_n^2(B/c)^2$ and $\mu_p^2(B/c)^2$ must be much greater than unity (or greater than 10^8, if practical units are employed). For example, $\mu_p \simeq 400$ cm^2/V s for pure GaAs at room temperature, so that we would require $B \gg 250$ kG in order to have saturation in this case.

To find the functional forms of $\rho_0/\Delta\rho$ and $R_0/\Delta R$ we return to Eqs. (A50) and (A52), and use Eqs. (A51) and (A53):

$$\sigma_{B\to\infty} = \sigma_\infty + \frac{1}{\sigma_\infty R_\infty^2} \frac{c^2}{B^2}, \quad (A54)$$

$$R_{B\to\infty} = \frac{R_\infty}{1 + \frac{1}{\sigma_\infty^2 R_\infty^2} \frac{c^2}{B^2}}. \quad (A55)$$

Then it can be shown that

$$\frac{\rho_0}{\Delta\rho_{B\to\infty}} \equiv \frac{\rho_0}{\rho - \rho_0} = \frac{\sigma}{\sigma_0 - \sigma} = \frac{\sigma_\infty\left(1 + \frac{1}{\sigma_\infty^2 R_\infty^2}\frac{c^2}{B^2}\right)}{(\sigma_0 - \sigma_\infty)\left[1 - \frac{1}{\sigma_\infty(\sigma_0 - \sigma_\infty)R_\infty^2}\frac{c^2}{B^2}\right]}$$

$$\simeq \frac{\sigma_\infty}{\sigma_0 - \sigma_\infty} + \frac{\sigma_0}{\sigma_\infty(\sigma_0 - \sigma_\infty)^2 R_\infty^2}\frac{c^2}{B^2}. \quad (A56)$$

Similarly, we find that

$$-\frac{R_0}{\Delta R_{B\to\infty}} \equiv \frac{R_0}{R_0 - R} = \frac{R_0}{R_0 - R_\infty} - \frac{R_0}{\sigma_\infty^2 R_\infty(R_0 - R_\infty)^2}\frac{c^2}{B^2}. \quad (A57)$$

These expressions can also be written in the functional form introduced earlier:

$$\frac{c^2}{B^2} + \frac{\sigma_\infty^2(\sigma_0 - \sigma_\infty)R_\infty^2}{\sigma_0} = \frac{\sigma_\infty(\sigma_0 - \sigma_\infty)^2 R_\infty^2}{\sigma_0} \frac{\rho_0}{\Delta\rho_{B\to\infty}}, \quad (A58)$$

$$\frac{c^2}{B^2} - \sigma_\infty^2 R_\infty(R_0 - R_\infty) = \frac{\sigma_\infty^2 R_\infty(R_0 - R_\infty)^2}{R_0} \frac{R_0}{\Delta R_{B\to\infty}}, \quad (A59)$$

where σ_0, R_0, σ_∞, and R_∞ are given by Eqs. (A39), (A40), (A51), and (A53), respectively. Note that the *forms* of Eqs. (A58) and (A59) are identical to those of the low-field case [Eqs. (A44) and (A45)]. However, the slopes and intercepts are, or course, different.

c. *Energy-Independent τ, Arbitrary B*

When τ is *independent* of energy we can write Eqs. (A15) and (A34) as

$$\sigma_{nxx} = \frac{ne^2}{m_n^*} \frac{\tau_n}{1 + \omega_{cn}^2 \tau_n^2} = \frac{\sigma_n}{1 + \sigma_n^2 R_n^2 (B/c)^2}, \quad (A60a)$$

$$\sigma_{nxy} = -\frac{ne^2}{m_n^*} \frac{\omega_{cn}\tau_n^2}{1 + \omega_{cn}^2 \tau_n^2} = \frac{R_n \sigma_n^2 (B/c)}{1 + \sigma_n^2 R_n^2 (B/c)^2}, \quad (A60b)$$

$$\sigma_{pxx} = \frac{pe^2}{m_p^*} \frac{\tau_p}{1 + \omega_{cp}^2 \tau_p^2} = \frac{\sigma_p}{1 + \sigma_p^2 R_p^2 (B/c)^2}, \quad (A60c)$$

$$\sigma_{pxy} = \frac{pe^2}{m_p^*} \frac{\omega_{cp}\tau_p^2}{1 + \omega_{cp}^2 \tau_p^2} = \frac{R_p \sigma_p^2 (B/c)}{1 + \sigma_p^2 R_p^2 (B/c)^2}. \quad (A60d)$$

Again, it is but an algebraic exercise to show that Eqs. (A37) and (A38) now become (Putley, 1968, p. 88; Bube, 1974, p. 373)

$$\sigma = \frac{(\sigma_n + \sigma_p)^2 + \sigma_n^2 \sigma_p^2 (R_n + R_p)^2 (B/c)^2}{(\sigma_n + \sigma_p) + \sigma_n \sigma_p (\sigma_n R_n^2 + \sigma_p R_p^2)(B/c)^2}, \quad (A61)$$

$$R = \frac{\sigma_n^2 R_n + \sigma_p^2 R_p + \sigma_n^2 \sigma_p^2 R_n R_p (R_n + R_p)(B/c)^2}{(\sigma_n + \sigma_p)^2 + \sigma_n^2 \sigma_p^2 (R_n + R_p)^2 (B/c)^2}. \quad (A62)$$

Equations (A61) and (A62) hold for *arbitrary* magnetic field strength as long as $kT \gg \hbar\omega_c$. It is interesting to compare the $B = 0$ and $B = \infty$ values of σ and R in this case (energy-*independent* τ) with those in the energy-*dependent*-τ case. In fact, it can be easily shown that σ_0, R_0, σ_∞, and R_∞ [Eqs. (A39), (A40), (A51), and (A53), respectively] are exactly the same if we simply let $\gamma_i = r_i = 1$, i.e., their value when τ is independent of energy.

2. PROPERTIES OF SEMI-INSULATING GaAs

To proceed further with Eqs. (A61) and (A62) we can write them in a form similar to that used before:

$$\frac{c^2}{B^2} + \mu_n^2 Y = S_{\rho m}\frac{\rho_0}{\Delta\rho} = -S_{Rm}\frac{R_0}{\Delta R}, \quad \text{(A63)}$$

where $S_{\rho m}$ and S_{Rm} were defined by Eqs. (A46) and (A48), respectively, and

$$\mu_n^2 Y = \frac{\sigma_n^2 \sigma_p^2 (R_n + R_p)^2}{(\sigma_n + \sigma_p)^2} = \frac{r_{n0}^2 \mu_n^2 (c - \alpha)^2}{\alpha^2 (1 + bc)^2}. \quad \text{(A64)}$$

For this case (energy-independent τ) we can determine n, p, μ_n, and μ_p *exactly*, by simultaneously solving Eqs. (A46) ($S_{\rho m}$), (A48) (S_{Rm}), (A39) (σ_0), and (A40) (R_0). We give only the final result, again making use of the relations $\sigma_n = ne\mu_n$, $\sigma_p = pe\mu_p$, $R_n = -1/ne$, $R_p = 1/pe$:

$$\mu_n = \frac{1 + \beta^{-1}}{1 - b^{-1}}(-R_0\sigma_0), \quad \text{(A65)}$$

$$\mu_p = \mu_n/b, \quad \text{(A66)}$$

$$p = \frac{1}{e(-R_0)}\frac{(1 - b^{-1})}{(1 + b^{-1})}\frac{b(1 + \beta b^{-1})}{(1 + \beta)(1 + \beta^{-1})}, \quad \text{(A67)}$$

$$n = cp = \frac{1}{e(-R_0)}\frac{(1 - b^{-1})}{(1 + b^{-1})}\frac{(\beta + b^{-1})}{(1 + \beta)(1 + \beta^{-1})}, \quad \text{(A68)}$$

$$n_i = (np)^{1/2}, \quad \text{(A69)}$$

where

$$b = \tfrac{1}{2}[A + (A^2 - 4)^{1/2}], \quad \text{(A70)}$$

$$A = \frac{2 + T + T/\beta^2}{1 - T/\beta}, \quad \text{(A71)}$$

$$T = \frac{(R_0\sigma_0)^2}{S_{\rho m}}, \quad \text{(A72)}$$

and where $\beta \equiv S_\rho/S_R$.

We once again note that the form of Eq. (A63) (energy-*independent* τ) is identical to that of Eqs. (A44) and (A45) (low field, energy-*dependent* τ), and that of Eqs. (A58) and (A59) (high field, energy-*dependent* τ). We also note that the *slopes* ($S_{\rho m}$ and S_{Rm}) given in Eq. (A64) are the same as the *mixed-carrier* terms in the respective slopes of Eqs. (A44) and (A45) if we let $r_n = r_p = 1$. Thus, the close relationships between all of these equations are evident.

Some of these effects described above are illustrated in Fig. 13, which shows B^{-2} versus $\rho_0/\Delta\rho$, and B^{-2} versus $-R_0/\Delta R$ plots for two SI GaAs samples, one O doped, and the other Cr doped. The O-doped crystal has a resistivity of 4.5×10^7 Ω cm, which is too low to expect any appreciable mixed-conductivity effects (Look, 1982b). In fact, we can estimate $c = n/p = n^2/n_i^2 \simeq (eR_0)^{-2}/n_i^2 \simeq 10^2$. It is interesting to compare the experimental slopes, $S_\rho = 1.6 \times 10^7$ and $S_R = 7.2 \times 10^6$ cm^4/V^2 s^2, with those predicted, say from acoustic deformation potential (ADP) scattering. From Fig. 12 and the data in the caption for Fig. 13 we would predict $S_\rho = 0.273(4.3 \times 10^3)^2 = 5.0 \times 10^6$, and $S_R = [1.546 - 2(0.273)] (4.3 \times 10^3)^2 = 1.8 \times 10^7$ cm^4/V^2 s^2. Although both measured values are about a factor 3 different than the predicted values, we must remember that ADP scattering is certainly not the dominant mechanism in GaAs at room temperature. (The dominant mechanism, polar optical-phonon scattering, cannot be dealt with by the theory described in this paper.) However, a feature that seems to be common to all the scattering mechanisms is that S_ρ should begin falling, and S_R should begin rising rapidly when $c < \mu_p/\mu_n$ or $c < 0.1$ in this case. This feature should be evident in the results for the Cr-doped sample, also described in Fig. 13. This sample has a

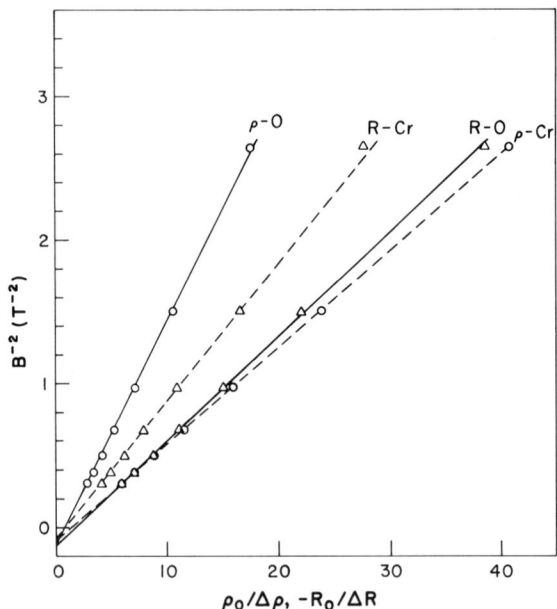

FIG. 13. Plots of B^{-2} versus $\rho_0/\Delta\rho$, and B^{-2} versus $-R_0/\Delta R$ for two SI GaAs crystals, one O-doped and the other Cr-doped. The relevant zero-field parameters are as follows: (1) O doped: $\rho_0 = 4.5 \times 10^7$ Ω cm, $(eR_0^{-1}) = 3.2 \times 10^7$ cm^{-3}, $R_0/\rho_0 = 4.3 \times 10^3$ cm^2/V s; (2) Cr doped: $\rho_0 = 9.1 \times 10^8$ Ω cm, $(eR_0)^{-1} = 4.1 \times 10^6$ cm^{-3}, $R_0/\rho_0 = 1.7 \times 10^3$ cm^2/V s.

resistivity of 9.1×10^8 Ω cm, and is expected to show strong mixed-conductivity effects, including $c = n/p < 1$, even though the Hall coefficient is negative. Thus, we might expect that S_R would be larger, and S_ρ smaller, than the corresponding values for the O-doped crystal, and indeed, such is the case. However, for truly quantitative comparisons of experiment and theory, it will be necessary to include the effects of all the scattering mechanisms within a framework that treats polar optical-phonon scattering properly (Ehrenreich, 1960; Rode, 1970; Nag, 1980; Meyer and Bartoli, 1981).

A final precaution, which is evident from a glance at Fig. 12, is that one should not assume that just because mixed-conductivity effects are strong, single-carrier effects can be ignored, and that therefore Eqs. (A65)–(A72) can be used to exactly calculate n, p, μ_n, and μ_p. For example, in Fig. 12 it is seen that $S_{Rs} > S_{Rm}$ for *all* values of c, and although this condition does not hold for every scattering mechanism, it must be kept in mind.

d. Other Approaches to Mixed-Conductivity Phenomena

All of the approaches to GaAs mixed-conductivity problems discussed so far have drawbacks. The "pure" mixed-conductivity problem can be solved exactly, but, unfortunately, most semi-insulating samples also exhibit appreciable single-carrier magnetic field effects that cannot be easily separated. Inclusion of the single-carrier effects simply results in too many unknown parameters [cf. Equations (A42) and (A43)]. A somewhat different approach, suggested recently (Look, 1980), is to completely eliminate dealing with the ζ's and β's. In this approach, we start by measuring σ and R at $B = 0$ only, giving Eqs. (A39) and (A40), which are repeated below:

$$\sigma_0 = e(n\mu_n + p\mu_p), \tag{A39}$$

$$R_0 = \frac{r_p p \mu_p^2 - r_n n \mu_n^2}{e(n\mu_n + p\mu_p)^2}. \tag{A40}$$

If we ignore r_n and r_p for the moment, then two more equations (four, total) are necessary in order to determine the four unknowns, n, p, μ_n, and μ_p. One of these equations comes to mind immediately:

$$np = n_i^2, \tag{A73}$$

where $n_i = 2.6 \times 10^6$ cm^{-3} at 296°K (Look, 1980). A fourth equation may be obtained by exploiting an obvious connection between μ_p and μ_n since holes and electrons experience similar scattering mechanisms. That is, from a Matthiessen's-rule point of view, we can write $\mu_n^{-1} = \mu_{nL}^{-1} + \mu_{nI}^{-1} = \mu_{nL}^{-1} + C_n N_I$, and $\mu_p^{-1} = \mu_{pL}^{-1} + C_p N_I$, where "L" denotes a lattice scattering contribution and "I" an ionized impurity contribution. Here, μ_{nL} and μ_{pL} are well known for many materials, and the C_n and C_p can be treated as unknown constants. (We ignore the weak dependence of C_n and C_p on n and N_1, which would arise from screening effects.) By eliminating N_1 in these equations it is possible to write $\mu_p^{-1} = D + E\mu_n^{-1}$, where D and E can be determined, in principle, from experimental data. Strictly speaking, however, it is necessary to have measurements of μ_n and μ_p *on the same sample*, but such data are scarce. As a second resort, we can use curves of μ_n versus n, and μ_p versus p from various literature sources (Sze, 1969; Wiley, 1975), which will yield $\mu_p = f(\mu_n)$ if relationships between n and p, and N_1 are assumed. From some of these data, as well as some of our own, we have deduced the following *tentative* relationship:

$$\mu_p^{-1} \simeq 9 \times 10^{-4} + 13\mu_n^{-1}. \tag{A74}$$

We must emphasize that Eq. (A74) is very approximate, applies only to a restricted range, and should be used with *extreme caution*. However, for our purposes here it is probably adequate. Equations (A39), (A40), (A73), and (A74) can now be solved to yield n, p, μ_n, and μ_p if $r_n = r_p = 1$ or if r_n and r_p are known, say, from a self-consistent theoretical study, which makes use of the actual data. The above equations and analysis are discussed further in Section 3.

It should be mentioned that very recently some theoretical relationships between μ_p and μ_n have been derived for GaAs (Walukiewicz et al., 1982; Hrivnak et al., 1982). One of these results (Walukiewicz et al., 1982) is only numerical at this point, although it would be useful to fit the numerical data with an analytical expression such as Eq. (A74). The other (Hrivnak et al., 1982) has the form of Eq. (A74), but with different constants.

We will mention two other possibilities for obtaining the elusive fourth equation. One is

$$R_\infty = 1/e(p - n), \tag{A53}$$

which is excellent if a large magnetic field is available. Unfortunately, Eq. (A53) is valid only if both $\mu_n B \gg 10^8$ and $\mu_p B \gg 10^8$, where the μ's are in cm^2/V s, and B is in G. For $\mu_p = 400$ cm^2/V s, the demand is $B \gg 250$ kG, well beyond the capabilities of the magnets available in most laboratories.

The final possibility for the fourth equation arises from the thermopower coefficient, $\Theta \equiv E_x/(dT/dX)$. The relevant theoretical expression at $B = 0$ is

$$\Theta_0 = -\frac{k}{e} \frac{[\chi_n + \ln(N_C/n)]n\mu_n - [\chi_p + \ln(N_V/p)]p\mu_p}{(n\mu_n + p\mu_p)}, \tag{A75}$$

where the Boltzmann approximation is assumed, and $\chi_i \equiv \Gamma(-s_i + \frac{7}{2})/\Gamma(-s_i + \frac{5}{2}) = \frac{5}{2} - s_i$. Here s_i has been defined as before: $\tau_n = a\varepsilon^{-s_i}$. Note that $\chi = 2, 3,$ and 4, for $s_i = \frac{1}{2}, -\frac{1}{2},$ and $-\frac{3}{2}$, respectively. The other terms, $\ln(N_C/n)$ and $\ln(N_V/p)$, are much larger than the χ's, for semi-insulating samples. For example, if $n \sim p \sim 10^6$ cm^{-3}, typical for SI GaAs, $\ln(N_C/n) \simeq 27$, and

$\ln(N_V/p) \simeq 30$, at room temperature. Thus, it is fortunately not necessary to know the scattering factors, χ_n and χ_p, accurately.

To summarize Section 18d, we can determine our four basic electrical parameters, n, p, μ_n, and μ_p, in the presence of both single- and mixed-carrier magnetic field effects, from any four of the following equations:

1. $$\sigma_0 = e(n\mu_n + p\mu_p), \tag{A39}$$

2. $$R_0 = \frac{r_p p \mu_p^2 - r_n n \mu_n^2}{e(n\mu_n + p\mu_p)^2}, \tag{A40}$$

3. $$n_i^2 = np, \tag{A73}$$

4. (a) $\mu_p^{-1} = f(\mu_n^{-1}) \simeq 9 \times 10^{-4} + 13\mu_n^{-1}$ (tentative), (A74)

or

(b) $R_\infty = 1/e(p - n),$ (A53)

or

(c) $$\Theta_0 = -\frac{k}{e} \frac{[\chi_n + \ln(N_C/n)]n\mu_n - [\chi_p + \ln(N_V/p)]p\mu_p}{(n\mu_n + p\mu_p)}. \tag{A75}$$

For *exact* solutions of n, p, μ_n, and μ_p from these equations, it is necessary to assume that $r_n = r_p = 1$, and $\chi_n \ll \ln(N_C/n)$, $\chi_p \ll \ln(N_V/p)$, unless these parameters are explicitly known. If *mixed-carrier* magnetic field effects are dominant (τ independent of energy), then Eqs. (A65)–(A72) will give n, p, μ_n, and μ_p exactly. However, as discussed earlier, mixed- and single-carrier effects are often of comparable magnitude, thus invalidating Eqs. (A65)–(A72). Further information on n, p, μ_n, and μ_p can be obtained from the low field magnetoresistance and magneto-Hall expressions, Eqs. (A42) and (A43), respectively, but only if the ξ and β parameters are known. Thus, the approach suggested above seems the best hope for solving mixed-conductivity problems. It should also be mentioned that one is obviously not restricted to using only *one* of the final three expressions listed above [Eqs. (A74), (A53), and (A75)]. For example, the simultaneous solutions of Eqs. (A39), (A40), (A73), and (A74) are double valued, but a thermopower measurement, Eq. (A75), can usually indicate immediately which of the two possible solutions is correct.

19. Units

So far we have employed gaussian units, which are more common in theoretical discussions. However, experimental results are generally discussed in terms of practical units. Since conversions between different systems of units can sometimes be confusing we include here an example. Consider R. In gaussian units E and B are in statvolt/cm (for B this is called "gauss"), j in statamp/cm^2. Thus,

$$R = \frac{c\left(\dfrac{cm}{s}\right) E_y\left(\dfrac{statvolt}{cm}\right)}{\underbrace{j_x\left(\dfrac{statamp}{cm^2}\right) B\left(\dfrac{statvolt}{cm}\right)}_{G}} \frac{cm^3}{statcoul}, \qquad (A76)$$

where the relevant units for the measurement parameters are indicated in parentheses. To convert to practical units we write

$$R = \frac{c\left(\dfrac{cm}{s}\right)(300)^{-1} E_y\left(\dfrac{V}{cm}\right)}{3 \times 10^9 j_x\left(\dfrac{A}{cm^2}\right) B(G)} \frac{cm^3}{statcoul} \times \frac{3 \times 10^9 \text{ statcoul}}{C}$$

$$= 10^8 \frac{E_y\left(\dfrac{V}{cm}\right)}{j_x\left(\dfrac{A}{cm^2}\right) B(G)} \frac{cm^3}{C} \qquad (A77)$$

since $c = 3 \times 10^{10}$ cm s^{-1}. Next consider μ. Theoretical expressions for μ are often conveniently evaluated in gaussian units, cm^2/statvolt s. To convert this to practical units, cm^2/V s, divide by 300. A good discussion of units is given in Jackson (1962).

Appendix B. Semiconductor Statistics

20. Derivation of Occupation Factors

The maximum probability method of deriving energy distribution functions is a familiar approach described in many textbooks on solid state physics (Nichols and Vernon, 1966; Bube, 1974, p.98). The heart of the technique is in finding the number of distinguishable ways that a given number of particles can be distributed among a given number of boxes (Tolman, 1938). The particles in this case are electrons and the boxes are the "elementary" eigenstates of the system. Here a "box" could be a band state, denoted by $k_x, k_y,$ and k_z quantum numbers, or an atomic impurity state, denoted by its own appropriate quantum numbers. Now an important assumption is that an electron in one box must interact only weakly with the electrons in other boxes. If this is the case, then the total Hamiltonian for the system will simply be the sum of the individual Hamiltonians, and the grand eigenstate can be written as a product of the individual eigenstates, properly symmetrized, of course. It should be noted here that, although the interaction between boxes must be weak, the interactions of electrons within a box may be strong indeed. We simply require that the eigenstate and eigenvalue for the box as a whole be known.

The available electrons in the system will come from the valence band and donor centers. These electrons can be distributed among states in the valence band, conduction band, donor centers, and acceptor centers. Each *distinguishable* arrangement of the electrons among all of these states results in a different grand eigenstate, and it is assumed that every grand eigenstate that leads to the same total energy is equally likely to occur. The maximum probability principle then states that the electrons will distribute themselves among the possible energy cells of the system in such a way that the maximum number of grand eigenstates can be formed, consistent with a given total energy.

We begin with the conduction band and divide it into "energy cells" of a width $\Delta\varepsilon_C$, large enough to encompass many states but smaller than our energy measurement resolution. Consider cell i (at energy ε_{Ci}) that contains n_{Ci} electrons distributed among N_{Ci} states. We have N_{Ci} ways to put in the first electron, $N_{Ci} - 1$ ways to put in the second, etc., or a total of $N_{Ci}!/(N_{Ci} - n_{Ci})!$ ways to put in all n_{Ci} electrons. However, permutations among the n_{Ci} electrons themselves cannot be counted separately since they do not result in new states; therefore, the number of *distinguishable* ways to distribute n_{Ci} electrons in N_{Ci} states is

$$W_{Ci} = N_{Ci}!/n_{Ci}!(N_{Ci} - n_{Ci})!. \tag{B1}$$

The number of distinguishable ways to distribute *all* of the conduction electrons in their various energy cells is then simply the product

$$W_C = \prod_i W_{Ci} = \prod_i \frac{N_{Ci}!}{n_{Ci}!(N_{Ci} - n_{Ci})!}. \tag{B2}$$

The same analysis can be carried out for the valence band, giving

$$W_V = \prod_j \frac{N_{Vj}!}{n_{Vj}!(N_{Vj} - n_{Vj})!}. \tag{B3}$$

Now the inclusion of impurities is relatively simple if we assume that each impurity has only one bound state within the band gap. (This condition will be relaxed later.) That is, suppose that if a particular impurity atom contains just n_a electrons, it takes *more* than band-gap energy (E_G) to excite any one of them to the conduction band, whereas if this same atom contains $n_a + 1$ electrons, it takes an energy of only E_{Ik} to excite one of them, where $E_{Ik} < E_G$. To complete this picture, we must further suppose that any atoms having a larger number of electrons (i.e., $n_a + 2$, $n_a + 3$, etc.) are *resonant* in energy with the conduction band, and therefore are not stable against losing these electrons by autoionization. Thus, at temperatures less than E_G/k (i.e., $T < 15,000°K$ in GaAs!) our model impurity atom will contain either n_a or $n_a + 1$ electrons. Let the total number of these "type-k" impurity atoms (or defect sites) be N_{Ik}, with n_{Ik} of these containing $n_a + 1$ electrons, and the rest containing n_a electrons. Further, let the energy of an atom with $n_a + 1$ electrons be ε_{Ik}. Ignoring degeneracy, the number of distinguishable permutations is, as before, just $N_{Ik}!/n_{Ik}!(N_{Ik} - n_{Ik})!$. However, there are now extra possible permutations because the electrons (either n_a or $n_a + 1$) on each atom can be permuted in g ways, where g is the degeneracy of that particular state. (For example, if there are three equivalent electrons in an orbital that can hold six equivalent electrons, then $g = 6!/3!3! = 20$.) Let g_1 be the degeneracy of the state containing $n_a + 1$ electrons and g_0 the degeneracy of the state containing n_a electrons. Then it is clear that we will have $g_1^{n_{Ik}} g_0^{(N_{Ik} - n_{Ik})}$ *extra* permutations so that

$$W_{Ik} = g_{1k}^{n_{Ik}} g_{0k}^{(N_{Ik} - n_{Ik})} N_{Ik}!/n_{Ik}!(N_{Ik} - n_{Ik})!. \tag{B4}$$

Again, if there are several donor or acceptor impurities (or defects), designated by $k = 1, 2, \ldots,$ then

$$W_I = \prod_k W_{Ik}. \tag{B5}$$

The total number of distinguishable ways of arranging all the electrons in our system is simply $W = W_C \cdot W_V \cdot W_I$. Following the usual procedure we assume that each permutation that results in the same total energy is equally likely so that the most probable distribution of the electrons among the various energy cells is just that one that maximizes W. It is more convenient to work with $\ln W$ since then, for large N, we can apply Stirling's approximation $(\ln N! \simeq N \ln N - N)$. The result is

$$\ln W = \sum_{i, \text{cond.-band cells}} N_{Ci} \ln N_{Ci} - n_{Ci} \ln n_{Ci} - (N_{Ci} - n_{Ci}) \ln(N_{Ci} - n_{Ci})$$

$$+ \sum_{j, \text{val.-band cells}} N_{Vj} \ln N_{Vj} - n_{Vj} \ln n_{Vj} - (N_{Vj} - n_{Vj}) \ln(N_{Vj} - n_{Vj})$$

$$+ \sum_{k, \text{impurities}} N_{Ik} \ln N_{Ik} - n_{Ik} \ln n_{Ik} - (N_{Ik} - n_{Ik}) \ln(N_{Ik} - n_{Ik})$$

$$+ n_{Ik} \ln g_{1k} + (N_{Ik} - n_{Ik}) \ln g_{0k}. \tag{B6}$$

There are two constraints that must be obeyed while maximizing $\ln W$: conservation of the number of electrons N_e and conservation of total energy ε_t.

$$F_1(n) \equiv \sum_i n_{Ci} + \sum_j n_{Vj} + \sum_k n_{Ik} = N_e, \tag{B7}$$

2. PROPERTIES OF SEMI-INSULATING GaAs

$$F_2(n, \varepsilon) \equiv \sum_i n_{Ci}\varepsilon_{Ci} + \sum_j n_{Vj}\varepsilon_{Vj} + \sum_k n_{Ik}\varepsilon_{Ik} = \varepsilon_t. \tag{B8}$$

To maximize ln W, subject to the above constraints, it is convenient to use the Lagrangian multiplier method:

$$\frac{\partial}{\partial n_l}\{\ln W + \alpha[N_e - F_1(n)] + \beta[\varepsilon_t - F_2(n, \varepsilon)]\} = 0, \tag{B9}$$

where α and β are constants. By letting n_l be n_{Cl}, n_{Vl}, and n_{Il}, in succession, we can generate three equations:

$$\ln(N_{Cl} - n_{Cl}) - \ln n_{Cl} - \alpha - \beta\varepsilon_{Cl} = 0, \tag{B10a}$$

$$\ln(N_{Vl} - n_{Vl}) - \ln \overline{n_{Vl}} - \alpha - \beta\varepsilon_{Vl} = 0, \tag{B10b}$$

$$\ln(N_{Il} - n_{Il}) - \ln n_{Il} + \ln g_{1l} - \ln g_{0l} - \alpha - \beta\varepsilon_{Il} = 0, \tag{B10c}$$

with solutions

$$n_{Cl} = \frac{N_{Cl}}{1 + e^{(\alpha + \beta\varepsilon_{Cl})}}, \tag{B11a}$$

$$n_{Vl} = \frac{N_{Vl}}{1 + e^{(\alpha + \beta\varepsilon_{Vl})}} = N_{Vl} - \frac{N_{Vl}}{1 + e^{-(\alpha + \beta\varepsilon_{Vl})}}, \tag{B11b}$$

$$n_{Il} = \frac{N_{Il}}{1 + (g_0/g_1)e^{(\alpha + \beta\varepsilon_{Il})}}. \tag{B11c}$$

Thus, the bound-state degeneracy factor, $K \equiv g_0/g_1$, falls out in a very natural way. It arises simply because of the assumption that the state with $n_a + 2$ electrons is at too high an energy to be stable (see Look, 1981, for more detail).

By using standard thermodynamic arguments, it can be shown that $\beta = (kT)^{-1}$ and $\alpha = -\varepsilon_F/kT$, where ε_F is the electrochemical potential, or Fermi energy. This procedure is carried out in many textbooks (Bube, 1974, p. 102) and will not be repeated here.

The total numbers of electrons in the conduction and valence bands respectively, are now easily derived. From this point on, all numbers of particles will be normalized to unit volume so that they become particle densities. From Eq. (B11a) we get

$$n \equiv \sum_{l,\text{ cond. band}} n_{Cl} = \sum_l \frac{\phi_C(\varepsilon_l)\Delta\varepsilon}{1 + \exp(\varepsilon_l - \varepsilon_F)/kT}$$

$$\to \int_{\varepsilon_C^{\min}}^{\varepsilon_C^{\max}} \frac{\phi_C(\varepsilon)\,d\varepsilon}{1 + \exp(\varepsilon - \varepsilon_F)/kT} = \frac{4\pi(2m_n^*)^{3/2}}{h^3} \int_0^\infty \frac{\varepsilon^{1/2}\,d\varepsilon}{1 + \exp(\varepsilon - \varepsilon_F)/kT}. \tag{B12}$$

The density of states function given here follows from the assumption of a parabolic energy band and includes spin degeneracy (Bube, 1974, p. 172). (Note that the Fermi function effectively cuts off the integrand at a few kT above ε_F, so that for $kT \ll \varepsilon_C^{max}$, the upper limit may be extended to infinity.) Also we have set $\varepsilon_C^{min} = 0$.

The occupation of the valence band may be more easily described in terms of unoccupied states (holes) than occupied states. Let p_{Vl} be the density of holes in energy cell l. Since $n_{Vl} = N_{Vl} - p_{Vl}$, the definition of p_{Vl} is obvious from Eq. (B11b). As above, we change from a sum to an integral:

$$p \equiv \sum_{l,\text{val. band}} p_{Vl} = \sum_{l} \frac{N_{Vl}}{1 + \exp(\varepsilon_F - \varepsilon_l)/kT} \rightarrow \int_{\varepsilon_V^{min}}^{\varepsilon_V^{max}} \frac{\phi_V(\varepsilon)\,d\varepsilon}{1 + \exp(\varepsilon_F - \varepsilon)/kT}$$

$$= \frac{4\pi(2m_p^*)^{3/2}}{h^3} \int_{-\infty}^{\varepsilon_G} \frac{(-\varepsilon + \varepsilon_G)^{1/2}\,d\varepsilon}{1 + \exp(\varepsilon_F - \varepsilon)/kT}$$

$$= \frac{4\pi(2m_p^*)^{3/2}}{h^3} \int_0^\infty \frac{x^{1/2}\,dx}{1 + \exp(x + \varepsilon_F - \varepsilon_G)/kT}, \tag{B13}$$

where we are again assuming a parabolic band and have included spin degeneracy. Here, $\varepsilon_G \equiv \varepsilon_V^{max}$.

If ε_F is at least a few kT from either band edge, then we can ignore the "1" in the integrands of Eqs. (B12) and (B13) (Boltzmann approximation) and explicitly solve the respective integrals. The final results are

$$n = \frac{2(2\pi m_n^* kT)^{3/2}}{h^3} e^{\varepsilon_F/kT} \equiv N_C e^{\varepsilon_F/kT}, \tag{B14}$$

$$p = \frac{2(2\pi m_p^* kT)^{3/2}}{h^3} e^{(\varepsilon_G - \varepsilon_F)/kT} \equiv N_V e^{(\varepsilon_G - \varepsilon_F)/kT}, \tag{B15}$$

where energies here are referred to the *conduction* band edge.

To solve for the Fermi energy we must return to Eq. (B7). First consider the case $N_{Ik} = 0$, i.e., no states in the band gap. Then the total number of electrons in the system is N_{Vt} (the total valence-band density of states) and thus $n + (N_{Vt} - p) = N_{Vt}$, or $p = n \equiv n_i$, the intrinsic concentration. Thus,

$$n_i = \sqrt{pn} = \sqrt{N_C N_V} e^{\varepsilon_G/2kT}. \tag{B16}$$

Note that this result is independent of ε_F in the Boltzmann approximation, used here.

We next consider the case when impurities are present. The total number of electrons available for distribution among the various energy cells is then $N_{Vt} + \sum_i N_{Di}$, where the subscript D denotes a donor impurity or defect.

2. PROPERTIES OF SEMI-INSULATING GaAs

(Here we are dealing only with *single* donors and acceptors.) Equation (B7) now becomes

$$n + (N_{Vt} - p) + \sum_i n_{Di} + \sum_j n_{Aj} = N_{Vt} + \sum_i N_{Di}.$$

This can be rewritten

$$n + \sum_j n_{Aj} = p + \sum_i (N_{Di} - n_{Di}), \quad (B17)$$

which is simply an expression of charge balance, since each occupied acceptor has a single negative charge and each unoccupied donor has a single positive charge. Equation (B17) can now be solved for ε_F giving a value for n or p [Eq. (B14) or (B15)], which can be compared with experimental values.

We next consider the most general case, including excited states. Let n_{klm} denote the number of centers in the mth excited state of the lth charge state of the kth impurity, or defect, or energy cell in a band. Here we are assuming that the zeroth charge state is resonant with the valence band so that an energy greater than the band gap would be required to excite an electron from a center in this charge state into the conduction band. Let l_k be the number of charge states within the band gap for the kth center; then $l = 0, 1, 2, \ldots, l_k$. Here l denotes the number of *ionizable* electrons. The $(l_k + 1)$th charge state is presumed to be resonant with the conduction band, and thus unstable. For a given l, assume that there are m_{kl} excited states, i.e., $m_l = 0, 1, 2, \cdots m_{kl}$. Finally, let the degeneracy of the klmth state be g_{klm}. Each excited state is now a separate box in this problem, and from the above considerations it should be clear that

$$W = \prod_k \prod_{l=0}^{l_k} \prod_{m=0}^{m_{kl}} \frac{g_{klm}^{n_{klm}} N_k!}{n_{klm}!}. \quad (B18)$$

The constraints are

$$\sum_{l,m} n_{klm} = N_k \quad \text{for each } k, \quad (B19)$$

$$\sum_{k,l,m} l n_{klm} = N_e, \quad (B20)$$

$$\sum_{k,l,m} n_{klm} \varepsilon_{klm} = \varepsilon_t. \quad (B21)$$

Equation (B20) incorporates the fact that each impurity in a state klm contains l ionizable electrons. Let the Lagrange multipliers be γ_k, $k = 1, 2, \ldots$ for Eq. (B19), α for Eq. (B20), and β for Eq. (B21). Then the derivative with respect to n_{klm} of the total Lagrange function will give

$$\ln g_{klm} - 1 - \ln n_{klm} - \gamma_k - l\alpha - \beta \varepsilon_{klm} = 0. \quad (B22)$$

When $l = 0$ and $m = 0$ we get

$$\ln g_{k00} - 1 - \ln n_{k00} - \gamma_k - \beta\varepsilon_{k00} = 0. \tag{B23}$$

Thus, by combining Eqs. (B22) and (B23) we get

$$n_{klm} = \frac{g_{klm}}{g_{k00}} n_{k00} e^{-l\alpha - \beta(\varepsilon_{klm} - \varepsilon_{k00})} = \frac{g_{klm}}{g_{k00}} n_{k00} e^{l\eta_F - (\eta_{klm} - \eta_{k00})}, \tag{B24}$$

where again $\eta \equiv \varepsilon/kT$. Equation (B19) can be written

$$N_k = \sum_{l,m} n_{klm} = \frac{n_{k00}}{g_{k00}} \sum_{l,m} g_{klm} e^{l\eta_F - (\eta_{klm} - \eta_{k00})} \tag{B25}$$

so that the final result is

$$n_{klm} = \frac{N_k}{1 + \sum_{l',m' \neq l,m} (g_{kl'm'}/g_{klm}) e^{\eta_{klm} - \eta_{kl'm'} - (l-l')\eta_F}}. \tag{B26}$$

The restriction on the summation is that $l' \neq l$, and $m' \neq m$ *at the same time*. (Note that $l' = l$ is permissible as long as $m' \neq m$, i.e., as long as excited states exist.) Also, we should remember that ε_{klm} is the energy required to place l electrons from the valence band into the mth excited state on one of the centers in group k.

We may now also generalize the charge-balance equation. In Eq. (B20), the left-hand side (LHS) becomes

$$\text{LHS} = \sum_{l,m} l n_{klm} = n + (N_{Vt} - p) + \sum_{k,l,m \text{ donors}} l n_{klm} + \sum_{k,l,m \text{ acceptors}} l n_{klm}, \tag{B27}$$

and the right-hand side (RHS) is just the total number (density) of electrons available for distribution:

$$\text{RHS} = N_e = N_{Vt} + \sum_{k \text{ donors}} l_k N_k. \tag{B28}$$

Thus, Eq. (B20) can be rearranged as follows:

$$n + \sum_{k,l,m \text{ acceptors}} l n_{klm} = p + \sum_{k \text{ donors}} \left(l_k N_k - \sum_{l,m} l n_{klm} \right). \tag{B29}$$

Equation (B29) is a generalization of the simple charge-balance equation represented by Eq. (B17). Inclusion of centers that can have *both* donor and acceptor states results in a slight modification of Eq. (B29), and will be considered in the next section.

21. Applications

In this section we will apply Eqs. (B26) and (B29) in order to find the temperature dependences of n or p for several cases of interest. In all cases we

2. PROPERTIES OF SEMI-INSULATING GaAs

will assume that the Fermi level ε_F is more than a few kT from either band edge; then Eqs. (B14) and (B15) apply for n and p, respectively. We will also assume there is only one, or at most two, centers of *significant* concentration within a few kT of ε_F. Thus, all *other* centers will be either completely full or completely empty, or, for multi-charge-state centers, in the same occupation state.

a. One Single-Charge-State Donor

By this notation we mean that there is only *one* center of significant concentration within a few kT of ε_F, and that this center is a donor that can exist in only two occupation states, or charge states. In general, one of these charge states will be neutral. We will assume that all other donors and acceptors in the system are also single-charge-state centers, i.e., $l_k = 1$ and $l = 0$ or 1, and that there are no excited states of consequence within the band gap, i.e., $m = 0$. Any excited states within kT or so of the ground state can, of course, be considered degenerate, and lumped together.

Let the donor center of interest be designated with a subscript D. Then Eq. (B29) becomes

$$n + \sum_{k;l=0,1 \text{ acceptors}} ln_{kl} = p + N_D - \sum_{l=0,1} ln_{Dl} + \sum_{k \neq D \text{ donors}} \left(N_k - \sum_{l=0,1} ln_{kl} \right)$$

or

$$n + \sum_{k \text{ acceptors}} n_{k1} = p + N_D - n_{D1} + \sum_{k \neq D \text{ donors}} (N_k - n_{k1}). \quad \text{(B30)}$$

Now from the disscussion above it is clear that $n_{k1} = N_k$ for *all* centers having $\varepsilon_k < \varepsilon_F$, i.e., centers *below* the Fermi level, because such centers will be occupied with an electron, be they donors or acceptors. Conversely, $N_k - n_{k1} = N_k$ for all centers lying *above* ε_F, since $n_{k1} = 0$ in this case. We will define

$$N_{AS} \equiv \sum_{k \text{ acceptors } (\varepsilon < \varepsilon_F)} n_{k1}, \quad \text{(B31a)}$$

$$N_{DS} \equiv \sum_{k \text{ donors } (\varepsilon > \varepsilon_F)} (N_k - n_{k1}), \quad \text{(B31b)}$$

where the subscript S stands for shallow, or closer to the appropriate band than the center of interest. We now need only an expression for n_{D1}, which can be obtained from Eq. (B26):

$$n_{D1} = \frac{N_D}{1 + \sum_{l' \neq 1}(g_{Dl'}/g_{D1})e^{n_{D1} - n_{Dl'} - (1-l')n_F}}$$

$$= \frac{N_D}{1 + (g_{D0}/g_{D1})e^{n_{D1} - n_{D0} - n_F}} = \frac{N_D}{1 + (g_{D0}/g_{D1})e^{(\varepsilon_{D1} - \varepsilon_{D0} - \varepsilon_F)/kT}}, \quad \text{(B32)}$$

since l' can only be 0 or 1 (single-charge state). The unoccupied-state energy, ε_{D0}, must vanish since it takes no energy to transfer the "final" electron in this state from the valence band. Therefore,

$$N_D - n_{D1} = N_D - \frac{N_D}{1 + (g_{D0}/g_{D1})e^{(\varepsilon_D - \varepsilon_F)/kT}} = \frac{N_D}{1 + (g_{D1}/g_{D0})e^{(\varepsilon_F - \varepsilon_D)/kT}}, \quad (B33)$$

where we have dropped the subscript 1 on the energy. Again, ε_D is the energy necessary to transfer an electron from the valence band to the *unoccupied* state, making it occupied. Note, however, that we can reference ε_D and ε_F to any fixed energy that we choose, since only the *difference* appears in Eq. (B33). For donors, it is often convenient to use the conduction band edge as a reference.

Finally, we will assume that ε_F is much closer to the conduction band than the valence band, so that $n \gg p$. Equation (B30) then becomes

$$n + N_{AS} = \frac{N_D}{1 + (g_{D1}/g_{D0})e^{(\varepsilon_F - \varepsilon_D)/kT}} + N_{DS}. \quad (B34)$$

By referencing ε_F and ε_D to the conduction band edge, we can use Eq. (B14) to write $n = N_C \exp(\varepsilon_F/kT)$. Also we define what we will call a "modified" density of states function:

$$\phi_{DC} = \frac{g_{D0}}{g_{D1}} N_C e^{\varepsilon_D/kT} = \frac{g_{D0}}{g_{D1}} N'_C T^{3/2} e^{-E_{DC}/kT}, \quad (B35)$$

where $N'_C = 2(2\pi m_n^* k)^{3/2}/h^3$ and $E_{DC} = \varepsilon_C - \varepsilon_D = -\varepsilon_D$, the donor ionization energy with respect to the conduction band (a positive quantity). Thus, finally, Eq. (B33) becomes

$$(n + \phi_{DC})[n + (N_{AS} - N_{DS})] = \phi_{DC} N_D. \quad (B36)$$

This quadratic equation can be solved for n:

$$n = \frac{1}{2}[\phi_{DC} + (N_{AS} - N_{DS})]\left\{\left[1 + \frac{4\phi_{DC}[N_D - (N_{AS} - N_{DS})]}{[\phi_{DC} + (N_{AS} - N_{DS})]^2}\right]^{1/2} - 1\right\}. \quad (B37)$$

b. *One Single-Charge-State Acceptor*

As before, we will assume that $l = 0$ or 1 for all donors and acceptors in the system, including the dominant acceptor, designated by subscript A. Then Eq. (B29) becomes

$$n + n_{A1} + \sum_{k \neq A \text{ acceptors}} n_{k1} = p + \sum_{k \text{ donors}} (N_k - n_{k1}). \quad (B38)$$

From Eq. (B26),

$$n_{A1} = \frac{N_A}{1 + \sum_{l' \neq 1}(g_{Al'}/g_{A1})e^{\eta_{A1}-\eta_{Al'}-(1-l')\eta_F}}$$

$$= \frac{N_A}{1 + (g_{A0}/g_{A1})e^{(\varepsilon_A-\varepsilon_F)/kT}}, \quad (B39)$$

where, as before, $\varepsilon_{A0} = 0$, and we have defined $\varepsilon_{A1} \equiv \varepsilon_A$. By using Eq. (B15) we can write

$$n_{A1} = \frac{N_A}{1 + (g_{A0}/g_{A1})e^{\varepsilon_A/kT}(p/N_V)e^{-\varepsilon_G/kT}} = \frac{N_A}{1 + p/\phi_{AV}}, \quad (B40)$$

where

$$\phi_{AV} = \frac{g_{A1}}{g_{A0}} N_V e^{(\varepsilon_G-\varepsilon_A)/kT} = \frac{g_{A1}}{g_{A0}} N'_V T^{3/2} e^{-E_{AV}/kT}. \quad (B41)$$

Here $N'_V = 2(2\pi m_p^* k)^{3/2}/h^3$, and $E_{AV} = \varepsilon_A - \varepsilon_G$, the acceptor ionization energy with respect to the valence band. By making use of the definitions in Eq. (B31), and assuming that $p \gg n$, Eq. (B38) becomes

$$\frac{N_A}{1 + p/\phi_{AV}} + N_{AS} = p + N_{DS}, \quad (B42)$$

which can be written as

$$(p + \phi_{AV})[p + (N_{DS} - N_{AS})] = \phi_{AV} N_A. \quad (B43)$$

This equation has precisely the same form as its donor analog, Eq. (B36), except for the inversion of the degeneracy factor in ϕ_{AV}.

It is interesting to suppose that our dominant acceptor here is much closer to the *conduction* band than the valence band. Then $n \gg p$, and we can write Eq. (B38) as

$$n + \frac{N_A}{1 + n_i^2/n\phi_{AV}} + N_{AS} = N_{DS}$$

or

$$n + \frac{N_A}{1 + \phi_{AC}/n} + N_{AS} = N_{DS}, \quad (B44)$$

where

$$\phi_{AC} \equiv n_i^2/\phi_{AV} = \frac{g_{A0}}{g_{A1}} N'_C T^{3/2} e^{-E_{AC}/kT}. \quad (B45)$$

Here the subscript AC denotes an acceptor level referenced to the conduction band. Equation (B44) can be written as

$$(n + \phi_{AC})[(N_{DS} - N_{AS}) - n] = nN_A. \quad (B46)$$

This form is obviously different than that of Eq. (B36) or Eq. (B43). Thus, we have a hope of determining whether a particular center, say near the conduction band, is a donor or acceptor; i.e., the n versus T behavior should be different in the two cases. Unfortunately, this difference is not always recognizable, as was discussed in Section 5, in connection with Fig. 3. For the data of Fig. 3, the donor fit [Eq. (B36)] is somewhat better than the acceptor fit [Eq. (B46)], but not significantly better. The differences in the two fits would become more appreciable at lower temperatures, but then impurity conduction effects begin to affect the measurements. One point, which should be clear from the above discussion, is that it is often difficult to distinguish between donor and acceptor behavior. Usually, it is assumed that an n-type sample must have a dominant donor near ε_F, and a p-type sample, a dominant acceptor. However, there are certainly many counterexamples in the literature (Lorenz et al., 1964). Mobility measurements can sometimes help since an occupied acceptor is charged (negatively), while an occupied donor is neutral (if both are single-charge-state centers). Thus, a decrease in temperature should increase the concentration of ionized scattering centers in the acceptor case, and decrease this concentration in the donor case.

c. One Donor, One Acceptor

For this case, we assume that a single-charge-state donor and a single-charge-state acceptor are both within a few kT of ε_F, and are of comparable concentration. Then Eq. (B29) becomes

$$n + n_{A1} + \sum_{k \neq A \text{ acceptors}} n_{k1} = p + N_D - n_{D1} + \sum_{k \neq D \text{ donors}} (N_k - n_{k1}). \quad (B47)$$

From Eqs. (B31), (B33), (B35), (B44), and (B45), it should be clear that we can write Eq. (B47) as

$$n + \frac{N_A}{1 + \phi_{AC}/n} + N_{AS} = \frac{n_i^2}{n} + \frac{N_D}{1 + n/\phi_{DC}} + N_{DS}. \quad (B48)$$

This equation leads to a quartic in n, but for certain cases (e.g., SI GaAs), we can neglect n and n_i^2/n with respect to N_{AS} and N_{DS}. That is, the carrier concentrations are typically only 10^6–10^7 cm^{-3} in SI GaAs, whereas the impurity concentrations are typically 10^{16} cm^{-3}. In this approximation, Eq. (B48) becomes

$$n = \frac{1}{2B} \{C\phi_{DC} + A\phi_{AC} \pm [C^2\phi_{DC}^2 + A^2\phi_{AC}^2 + 2(2N_DN_A - AC)\phi_{AC}\phi_{DC}]^{1/2}\}, \quad (B49)$$

where $A \equiv N_{AS} - N_{DS}$, $B \equiv N_{DS} - N_{AS} - N_A$, and $C = N_A - N_D + N_{AS} - N_{DS}$. The one donor–one acceptor case is important in SI GaAs because of simultaneous O doping, and Cr doping.

d. One Double-Charge-State Center

A particularly important application of Eq. (B26) is the system with two charge states within the gap, i.e., two ionizable electrons. Then $l = 0, 1,$ or 2. We will not consider excited states, and thus will suppress m. The three occupation numbers can be written down immediately by referring to Eq. (B26):

$$n_{k0} = \frac{N_k}{1 + (g_{k1}/g_{k0})e^{\eta_F - \eta_{k1}} + (g_{k2}/g_{k0})e^{2\eta_F - \eta_{k2}}}, \qquad \text{(B50a)}$$

$$n_{k1} = \frac{N_k}{1 + (g_{k0}/g_{k1})e^{\eta_{k1} - \eta_F} + (g_{k2}/g_{k1})e^{\eta_{k1} - \eta_{k2} + \eta_F}}, \qquad \text{(B50b)}$$

$$n_{k2} = \frac{N_k}{1 + (g_{k0}/g_{k2})e^{\eta_{k2} - 2\eta_F} + (g_{k1}/g_{k2})e^{\eta_{k2} - \eta_{k1} - \eta_F}}. \qquad \text{(B50c)}$$

Here all energies are referred to the *valence* band edge. For example, ε_{k2} is the energy necessary to put two electrons from the valence band edge onto a center of group k. Thus, $\varepsilon_{k0} = \eta_{k0} = 0$, since the state with no ionizable electrons is resonant with the valence band. The average number of electrons per center can be shown to be (Look, 1981)

$$n_{avg} = \frac{n_{k1} + 2n_{k2}}{N_k} = \frac{2}{1 + \dfrac{1 + 2\dfrac{g_{k0}}{g_{k1}} \exp(\eta_{k1} - \eta_F)}{1 + 2\dfrac{g_{k2}}{g_{k1}} \exp(\eta_{k1} - \eta_{k2} + \eta_F)}}. \qquad \text{(B51)}$$

Equation (B51) is a generalization of a well-known formula due to Harvey Brooks (1955). The derivation was not published, but the formula is often quoted (Putley, 1968, p. 136) either with factors of 1 in front of the exponentials, or factors of 4. The first case, which requires that $g_{k0}/g_{k1} = g_{k2}/g_{k1} = \frac{1}{2}$, can be easily understood in terms of an orbital singlet state, which can hold 0, 1, or 2 electrons ($g_0 = 1, g_1 = 2, g_2 = 1$). The second case requires $g_{k0}/g_{k1} = g_{k2}/g_{k1} = 2$, and can perhaps be represented by two singlet states at different energies ($g_0 = 2, g_1 = 1, g_2 = 2$).

Formulas such as Eq. (B51) become useful when dealing with states having several charge configurations within the band gap. For example, Cr in GaAs has charge states Cr^{4+}, Cr^{3+}, and Cr^{2+}, at standard pressure (Stauss *et al.*, 1980), and Cr^{1+} at higher pressures (Hennel *et al.*, 1981). The g_{kl} can be determined from the ground state symmetries. If excited states exist within a few kT of any of the ground states, then summations over "m" must be included in Eq. (B26).

Suppose our double-charge-state center is an acceptor. Then the charge balance expression [Eq. (B29)] becomes

$$n + n_{A1} + 2n_{A2} + N_{AS} = p + N_{DS} \tag{B52}$$

or

$$n + \frac{2N_A}{1 + \dfrac{1 + 2(g_{A0}/g_{A1})\exp[(\varepsilon_{A1} - \varepsilon_F)/kT]}{1 + 2(g_{A2}/g_{A1})\exp[(\varepsilon_{A1} - \varepsilon_{A2} + \varepsilon_F)/kT]}} + N_{AS} = p + N_{DS}. \tag{B53}$$

The usefulness of Eq. (B26) is obvious here, because, although we could probably have written down the charge-balance equations for the previous cases by inspection, we certainly could not have done so for the double-charge-state case.

Before leaving this example, we should caution the reader about a possible problem in the interpretation of the various energies. We *experimentally* measure *ionization* energies, or the energy it takes to add or subtract *one* electron. For any state capable of changing its occupation number only from zero to one (or vice versa), no problem exists, because ε_{A1} is not only the energy of that one electron (relative to the valence band), but is also the energy necessary to change the center from a zero-electron state to a one-electron state. However, the two-electron state cannot be interpreted so simply, because ε_{A2} is the energy necessary to go from the *zero*-electron state to the *two*-electron state, whereas the *experimentally* measured quantity will, in general, be the energy required to change the *one*-electron state to the two-electron state. Thus, the relevant *ionization* energy corresponding to the two-electron state is $E_{A2} = \varepsilon_{A2} - \varepsilon_{A1}$. (For the l-electron state, the ionization energy will be $E_l = \varepsilon_l - \varepsilon_{l-1}$.) Multielectron formulas such as Eq. (B53) can be written in terms of ionization energies, if desired, but, in any case, the distinction should be remembered when comparing with experimental data.

e. Degeneracy Factors

To determine the various state degeneracies, the g_{klm}, we, of course, have to know something about the states themselves. Sometimes symmetry considerations are sufficient to determine the g_{klm}, even when the various state energies are totally unknown. We will consider only two simple examples here.

Suppose a donor electron occupies an s-like state, e.g., an a_1 state for a center with tetrahedral symmetry. Such a state is orbitally nondegenerate, but may accomodate two electrons, with opposite spins. Thus, the zero-electron

2. PROPERTIES OF SEMI-INSULATING GaAs

state has $g_{D0} = 2!/2! = 1$, the one-electron state, $g_{D1} = 2!/1!1! = 2$, and the two-electron state, $g_{D2} = 2!/2! = 1$. The degeneracy factor in Eqs. (B33) and (B34) is therefore $g_{D1}/g_{D0} = 2$. This result is used quite commonly, but its justification is often "hand-waved," rather than put on a simple mathematical basis, as is done here. The problem with the hand waving is that more complicated situations may be intractable with the same reasoning.

We next consider an acceptor center near the valence band. Here the states can sometimes be described in terms of a deep s-like state (not important in the problem), and three p-like states ($p_+ = p_x + ip_y$, $p_- = p_x - ip_y$, and p_z). Suppose that these p-like states are degenerate (or nearly so, within a few kT), so that their six electrons (three spin-up, three spin-down) can be considered equivalent. Consider the acceptor formed by Cd on a Ga site in GaAs. This site has seven electrons, two in the deep s-like state and five in the p-like states. Therefore, the *unoccupied* state has degeneracy $g_{A0} = 6!/5!1! = 6$, and the occupied state (after accepting one electron) has $g_{A1} = 6!/6! = 1$. Thus, in Eq. (B41), the degeneracy factor is $g_{A1}/g_{A0} = \frac{1}{6}$.

A more realistic situation might be to assume that the p_z level is well below the p_+ and p_- levels, as is true for the GaAs valence bands (Teitler and Wallis, 1960). The p_+ and p_- states can then hold four electrons. The *unoccupied* state contains three electrons, giving $g_{A0} = 4!/3!1! = 4$, and the occupied state, 4 electrons, giving $g_{A1} = 4!/4! = 1$. In this case, $g_{A1}/g_{A0} = \frac{1}{4}$, certainly the value most commonly assumed for acceptors.

We next consider Cu substituting for Ga, leaving the p_+, p_- levels with only 2 electrons. This center could well be a double-charge-state acceptor, with $g_{A0} = 4!/2!2! = 6$, $g_{A1} = 4!/3!1! = 4$, and $g_{A2} = 4!/4! = 1$. For a given mixture of shallow donors, shallow acceptors, and temperature, only the $l = 0$ and $l = 1$ states might be important. Then $g_{A1}/g_{A0} = \frac{4}{6} = \frac{2}{3}$, a value that has also been invoked on several occasions. The general case here, for any ε_F and T, would be given by Eq. (B51), with $g_{A0}/g_{A1} = \frac{3}{2}$, and $g_{A2}/g_{A1} = \frac{1}{4}$:

$$n_{\text{avg}} = \frac{2}{1 + \dfrac{1 + 3\exp[(\varepsilon_{A1} - \varepsilon_F)/kT]}{1 + (\frac{1}{2})\exp[(\varepsilon_{A1} - \varepsilon_{A2} + \varepsilon_F)/kT]}}. \tag{B54}$$

The complexity of this example (Cu in GaAs) can be increased by supposing that a single-donor state is also possible, i.e., that one of the two (p_+, p_-) electrons can be excited by less than band-gap energy. Then the $l = 0$ state has one electron, and a degeneracy $g_{A0} = 4!/3!1! = 4$. Similarly, $g_{A1} = 4!/2!2! = 6$, $g_{A2} = 4$, and $g_{A3} = 1$. The charge-balance equation must include these additional electrons available for distribution. Thus, Eq. (B28) becomes

$$\text{RHS} = N_{Vt} + \sum_{k \text{ donors}} l_k N_k + \sum_{k \text{ acceptors}} l'_k N_k, \tag{B55}$$

where l'_k is the total number of ionizable electrons on a given acceptor *when in its neutral state*. For our example here, $l'_k = 1$. The equating of Eqs. (B27) and (B28) now gives

$$n + \sum_{k,l,m \text{ acceptors}} ln_{klm} = p + \sum_{k \text{ donors}} \left[l_k N_k - \sum_{l,m} ln_{klm} \right] + \sum_{k \text{ acceptors}} l'_k N_k, \quad (B56)$$

which can be written as a "generalized" charge-balance equation:

$$n + \sum_{k,l,m \text{ acceptors}} (l - l'_k) n_{klm} = p + \sum_{k \text{ donors}} \left[l_k N_k - \sum_{l,m} ln_{klm} \right]. \quad (B57)$$

Our example then gives

$$n - n_{\text{Cu}0} + n_{\text{Cu}2} + 2n_{\text{Cu}3} + N_{\text{AS}} = p + N_{\text{DS}}, \quad (B58)$$

where the various $n_{\text{Cu}l}$ can be determined from Eq. (B26) and the state degeneracies computed above. Note that Eq. (B58) is not really a "charge-balance" equation, as such, although it could be made into one by transposing the $n_{\text{Cu}0}$ term to the right-hand side.

Finally, by noting that $N_k = \sum_{l,m} n_{klm}$ we can write Eq. (B56) in its simplest form:

$$n = p + \sum_{k,l,m \text{ donors, acceptors}} (l_k - l) n_{klm}, \quad (B59)$$

where all centers, donors or acceptors, are treated equally. For example, an (unlikely) center with two acceptor states and three donor states would have $l_k = 3$, and $l = 0, 1, 2, 3, 4,$ or 5. Although we could formally write out Eq. (B59) for this center and, indeed, any number of others, the number of parameters involved (energies, degeneracies, and concentrations) would be too great for any sort of meaningful fit to n (or p) versus T data, unless many of the parameters were already known.

Appendix C. Derivation of TSC and PITS Equations

We wish to solve, under various conditions, the following differential equations for the electron and hole concentrations:

$$\frac{dn}{dt} = \sum_i e_{ni} n_i - n \sum_i \sigma_{ni} v_n (N_i - n_i), \quad (C1)$$

$$\frac{dp}{dt} = \sum_i e_{pi} (N_i - n_i) - p \sum_i \sigma_{pi} v_p n_i, \quad (C2)$$

where all symbols have been defined in Sections 7–10, and in the List of Symbols. As before, we assume that only one dominant electron (or hole)

2. PROPERTIES OF SEMI-INSULATING GaAs

trap is emitting and trapping carriers, while the rest are simply acting as recombination centers. We also assume that recombination between free electrons and free holes is negligible. Then it follows that

$$\frac{dn}{dt} = e_{ni}n_i - n\sigma_{ni}v_n(N_i - n_i) - n \sum_{\substack{j \neq i \text{ recomb. centers}}} \sigma_{nj}v_n(N_j - n_j), \quad \text{(C3)}$$

$$\frac{dp}{dt} = e_{pi}(N_i - n_i) - p\sigma_{pi}v_p n_i - p \sum_{\substack{j \neq i \text{ recomb. centers}}} \sigma_{pj}v_p n_j. \quad \text{(C4)}$$

We will further define the recombination times, τ_n and τ_p, as

$$\tau_n^{-1} = \sum_{\substack{j \neq i \text{ recomb. centers}}} \sigma_{nj}v_n(N_j - n_j), \quad \text{(C5)}$$

$$\tau_p^{-1} = \sum_{\substack{j \neq i \text{ recomb. centers}}} \sigma_{pj}v_p n_j. \quad \text{(C6)}$$

If $\tau_n^{-1} \gg \sigma_{ni}v_n(N_i - n_i)$ and $\tau_p^{-1} \gg \sigma_{pi}v_p n_i$, then Eqs. (C3) and (C4) become linear in n (or p) and n_1:

$$\frac{dn}{dt} = e_{ni}n_i - \frac{n}{\tau_n}, \quad \text{(C7)}$$

$$\frac{dp}{dt} = e_{pi}(N_i - n_i) - \frac{p}{\tau_p}. \quad \text{(C8)}$$

22. TSC Equation: Electrons

We suppose that only electrons are being emitted in appreciable quantities, and $n \gg p$ at all times. Then, if a constant heating rate, $a \equiv dT/dt$ is imposed upon the sample, Eq. (C7) becomes

$$\frac{dn}{dt} = a\frac{dn}{dT} = e_{ni}n_i - \frac{n}{\tau_n}, \quad \text{(C9)}$$

where e_{ni} is given by Eq. (31). This equation can be solved by use of the integrating factor $\exp(T/a\tau_n)$ to get

$$n = n_{i0}a^{-1}e^{-T/a\tau_n}\int_{T_0}^{T} e_{ni}e^{-Q+T'/a\tau_n}\,dT' + Ce^{-T/a\tau_n}, \quad \text{(C10)}$$

where C is a constant and

$$Q = \int_{T_0}^{T} \frac{e_{ni}}{a}\,dT'. \quad \text{(C11)}$$

A glance at the temperature dependence of e_{ni}, as shown in Eq. (31), quickly convinces one that Eq. (C10) cannot be solved analytically. However, progress can be made by moving the $\exp(-T/a\tau_n)$ factor inside the integral. Then

$$n = n_{t0}a^{-1}\int_{T_0}^{T} e_{ni}(T')e^{-Q(T')}e^{(T'-T)/a\tau_n}\,dT' + Ce^{-T/a\tau_n}. \quad \text{(C12)}$$

Now if τ_n^{-1} is large enough, specifically, if $\tau_n^{-1} \gg e_{ni}$, then the $\exp[(T' - T)/a\tau_n]$ term will dominate the temperature dependence and, in fact, will keep the integrand small until T' approaches T. Thus, we can evaluate the slowly varying part, $e_{ni}(T')\exp[-Q(T')]$ at $T' = T$ and pull it out of the integral. That is,

$$\int_{T_0}^{T} e_{ni}(T')e^{-Q(T')}e^{(T'-T)/a\tau_n}\,dT = e_{ni}(T)e^{-Q(T)}\int_{T_0}^{T} e^{(T'-T)/a\tau_n}\,dT$$

$$= e_{ni}(T)e^{-Q(T)}a\tau_n[1 - e^{-(T-T_0)/a\tau_n}]. \quad (C13)$$

Here the last term will be negligible if $\tau_n \ll 1$ s, since usually $a \simeq 1°$K/s. Also, we will be able to nelgect the last term in Eq. (C10) [i.e., $C\exp(-T/a\tau_n)$] in the spirit of the present approximations. The final result is therefore

$$n = n_{i0}\tau_n e_{ni}\exp\left(-\int_{T_0}^{T}\frac{e_{ni}}{a}\,dT'\right). \quad (C14)$$

The case for hole emission can be solved similarly.

23. PITS Equation: Electrons

We assume that a strong light pulse of duration t_p has been applied to the sample. Then the solution for $n_i(t)$ is given by Eq. (63) in the text:

$$n_i(t) = \frac{N_i}{1 + e_{ni}/\beta_{ni}}[1 - e^{-(e_{ni}+\beta_{ni})t_p}]e^{-e_{ni}(t-t_p)} \equiv C_{ni}(t_p)e^{-e_{ni}t} \quad (C15)$$

for times $t \geq t_p$. Equation (C7) can now be solved by use of the integrating factor $\exp(t/\tau_n)$:

$$\int_{t_p}^{t}\left(\frac{dn}{dt} + \frac{n}{\tau_n}\right)e^{t/\tau_n}\,dt = \int_{t_p}^{t} e_{ni}n_i e^{t/\tau_n}\,dt$$

$$= C_{ni}(t_p)e_{ni}\int_{t_p}^{t} e^{-e_{ni}t}e^{t/\tau_n}\,dt \quad (C16)$$

or

$$ne^{t/\tau_n} - n(t_p)e^{t_p/\tau_n} = C_{ni}(t_p)\frac{e_{ni}}{\tau_{ni}^{-1} - e_{ni}}[e^{-(e_{ni}-\tau_n^{-1})t}$$

$$- e^{-(e_{ni}-\tau_n^{-1})t_p}]. \quad (C17)$$

By using Eq. (C15) for $C_{ni}(t_p)$ we get

$$n(t) = \frac{e_{ni}}{(\tau_n^{-1} - e_{ni})}\frac{N_i}{(1 + e_{ni}/\beta_{ni})}$$

$$\times [1 - e^{-(e_{ni}+\beta_{ni})t_p}][e^{-e_{ni}(t-t_p)} - e^{-(t-t_p)/\tau_n}] + n(t_p)e^{-(t-t_p)/\tau_n}. \quad (C18)$$

Equation (C18) is identical to Eq. (65) in the text.

24. PITS Equation: Electrons and Holes

When both electrons and holes are important we must modify the equation for dn_i/dt:

$$\frac{dn_i}{dt} = -e_{ni}n_i + e_{pi}(N_i - n_i) + n(N_i - n_i)\sigma_{ni}v_n - pn_i\sigma_{pi}v_p$$

$$= -(e_{ni} + e_{pi} + n\sigma_{ni}v_n + p\sigma_{pi}v_p)n_i + (e_{pi} + n\sigma_{ni}v_n)N_i. \quad (C19)$$

As before, we are assuming that only one trap (trap i) is dominant, at least in the temperature range of interest. Now Eq. (C19) is obviously nonlinear in n, p, and n_i, but we can linearize it with reasonable assumptions, in the two different time regimes, as described below.

a. $t < t_p$

During the light pulse ($t = 0$ to t_p) we assume that the carrier concentrations are essentially constant and given by $n_l = I_0\alpha_n\tau_n$ and $p_l = I_0\alpha_p\tau_p$. This assumption is justified if the carrier concentrations rise and fall in times much less than t_p (i.e., if τ_n, $\tau_p \ll t_p$). Then Eq. (C19) becomes

$$\frac{dn_i}{dt} = -(e_{ni} + e_{pi} + \beta_{ni} + \beta_{pi})n_i + (e_{pi} + \beta_{ni})N_i$$

$$\equiv -e_{npi}n_i + (e_{pi} + \beta_{ni})N_i, \quad (C20)$$

where $\beta_{ni} \equiv I_0\alpha_n\tau_n\sigma_{ni}v_n$, $\beta_{pi} \equiv I_0\alpha_p\tau_p\sigma_{pi}v_p$, and $e_{npi} \equiv e_{ni} + e_{pi} + \beta_{ni} + \beta_{pi}$. By use of the integrating factor $\exp(e_{npi}t)$ it is easy to see that the solution for n_i is

$$n_i(t)e^{e_{npi}t} = N_i\frac{e_{pi} + \beta_{ni}}{e_{npi}}e^{e_{npi}t} + C = \frac{N_i}{1 + Z_i}e^{e_{npi}t} + C, \quad (C21)$$

where $Z_i \equiv (e_{ni} + \beta_{pi})/(e_{pi} + \beta_{ni})$. The constant C can be determined by noting that at $t = 0$, before the light is turned on, the carrier concentrations will have their equilibrium values, i.e., $n = n_0$ and $p = p_0$. At equilibrium, $dn_i/dt = 0$, so that Eq. (C20) becomes

$$n_i(0) = N_i/(1 + Z_{i0}) \quad (C22)$$

where $Z_{i0} \equiv (e_{ni} + p_0\sigma_{pi}v_p)/(e_{pi} + n_0\sigma_{ni}v_n)$. Thus,

$$C = N_i\left[\frac{1}{1 + Z_{i0}} - \frac{1}{1 + Z_i}\right] \quad (C23)$$

and

$$n_i(t) = \frac{N_i}{1 + Z_i} + N_i\left[\frac{1}{1 + Z_{i0}} - \frac{1}{1 + Z_i}\right]e^{-e_{npi}t}, \quad t < t_p. \quad (C24)$$

b. $t \geq t_p$

For $t \geq t_p$, we again wish to linearize Eq. (C19), but the task is now more difficult because n and p are no longer essentially constant, as they were with the light on. Nor is it clear that we can always neglect $(n\sigma_{ni}v_n + p\sigma_{pi}v_p)$ with respect to $(e_{ni} + e_{pi})$. To discuss this problem in more detail, we rewrite Eq. (30) in the text:

$$e_{ni} = (g_{i0}/g_{i1})N_C\sigma_{ni}v_n e^{-(\varepsilon_C - \varepsilon_i)/kT}. \tag{C25}$$

By setting Eq. (C2) equal to zero, at equilibrium, it is also easy to get the corresponding equation for e_{pi}:

$$e_{pi} = (g_{i1}/g_{i0})N_V\sigma_{pi}v_p e^{-(\varepsilon_i - \varepsilon_V)/kT}. \tag{C26}$$

From Eqs. (B14) and (B15) we remember that

$$n_0 = N_C e^{-(\varepsilon_C - \varepsilon_F)/kT}, \tag{C27}$$

$$p_0 = N_V e^{-(\varepsilon_F - \varepsilon_V)/kT}, \tag{C28}$$

in the Boltzmann approximation. If the *differences* in the magnitudes of e_{ni}, e_{pi}, $n_0\sigma_{ni}v_n$, and $p_0\sigma_{pi}v_p$ depend mainly upon the exponential terms, then the inequality $e_{ni} + e_{pi} \gg n_0\sigma_{ni}v_n + p_0\sigma_{pi}v_p$ will hold if $\varepsilon_C - \varepsilon_i < \varepsilon_C - \varepsilon_F$, $\varepsilon_F - \varepsilon_V$, or if $\varepsilon_i - \varepsilon_V < \varepsilon_C - \varepsilon_F$, $\varepsilon_F - \varepsilon_V$, i.e., if trap i is closer to *one* of the bands than is ε_F. This statement must be modified accordingly if $\sigma_{ni} \gg \sigma_{pi}$ or vice versa. From this discussion, it is clear that a problem can arise if both ε_F and the trap of interest are near midgap.

With this caveat, we will assume that within a time of order τ_n or τ_p after the pulse has been turned off, the carrier concentrations have decreased to the extent that $e_{ni} + e_{pi} \gg n\sigma_{ni}v_n + p\sigma_{pi}v_p$. As discussed above, this inequality may not hold for midgap traps (since n and p will not fall below their equilibrium values), but we will proceed with this assumption anyway, since it is necessary for obtaining an analytical solution of the problem. Then Eq. (C20) becomes, for $t \geq t_p$,

$$\frac{dn_i}{dt} = -(e_{ni} + e_{pi})n_i + e_{pi}N_i \tag{C29}$$

and, by use of the integrating factor $\exp(e_{ni} + e_{pi})t$, the solution is easily found:

$$n_i = \frac{N_i}{1 + e_{ni}/e_{pi}} + Ce^{-(e_{ni} + e_{pi})t}, \tag{C30}$$

where C, the constant of integration, can be related to $n_i(t_p)$:

$$C = \left[n_i(t_p) - \frac{N_i}{1 + e_{ni}/e_{pi}}\right]e^{(e_{ni} + e_{pi})t_p}. \tag{C31}$$

Finally, we can use Eqs. (C30) and (C31) to solve Eq. (C7) for n, by employing the integrating factor $\exp(t/\tau_n)$:

$$n = \frac{e_{ni}\tau_n N_i}{1 + e_{ni}/e_{pi}} + \frac{e_{ni}\tau_n}{1 - (e_{ni} + e_{pi})\tau_n}\left[n_i(t_p) - \frac{N_i}{1 + e_{ni}/e_{pi}}\right]e^{-(e_{ni}+e_{pi})(t-t_p)}$$
$$+ C_n e^{-(t-t_p)/\tau_n}, \qquad (C32)$$

where C_n is the constant of integration. Now $n_i(t_p)$ can be determined by setting $t = t_p$ in Eq. (C24), but we should also notice that, in the present approximation $(e_{ni} + e_{pi} \gg n_0\sigma_{ni}v_n + p_0\sigma_{pi}v_p)$, the quantity $1/(1 + Z_{i0}) \rightarrow 1/(1 + e_{ni}/e_{pi})$. Thus, Eq. (C32) becomes

$$n = \frac{e_{ni}\tau_n N_i}{1 + e_{ni}/e_{pi}} + \frac{e_{ni}\tau_n N_i}{1 - (e_{ni} + e_{pi})\tau_n}\left(\frac{1}{1 + Z_i} - \frac{1}{1 + e_{ni}/e_{pi}}\right)$$
$$+ (1 - e^{-e_{np}i t_p})e^{-(e_{ni}+e_{pi})(t-t_p)} + C_n e^{-(t-t_p)/\tau_n}. \qquad (C33)$$

This equation for n is repeated in the text as Eq. (80). The corresponding equation for p is given as Eq. (81), but will not be derived here since the same procedure as above is followed.

Acknowledgments

The author wishes to thank the group at the ET&D Laboratory, Ft. Monmouth, for the support that made much of this report possible. Special thanks go to J. Winter, H. Leupold, and L. Ross. The author is also grateful to S. Chaudhuri and A. C. Beer for critical readings of the manuscript and several suggested changes. Many other colleagues at the Avionics Laboratory, Wright-Patterson AFB, and at Wright State University have also made helpful comments regarding various aspects of the material. Finally, thanks are due to G. M. Martin for interesting and informative discussions, and permission to use several drawings from his published works.

References

Allen, G. A. (1968). *J. Phys. D* **1**, 593.
Ashby, A., Roberts, G. G., Ashen, D. J., and Mullin, J. B. (1976). *Solid State Commun.* **20**, 61.
Beer, A. C. (1963). "Galvanomagnetic Effects in Semiconductors" Academic Press, New York.
Betko, J., and Merinsky, K. (1979). *J. Appl. Phys.* **50**, 4212.
Blakemore, J. S. (1980). In *Semi-Insulating III–V Materials, Nottingham, 1980* (G. S. Rees, ed.), p. 29. Shiva Publ., Nantwich, UK.
Blakemore, J. S., Johnson, S. G., and Rahimi, S. (1982). In *Semi-Insulating III–V Compounds, Evian, 1982* (S. Makram-Ebeid and B. Tuck, eds.), p. 172. Shiva Publ., Nantwich, UK.
Blow, K. J., and Inkson, J. C. (1980). *J. Phys. C* **13**, 359.
Brooks, H. (1955). *Adv. Electron. Electron Phys.* **7**, 85.
Bube, R. H. (1960). "Photoconductivity of Solids," p. 294. Wiley, New York.

Bube, R. H. (1974). "Electronic Properties of Crystalline Solids: An Introduction to Fundamentals," Chap. 10. Academic Press, New York. [Note that Bube's Eq. (6.20), p. 172, does not include spin degeneracy.]
Castagne, M., Bonnate, J., Romestau, J., and Fillard, J. P. (1980). *J. Phys. C* **13**, 5555.
Chantre, A., Vincent, G., and Bois, D. (1981). *Phys. Rev. B* **23**, 5335.
Clark, M. G. (1980). *J. Phys. C* **13**, 2311.
Cronin, G. R., and Haisty, R. W. (1964). *J. Electrochem. Soc.* **III**, 874.
Debney, B. T., and Jay, P. R. (1980). *Solid State Electron.* **23**, 773.
Deveaud, B., and Toulouse, B. (1980). In *Semi-Insulating III–V Materials, Nottingham, 1980* (G. S. Rees ed.), p. 241. Shiva Publ., Nantwich, UK.
Ehrenreich, H. (1960). *Phys. Rev.* **120**, 1951.
Fairman, R. D., and Oliver, J. R. (1980). In *Semi-Insulating III–V Materials, Nottingham, 1980* (G. S. Rees, ed.), p. 83. Shiva Publ., Nantwich, UK.
Fairman, R. D., Morin, F. J., and Oliver, J. R. (1979). *Conf. Ser. Inst. Phys.* No. 45, p. 134.
Fazzio, A., Brescansin, L. M., Caldas, M. J., and Leite, J. R. (1979). *J. Phys. C* **12**, L831.
Fillard, J. P., Gasiot, J., and Manifacier, J. C. (1978). *Phys. Rev. B* **18**, 4497.
Ford, L. (1955). "Differential Equations." McGraw-Hill, New York.
Haloulos, S. G., Papastamatiou, M. J., Kalkanis, G. T., Nomicos, C. D., Euthymiou, P. C., and Papaioannou, G. J. (1980). *Solid State Commun.* **34**, 245.
Hemenger, P. M. (1973). *Rev. Sci. Instrum.* **44**, 698.
Hemstreet, L. A. (1980). *Phys. Rev. B* **22**, 4590.
Hemstreet, L. A., and Dimmock, J. O. (1979). *Phys. Rev. B* **20**, 1527.
Hennel, A. M., Szuszkiewicz, W., Balkanski, M., Martinez, G., and Clerjaud, B. (1981). *Phys. Rev. B* **23**, 3933.
Holeman, B. R., and Hilsum, C. (1961). *J. Phys. Chem. Solids* **22**, 19.
Hrivnak, L., Betko, J., and Merinsky, K. (1982). In *Semi-Insulating III–V Materials, Evian, 1982* (S. Makram-Ebeid and B. Tuck, eds.), p. 139. Shiva Publ., Nantwich, UK.
Huber, A. M., Linh, N. T., Debrun, J. C., Valladon, M., Martin, G. M., Mitonneau, A., and Mircea, A. (1979). *J. Appl. Phys.* **50**, 4022.
Hurtes, Ch., Hollan, L., and Boulou, M. (1978a). *Conf. Ser. Inst. Phys.* No. 45, p. 342.
Hurtes, Ch., Boulou, M., Mitonneau, A., and Bois, D. (1978b). *Appl. Phys. Lett.* **32**, 821.
Inoue, T., and Ohyama, M. (1970). *Solid State Commun.* **8**, 1309.
Ippolitova, G. K., Omelianovski, E., and Pervova, L. Ya. (1976). *Sov. Phys.—Semicond.* (*Eng. Transl.*) **9**, 864.
Jackson, J. D. (1962). "Classical Electrodynamics," p. 611. Wiley, New York.
Jesper, T., Hamilton, B., and Peaker, A. R. (1980). In *Semi-Insulating III–V Materials, Nottingham, 1980* (G. S. Rees, ed.), p. 233. Shiva Publ., Nantwich, UK.
Kang, K. N., Cristoloveanu, S., and Chovet, A. (1982). In *Semi-Insulating III–V Materials, Evian, 1982* (S. Makram-Ebeid and B. Tuck, eds.), p. 113. Shiva Publ., Nantwich, UK.
Kaufmann U., and Schneider, J. (1976). *Solid State Commun.* **20**, 143.
Kirchner, P. D., Schaff, W. J., Maracas, G. N., Eastman, L. F., Chappell, T. I., and Ransom, C. M. (1981). *J. Appl. Phys.* **52**, 6462.
Kocot, K., Rao, R. A., and Pearson, G. L. (1979). *Phys. Rev. B* **19**, 2059.
Koschel, W. H., Bishop, S. G., and McCombe, B. D. (1976). *Solid State Commun.* **19**, 521.
Krebs, J. J., and Stauss, G. H. (1977). *Phys. Rev. B* **15**, 17.
Krebs, J. J., and Stauss, G. H. (1979). *Phys. Rev. B* **20**, 795.
Lang, D. V. (1974). *J. Appl. Phys.* **45**, 3023.
Lang, D. V., and Logan, R. A. (1975). *J. Electron. Mater.* **4**, 1053.
Lee, C. P. (1982). In *Semi-Insulating III–V Materials, Evian, 1982* (S. Makram-Ebeid and B. Tuck, eds.), p. 324. Shiva Publ., Nantwich, UK.

Li, S. S., and Huang, C. I. (1972). *J. Appl. Phys.* **43**, 1757.
Lin, A. L., and Bube, R. H. (1976). *J. Appl. Phys.* **47**, 1859.
Lindquist, P. F. (1977). *J. Appl. Phys.* **48**, 1262.
Ll'in, N. P., and Masterov, V. F. (1977). *Sov. Phys.—Semicond. (Eng. Transl.)* **11**, 864.
Look, D. C. (1975). *J. Phys. Chem. Solids* **36**, 1311.
Look, D. C. (1977a). *Solid State Commun.* **24**, 825.
Look, D. C. (1977b). *Phys. Rev. B* **16**, 5460.
Look, D. C. (1980). In *Semi-Insulating III-V Materials, Nottingham, 1980* (G. S. Rees, ed.), p. 183. Shiva Publ., Nantwich, UK.
Look, D. C. (1981). *Phys. Rev. B* **24**, 5852.
Look, D. C., Walters, D. C., and Meyer, J. R. (1982a). *Solid State Commun.* **42**, 745.
Look, D. C. (1982b). *Phys. Rev. B* **25**, 2920.
Look, D. C. (1982c). In *Semi-Insulating III-V Materials, Evian, 1982* (S. Makram-Ebeid and B. Tuck, eds.), p. 372. Shiva Publ., Nantwich, UK.
Look, D. C., and Farmer, J. W. (1981). *J. Phys. E* **14**, 472.
Look, D. C., Chaudhuri, S., and Eaves, L. (1982). *Phys. Rev. Lett.* **49**, 1728.
Look, D. C., Farmer, J. W., and Ely, R. N. (1980). *Rev. Sci. Instrum.* **51**, 968.
Lorenz, M. R., Segall, B., and Woodbury, H. H. (1964). *Phys. Rev.* **134**, A751.
Lucovsky, G. (1965). *Solid State Commun.* **3**, 299.
Makram-Ebeid, S. (1980). *Appl. Phys. Lett.* **37**, 464.
Martin, G. M. (1981). *Appl. Phys. Lett.* **39**, 747.
Martin, G. M., and Bois, D. (1978). *Proc. Electrochem. Soc.* **78**, 32.
Martin, G. M., Verheijke, M. L., Jansen, J. A. J., and Poiblaud, G. (1979). *J. Appl. Phys.* **50**, 467.
Martin, G. M., Mitonneau, A., Pons, D., Mircea, A., and Woodward, D. W. (1980a). *J. Phys. C* **13**, 3855.
Martin, G. M., Farges, J. P., Jacob, G., and Hallais, J. P. (1980b). *J. Appl. Phys.* **51**, 2840.
Martinez, G., Hennel, A. M., Szuszkiewicz, W., Balkanski, M., and Clerjaud, B. (1981). *Phys. Rev. B* **23**, 3920.
Masut, R., and Penchina, C. M. (1981). *Bull. Am. Phys. Soc.* **26**, 287.
Meyer, J. R., and Bartoli, F. J. (1981). *Phys. Rev. B* **23**, 5413.
Milnes, A. G. (1973). "Deep Impurities in Semiconductors," p. 227. Wiley, New York.
Mitonneau, A., Mircea, A., Martin, G. M., and Pons, D. (1979). *Rev. Phys. Appl.* **14**, 853.
Nag, B. R. (1980). "Electron Transport in Compound Semiconductors." Springer-Verlag, Berlin and New York.
Nam, S. B. (1980). *Bull. Am. Phys. Soc.* **25**, 438.
Neumark, G. F., and Kosai, K. (1983). In "Semiconductors and Semimetals" (R. K. Willardson and A. C. Beer, eds.), this volume. Academic Press, New York.
Nichols, K. G., and Vernon, E. V. (1966). "Transistor Theory," Chap. 3. Chapman & Hall, London.
Noras, J. M. (1980). In *Semi-Insulating III-V Materials, Nottingham, 1980* (G. S. Rees, ed.), p. 292. Shiva Publ., Nantwich, UK.
Pankove, J. I. (1971). "Optical Processes in Semiconductors," p. 93. Dover, New York.
Philadelpheus, A. Th., and Euthymiou, P. C. (1974). *J. Appl. Phys.* **45**, 955.
Podor, B. (1983). *Phys. Rev. B* **27**, 2551 (1983).
Pons, D. (1980). *Appl. Phys. Lett.* **37**, 413.
Putley, E. H. (1968). "The Hall Effect and Semiconductor Physics." Dover, New York.
Ridley, B. K. (1977). *J. Phys. C* **10**, 1589.
Ridley, B. K. (1980). *J. Phys. C* **13**, 2015.
Robert, J. L., Pistoulet, B., Raymond, A., Dusseau, J. M., and Martin, G. M. (1979). *J. Appl. Phys.* **50**, 349.

Rode, D. L. (1970). *Phys. Rev. B* **2**, 1012.
Stauss, G. H., Krebs, J. J., Lee, S. H., and Swiggard E. M. (1980). *Phys. Rev. B* **22**, 3141.
Stillman, G. E., Wolfe, C. M., and Dimmock, J. O. (1970). *J. Phys. Chem. Solids* **31**, 1199.
Szawelska, H. R., and Allen, J. W. (1979). *J. Phys. C* **12**, 3359.
Sze, S. M. (1969). "Physics of Semiconductor Devices," p. 40. Wiley, New York.
Teitler, S., and Wallis, R. F. (1960). *J. Phys. Chem. Solids* **16**, 71.
Tolman, R. C. (1938). "The Principles of Statistical Mechanics," Chap. 10. Oxford, London.
van der Pauw, L. J. (1958). *Philips Res. Reps.* **13**, 1.
Van Vechten, J. A., and Thurmond, C. D. (1976). *Phys. Rev. B* **14**, 3539.
Walukiewicz, W., Pawlowicz, L., Lagowski, J., and Gatos, H. C. (1982). In *Semi-Insulating III–V Materials, Evian, 1982* (S. Makram-Ebeid and B. Tuck, eds.), p. 121. Shiva Publ., Nantwich, UK.
White, A. M. (1980). In *Semi-Insulating III–V Materials, Nottingham, 1980* (G. S. Rees, ed.), p. 3. Shiva Publ., Nantwich, UK.
White, A. M., Day, B., and Grant, A. J. (1979). *J. Phys. C* **12**, 4833.
Wieder, H. H. (1979). "Laboratory Notes on Electrical and Galvanomagnetic Measurements." Elsevier, New York.
Wiley, J. D. (1975). *In* "Semiconductors and Semimetals" (R. K. Willardson and A. C. Beer, eds.), Vol. 10. Academic Press, New York.
Winter, J. J., Leupold, H. A., Ross, R. L., and Ballato, A. (1982). In *Semi-Insulating III–V Materials, Evian, 1982* (S. Makram-Ebeid and B. Tuck, eds.), p. 134. Shiva Publ., Nantwich, UK.
Yu, P. W., and Walters, D. C. (1982). *Appl. Phys. Lett.* **41**, 863.
Zucca, R. (1977). *J. Appl. Phys.* **48**, 1977.
Zucca, R. (1980). In *Semi-Insulating III–V Materials, Nottingham, 1980* (G. S. Rees, ed.), p. 358. Shiva Publ., Nantwich, UK.

CHAPTER 3

Associated Solution Model for Ga–In–Sb and Hg–Cd–Te

R. F. Brebrick, Ching-Hua Su, and Pok-Kai Liao

MATERIALS SCIENCE AND METALLURGY PROGRAM
COLLEGE OF ENGINEERING
MARQUETTE UNIVERSITY
MILWAUKEE, WISCONSIN

	List of Symbols.	172
I.	Introduction.	173
II.	Thermodynamic Equations for the Liquidus Surface of $(A_{1-u}B_u)C(s)$.	178
III.	Solution Thermodynamics.	181
IV.	Associated Solution Model for the Liquid Phase.	186
	1. *Relation between Species Mole Fractions and Composition*	186
	2. *Interaction Terms.*	188
	3. *Development of the Liquid Model for the A–C Binary*	191
	4. *Constraints on Interaction Coefficients in the A–C and B–C Binaries*	196
V.	Ga–In–Sb Ternary.	197
	5. *Ga–In Binary.*	197
	6. *In–Sb Binary.*	198
	7. *Ga–Sb Binary.*	205
	8. *Ga–In–Sb Ternary.*	209
VI.	Hg–Cd–Te Ternary.	214
	9. *Hg–Cd Binary.*	217
	10. *Hg–Te and Cd–Te Binaries.*	217
	11. *Hg–Cd–Te Ternary.*	221
VII.	Summary.	230
	Appendix A	231
	Appendix B	234
	Appendix C	242
	References.	251

List of Symbols

A, B, C	Thermodynamic components. These are the elements Ga, In, and Sb, respectively, in the first application made of the theory and the elements Hg, Cd, and Te, respectively, in the second application.		in the A–C system ($z \equiv y_4$) or of bc in the B–C system ($z \equiv y_5$).
		z^*	Value of z at 50 at. % and the melting point of AC(s) in the A–C system, or value of z at 50 at. % and the melting point of BC(s) in the B–C system.
$x_j, j = $ A, B, C or 1, 2, 3	Atom fraction of the component j in the liquid.		
N_j	Number of gram-atoms of component j in the liquid.	$(A_{1-u}B_u)_{2-y}C_y(s)$	Formula for the solid-solution phase formed in the A–B–C system. The parameter y is assumed to always be close to unity.
a, b, c, ac, bc	Species assumed present in the liquid phase. Except in Part III, these are numbered consecutively 1 through 5 and are, respectively, Ga, In, Sb, GaSb, and InSb in the first application and, respectively, Hg, Cd, Te, HgTe, and CdTe in the second application. In Part III the A–C system with species a, c, and ac is treated. In order to utilize the summation symbol, these components are numbered 1 and 2 here and the species are numbered 1, 2, and 3.	R	Gas constant.
		T	Temperature in °K.
		T_{AC}	Maximum (congruent) melting point of AC(s).
		T_{BC}	Maximum melting point of BC(s).
		P	Total partial pressure in atmospheres.
		P_k	Partial pressure in atmospheres of k where k is Hg, Cd, or 2 (for Te$_2$).
		P_k°	Partial pressure in atmospheres of k over its pure liquid phase.
		ΔG_M	Gibbs energy of mixing for the liquid, i.e., change in the Gibbs energy upon forming the liquid solution at constant T and P from its pure liquid components A(l), B(l), and C(l).
n_j	Number of moles of species j in the liquid.		
$y_j, j = 1, \ldots, 5$	Mole fraction of species j in the liquid.	ΔG_M (gram-atom components)	Gibbs energy of mixing of the liquid for 1 gram-at. of components.
$y_j, j = 1, \ldots, 3$	In Part III, only these are the mole fraction of species a, c, and ac, respectively.	ΔG_M (mole species)	Gibbs energy of mixing of the liquid for 1 mol of species.
z	Mole fraction of ac	$\Delta G_m^x \equiv \Delta G_M$ (mole	Excess Gibbs energy

Symbol	Definition
species) $- RT\sum_{j=1}^{5} y_j \ln y_j$	of mixing per mole of species.
$\Delta H_M, \Delta S_M$	Enthalpy and entropy of mixing for the liquid.
ΔC_p	Change in constant pressure heat capacity upon forming the liquid solution from its pure liquid components.
$\mu_j^{\circ,l}, j = A, B, C$ or $1, 2, 3$	Chemical potential of pure liquid component j.
$\mu_j, j = 1, \ldots, 5$	Chemical potential of species j in the liquid.
$\bar{\mu}_j = \mu_j - \mu_j^{\circ,l}$, $j = 1, 2, 3$	Relative chemical potential of monatomic species j and of component j in the liquid.
$\bar{\mu}_4 = \mu_4 - \mu_A^{\circ,l} - \mu_C^{\circ,l}$	Relative chemical potential of species number 4, ac, in the liquid.
$\bar{\mu}_5 = \mu_5 - \mu_B^{\circ,l} - \mu_C^{\circ,l}$	Relative chemical potential of species number 5, bc, in the liquid.
$\bar{\mu}_j^x = \bar{\mu} - RT \ln y_j$, $j = 1, \ldots, 5$	Relative, excess chemical potential of species j in the liquid.
$\gamma_j = \exp(\bar{\mu}_j^x / RT)$	Activity coefficient of species j in the liquid.
$\mu_j^s, j = A, B, C$	Chemical potential of component j in the solid solution.
$\Delta G_f^\circ(AC(s))$, $\Delta H_f^\circ(AC(s))$, $\Delta S_f^\circ(AC(s))$	Standard Gibbs energy, enthalpy, and entropy of formation of AC(s) from A(l) and C(l). Similar quantities are defined for BC(s).
$\Delta G_4^\circ, \Delta H_4^\circ, \Delta S_4^\circ$	Standard Gibbs energy, enthalpy, and entropy of dissociation of the species ac into A(l) and C(l).
$\Delta G_5^\circ, \Delta H_5^\circ, \Delta S_5^\circ$	Standard Gibbs energy, enthalpy, and entropy of dissociation of the species bc into B(l) and C(l).
$\Delta H_{AC}^m, \Delta H_{BC}^m$	Enthalpy of melting of AC(s) and BC(s), respectively.
κ_4, κ_5	Effective equilibrium constants (including the activity coefficient ratio) for the dissociation of, respectively, ac and bc. $y_1 y_3 / y_4 = \kappa_4$; $y_2 y_3 / y_5 = \kappa_5$.
$\bar{h}_j(CO), \bar{s}_j(CO)$	Relative partial molar enthalpy and entropy of *component* j in the liquid.
$\bar{h}_j(\text{species})$, $\bar{s}_j(\text{species})$	Same quantities but for *species* j in the liquid.
$\alpha_{ij}, \beta_{ij}, \gamma_{ijk}$	Linearly temperature-dependent interaction coefficients appearing in the equation for the excess Gibbs energy of mixing that thermodynamically defines the liquid solution model.
$\Omega_A = \omega_A - \nu_A T$, $\Omega_S = \omega_S - \nu_S T$, $\Omega_R = \omega_R - \nu_R T$	The Ω_j are the convenient linear combinations of the α_{ij} coefficients.
$\beta_{ij} = B_{ij} - C_{ij} T$	Temperature dependence of the β_{ij} coefficients.

I. Introduction

In this chapter we show a large part of the experimental phase diagram and thermodynamic data for the Ga–In–Sb and Hg–Cd–Te systems along with calculated values for these data obtained from a simultaneous, quantitatively good, and thermodynamically self-consistent fit to the experimental data.

A prominent feature of the phase diagram for both systems is the existence of a complete range of solid solutions between congruently melting, narrow-homogeneity-range binary compounds, GaSb and InSb in the first system, and HgTe and CdTe in the second. Because one component of the solid solution, Sb or Te, is always near 50 at.%, it is possible to divide the analysis into two independent parts, as is shown later. We confine our considerations to the part concerning the liquidus surface. The analysis then requires a thermodynamic model for the liquid phase and a thermodynamic characterization of the solid solution phase to the extent that the chemical potentials of the thermodynamic components, GaSb and InSb, or HgTe and CdTe, must be given. It is then possible to obtain the properties of the liquid phase, the liquidus surface, the tie-lines connecting the compositions of liquid and solid solution coexisting at equilibrium, partial pressures, etc., but not the solidus surface of the solid solution or any description of the thermodynamic behavior of the solid solution for compositions interior to its stable composition range. The latter would require a specification of the composition dependence of the chemical potentials of all three components in the solid-solution phase and would involve the defect chemistry of the solid-solution phase and its binary compound end members. This second part of the analysis can be carried out independently. However, the chemical potentials on the solidus surface, which are calculated here, are required to determine this surface.

There are three aspects of the analysis given here that are important enough to emphasize:

(1) First, all of what we believe to be the reliable phase diagram and thermodynamic data are considered simultaneously. If a particular set of data is not considered in fixing the values of the adjustable parameters of the liquid model, it is still required that the final values for these adjustable parameters lead to a good fit to this data as well as all the rest.

(2) Second, the congruent melting of the binary compounds introduce pinning points for any proposed liquid model (Brebrick et al., 1981). For instance, at the 1092°C melting point of CdTe(s), the enthalpy of mixing of the 50-at.% liquid must equal the sum of the enthalpy of formation of CdTe(s) from its pure liquid elements and the enthalpy of fusion. A similar constraint binds the entropy of mixing of the 50-at.% Cd–Te liquid at 1092°C. Moreover, the fact that the enthalpy of mixing of the 50-at.% Cd–Te liquid at 1092°C is about −9 kcal/gram-at. shows that strong cohesive forces remain after the destruction of long-range order and that a liquid model with weak interactions is inappropriate. In contrast, the enthalpy of mixing for a number of 50. at % III–V liquids (Brebrick, 1977) is between zero and −1.5 kcal/gram-at.

(3) Third, the basic liquidus equations used here are *not* equivalent to

the commonly used simplification of the equation given by Vieland (1963), in which the constant pressure heat capacity of CdTe(s) and of CdTe(l) would be assumed to be equal. Such an assumption introduces an error into the calculation of liquidus temperatures that can be significant (Brebrick *et al.*, 1981; Tung *et al.*, 1981a).

Many models for the liquid phase have been used in the analysis of phase diagrams. Some of these are discussed briefly here to provide a basis for the choice of model we made and to cite earlier work. Most of the relevant points can be made by considering models for two component liquids. A number of simple models are special cases of one that is defined by giving the excess Gibbs energy of mixing as

$$\Delta G_M^x = Wx(1-x)[1 + a(x - \tfrac{1}{2})] - VTx(1-x)[1 + b(x - \tfrac{1}{2})], \quad (1)$$

where W, a, V, and b are adjustable parameters independent of temperature T and atom fraction x. The excess Gibbs energy on the left of Eq. (1) is, of course, for the process

$(1 - x)$A (pure elemental liquid) + xB (pure elemental liquid)

$\rightarrow A_{1-x}B_x$ (liquid solution). $\quad (2)$

Thermodynamic definitions show that the first term of Eq. (1) is the enthalpy of mixing, ΔH_M, while the second term is the negative of the excess entropy of mixing, ΔS_M^x, multiplied by T. When all four parameters are zero, the liquid is ideal with a zero enthalpy and excess entropy of mixing. What has been called the quasiregular model, $a = b = 0$, has been used by Panish and Ilegems (1972) to fit the liquidus lines of a number of III–V binary compounds. The particular extension of this special case of Eq. (1) to a ternary liquid given by

$$\Delta G_M^x = \Omega_{12}x_1x_2 + \Omega_{23}x_2x_3 + \Omega_{13}x_1x_3, \quad (3)$$

where the interaction coefficients Ω_{ij} depend linearly on T, was then used in an analysis of ternary systems derived from two III–V systems with one element in common. However, the auxiliary conditions [referred to under aspect (2)] in the last paragraph were not considered. Subsequently it was shown for four III–V binary systems (Brebrick, 1977) that the quasiregular model was inadequate and the full subregular model defined by Eq. (1) was necessary if the auxiliary conditions were to be satisfied as well as a good fit obtained to the liquidus lines and the other thermodynamic data then available. Recently Kaufman *et al.* (1981) have used an equation equivalent to Eq. (1) for a number of III–V binary liquids. The parameters established by Brebrick (1977) for the In–As and Ga–As systems were used but those for the Ga–Sb and In–Sb systems were changed to obtain a better fit to new data (Ansara *et al.*, 1976) for the enthalpy of mixing of the liquid phase. The auxiliary conditions are satisfied with the new parameters to within small quantities comparable to the probable experimental error in the enthalpy and entropy of formation except for the entropy of formation of GaSb(s). The ternary liquid is described by extending Eq. (1) using the Kohler equation (Kohler, 1960). No quantitative measures of fit are quoted. As discussed further in Sections 6 and 7, it appears that these fits are not as good as those we obtain here and, in our opinion, are not good enough.

When $a = b = 0$, then Eq. (1) describes the Braggs–Williams approximation of the nearest-neighbor interaction model with an interaction coefficient, $W - VT$. Guggenheim (1952) provided another level of approximation called the quasichemical model in which the atomic distribution is no longer random over the lattice sites. This model used for the liquid phase leads to liquidus lines for a binary compound AC(s) that are symmetric about 50 at.% and so it is clearly inappropriate for the Cd–Te and Hg–Te systems. Stringfellow and Greene (1969) applied the quasichemical model to a number of III–V binary systems neglecting the auxiliary conditions. Later, Brebrick (1971b) gave quantitative measures of fit to the liquidus lines of InSb, GaSb, InAs, and GaAs using the quasichemical model. Except for GaAs, good fits could be obtained. The auxiliary conditions were not used in advance of the fit but only afterwards to obtain calculated values for the enthalpy and entropy of formation of the binary compounds. These agree fairly closely with the experimental values. No attempt was made to determine whether a closer agreement could be achieved without an excessive increase in the average deviation between observed and calculated liquidus temperatures. Recently Kikuchi (1981) has applied the pair approximation of his cluster variation method to the liquidus lines and surfaces of a number of III–V binaries and ternaries. In the pair approximation the model is identical to the quasichemical model.

For an increasing number of liquids it has been found that the isothermal composition dependence of one or more properties shows a pronounced change near a composition at which a solid forms at lower temperatures (Cutler, 1977; Chang and Smith, 1979). This had led to renewed interest in the associated solution model, which can duplicate such changes. The associated solution model assumes the existence of a number of species in the liquid. The identification of the appropriate compounds or associated species has been based on chemical bonding arguments or, more often, inferred from the existence of solid compounds at lower temperatures in the same or chemically related systems. The interactions among the species are represented in the excess Gibbs energy of mixing by a sum of terms quadratic in the species mole fractions, as indicated by conformal solution theory in the limit of similar species that approach identity (Longuet-Higgins, 1951; Prigogine, 1957), and in some cases by additional terms that are cubic in the species mole fractions.

3. SOLUTION MODEL FOR Ga–In–Sb AND Hg–Cd–Te

Jordan (1970) considered a binary liquid consisting of one molecular or associated species ac and two uncombined species a and c corresponding to the system components. The excess Gibbs energy of mixing was taken as a sum of terms quadratic in the species mole fractions, with each term containing a composition-independent factor called an interaction coefficient. In order to obtain analytical expressions for the partial molar quantities of the components, Jordan approximated the activity coefficients of species a and c by their limiting values at infinite dilution and then assumed that the coefficient for the interaction of a and ac and that for the interaction of c and ac were equal. The partial molar quantities are then functions of the atom fractions of the components rather than the mole fractions of the species, and the need to solve a transcendental equation for the species mole fractions is avoided. This so-called regular associated solution (RAS) model was thought to be widely applicable (Jordan, 1979) and was extended to ternary systems by Szapiro (1976, 1980). Using a generalization of this model, which allowed the interaction coefficients to depend linearly on T and the dissociation constant to depend exponentially on $1/T$, Tung et al. (1981a) analyzed the HgTe–CdTe–Te subsystem. It was found possible to simultaneously satisfy the auxiliary conditions and to obtain good fits to the phase diagram and partial pressure data then available. However, the model can be shown analytically to give a liquidus line for an equiatomic binary solid compound symmetric about 50 at.% and so cannot give a good fit to either the Cd–Te or Hg–Te binary. The stratagem of dividing the phase diagram in half and using different interaction coefficients for either half leads to undesirable discontinuities once it is desired to consider more than just the liquidus surface. Chang and Sharma (1979) and Kellogg and Larrain (1979) have used associated solution models, in which the excess Gibbs energy of mixing is a sum of terms quadratic and cubic in the species mole fractions, to analyze systems of metallurgical interest such as Fe–S and Cu–S. For these systems the solid compounds have extensive rather than narrow homogeneity ranges and so some features of the analysis are different than here. Moreover, the limiting case of complete association was considered, which simplifies the equations considerably.

The partial structure factors for binary (Bhatia and Thorton, 1970) and multicomponent (Bhatia and Ratti, 1977) liquids have been expressed in terms of fluctuation correlation factors, which at zero wave number are related to the thermodynamic properties. An associated solution model in the limits of nearly complete association or nearly complete dissociation has been used to illustrate the composition dependence of the composition-fluctuation factor at zero wave number, $S_{cc}(0)$. For a binary liquid this is inversely proportional to the second derivative of the Gibbs energy of mixing with respect to atom fraction.

On the basis of the above it seems apparent that the simple solution model defined by Eq. (1), the quasichemical model, and the RAS model are inadequate for the Hg–Cd–Te system. Therefore we were led to try the associated solution model described in detail below and have been successful in obtaining a good fit to the data for the Hg–Cd–Te system. Although the Ga–In–Sb system represents a smaller departure from the simplest picture of an ideal solution, there are extensive experimental data available and it is of interest to test the associated solution model in this case where the degree of association of molecular species might be expected to be small or intermediate. A generally good fit is obtained here also. Although the general model is the same as that used for the Hg–Cd–Te system, many more of the interaction coefficients can be set equal to zero. In fact only the minimum set allowing asymmetric liquidus lines in the Ga–Sb and In–Sb binaries need be taken to get a good fit. The application of this associated solution model has been described in less detail than given here by Liao et al. (1982) for Ga–In–Sb and by Tung et al. (1982) for Hg–Cd–Te. A slightly different optimum set of interaction coefficients is given here for the first ternary for reasons discussed at the end of Appendix B.

II. Thermodynamic Equations for the Liquidus Surface of $(A_{1-u}B_u)_{2-y}C_y(s)$

We wish here to obtain the thermodynamic equations defining the liquidus surface of a solid solution, $(A_{1-u}B_u)_{2-y}C_y(s)$. It is assumed that the A and B atoms occupy the sites of one sublattice of the structure and the C atoms the sites of a second sublattice. For the specific systems considered here Sb and Te play the role of C in the general formula above. It is also assumed that the composition variable y is confined to values near unity so that the site fractions of atomic point defects is always small compared to unity. This apparently is the case for the solid solutions in the two systems considered. Then it can be shown theoretically (Brebrick, 1979), as well as experimentally for $(Hg_{1-u}Cd_u)_{2-y}Te_y(s)$ (Schwartz et al., 1981; Tung et al., 1981b), that the sum of the chemical potentials of A and C and that of B and C in the solid are independent of the composition variable y:

$$\mu_A^s + \mu_C^s = \mu_{AC}^s(T, P, u), \tag{4}$$

$$\mu_B^s + \mu_C^s = \mu_{BC}^s(T, P, u). \tag{5}$$

Thus AC and BC can be chosen as thermodynamic components of the solid solution whose chemical potentials are independent of the C to A + B atom ratio. Relying on the relative insensitivity of the thermodynamic properties of condensed phases to the pressure P, we neglect this pressure dependence in

3. SOLUTION MODEL FOR Ga–In–Sb AND Hg–Cd–Te

the following. The chemical potentials above can be formally expressed in terms of activity coefficients Γ_{ij} as

$$\mu^s_{AC} = RT\ln[(1-u)\Gamma_{AC}] + \mu^{\circ,s}_{AC}, \tag{6}$$

$$\mu^s_{BC} = RT\ln u\Gamma_{BC} + \mu^{\circ,s}_{BC}. \tag{7}$$

When $u = 0$, then $\Gamma_{AC} = 1$ and the chemical potential μ^s_{AC} is that of the binary solid AC(s), $\mu^{\circ,s}_{AC}$. This, of course, is also the Gibbs energy of AC(s). Since the restriction that y be confined to values near unity is intended to apply to the end members of the solid solution, $\mu^{\circ,s}_{AC}$ as well as $\mu^{\circ,s}_{BC}$ are independent of y also. Similarly, when $u = 1$, then $\Gamma_{BC} = 1$ and μ^s_{BC} equals the Gibbs energy of BC(s), $\mu^{\circ,s}_{BC}$.

When the liquid and solid-solution phases coexist in equilibrium, the chemical potential of component A must be the same in the liquid and solid phases and similarly for B and C. Therefore the sum of the chemical potentials of A and C as well as that for B and C is the same in both phases, i.e.,

$$\bar{\mu}^l_A + \bar{\mu}^l_C = \mu^s_{AC} - \mu^{\circ,l}_A - \mu^{\circ,l}_C, \tag{8}$$

$$\bar{\mu}^l_B + \bar{\mu}^l_C = \mu^s_{BC} - \mu^{\circ,l}_B - \mu^{\circ,l}_C, \tag{9}$$

where $\mu^{\circ,l}_A$ is the chemical potential of the pure liquid element A, $\mu^{\circ,l}_B$ is that of B, and $\mu^{\circ,l}_C$ that of C. Equations (4) and (5) have been used in writing Eqs. (8) and (9), and the chemical potentials for the liquid are written as the sum of a relative chemical potential $\bar{\mu}^l_j$ and that of the corresponding pure element. It is at this stage, of course, that the analysis has been split into two parts. Adding the chemical potentials results in two equations in place of the original three. A description of the thermodynamics of the solid solution for compositions interior to its stability limits is thereby forgone. However, by adding the chemical potentials one takes advantage of the fact that μ^s_{AC} and μ^s_{BC} are stoichiometric invariants independent of y and the dependence of μ^s_{AC} and μ^s_{BC} on T and u can be expected to be relatively simple.

The chemical potentials of the solid solution components can be eliminated from the liquidus equations given by Eqs. (8) and (9) using Eqs. (6) and (7). Chemical potential differences occur that can be recognized as the Gibbs energy of formation of the binary compounds from their pure liquid elements. These are

$$\Delta G^0_f[AC(s)] = \mu^{\circ,s}_{AC} - \mu^{\circ,l}_A - \mu^{\circ,l}_C, \tag{10}$$

$$\Delta G^0_f[BC(s)] = \mu^{\circ,s}_{BC} - \mu^{\circ,l}_B - \mu^{\circ,l}_C. \tag{11}$$

Using these, the liquidus equations become

$$\bar{\mu}^l_A + \bar{\mu}^l_C = RT\ln[(1-u)\Gamma_{AC}] + \Delta G^0_f(AC(s)), \tag{12}$$

$$\bar{\mu}^l_B + \bar{\mu}^l_C = RT\ln u\Gamma_{BC} + \Delta G^0_f(BC(s)). \tag{13}$$

When $u = 0$, Eq. (12) describes the liquidus of AC(s) in the A–C binary, whereas when $u = 1$, Eq. (13) describes the liquidus of BC(s). The liquidus equations can be put into a number of different but equivalent forms, as shown by Brebrick et al. (1981), but Eqs. (12) and (13) prove to be most convenient here.

In one of these alternate forms the enthalpy of fusion and melting point of AC(s) enter one equation, replacing the Gibbs energy of formation, whereas the corresponding quantities for BC(s) enter the second equation. The difference in the constant pressure heat capacities of AC(s) and that of a liquid of the same composition also enter the first equation, whereas corresponding quantities for BC(s) enter the second. As mentioned in the Introduction, it has been a common procedure to set these heat capacity differences to zero. The error ensuing is discussed by Brebrick et al. (1981) and by Tung et al. (1981a) and some examples given. In particular this error can be serious when an associated solution model is used, unless the degree of association is either near zero or near one.

Application of Eqs. (12) and (13) requires in part the specification of the relative chemical potentials in the liquid. The forms adopted are discussed in the following sections. The solid-solution phase is characterized by assuming that its components mix like a quasiregular solution, i.e.,

$$RT \ln \Gamma_{AC} = (W_s - V_s T) u^2, \tag{14}$$

$$RT \ln \Gamma_{BC} = (W_s - V_s T)(1 - u)^2, \tag{15}$$

where W_s and V_s are composition- and temperature-independent parameters. There is experimental information for the Hg–Cd–Te solid solutions from partial pressure measurments (Schwartz et al., 1981; Tung et al., 1981b, 1982), which is consistent with the form of Eqs. (14) and (15) and with $W_s - 880 V_s$ equal to about 600 cal. Therefore in this case one has an almost ideal solid solution of HgTe and CdTe.

Finally there are four additional equations that we call auxiliary conditions that supplement the liquidus equations given by Eqs. (12) and (13). These follow immediately from the zero change of Gibbs energy upon the congruent melting of AC(s) and BC(s). For AC(s) these are

$$\Delta H_f^0(AC(s), T_{AC}) + \Delta H_{AC}^m = 2 \Delta H_M(x_A = x_C = \tfrac{1}{2}, T_{AC}), \tag{16}$$

$$\Delta S_f^0(AC(s), T_{AC}) + \Delta H_{AC}^m / T_{AC} = 2 \Delta S_M(x_A = x_C = \tfrac{1}{2}, T_{AC}). \tag{17}$$

Here ΔH_f^0 and ΔS_f^0 are the enthalpy and entropy of formation per mole of AC(s) from A(l) and C(l) and in Eqs. (16) and (17) the values at the melting point of AC(s), T_{AC} are required. These, of course, are related at all temperatures to the Gibbs energy of formation appearing in Eq. (12) by

$$\Delta G_f^0(AC(s)) = \Delta H_f^0 - T \Delta S_f^0. \tag{18}$$

The quantity ΔH_{AC}^m is the enthalpy of fusion of AC(s). The quantities ΔH_M and ΔS_M are, respectively, the enthalpy and entropy of mixing of the liquid solution from its pure liquid elements. In Eqs. (16) and (17) these are evaluated at $x_A = x_C = \frac{1}{2}$ and the melting point of AC(s). The quantities ΔH_M and ΔS_M are determined by the liquid model adopted. Completely analogous equations hold for BC(s). Here we impose the auxiliary conditions, Eq. (16) and (17) and their analogues for BC(s), prior to fitting the phase diagram and thermodynamic data. Four of the adjustable parameters appearing in the liquid model are thereby fixed. If the liquid phase is assumed to be ideal, then Eq. (16) states that the enthalpy of formation of AC(s) is equal and opposite to its enthalpy of fusion, whereas Eq. (17) states that the entropy of formation of AC(s) is the negative of its entropy of fusion plus twice the ideal, configurational entropy of mixing, 1.38 cal/°K gram-at.

III. Solution Thermodynamics

The associated solution model used here for the liquid phase is presented in Part IV. The model gives the various thermodynamic quantities in terms of the mole fractions of the chemical species assumed to be present. However, the experimental results are given in terms of the atom fractions of the system components, which are the chemical elements comprising the system. Once the species are identified, general relations between the two sets of quantities can be obtained, which are independent of the specific form assumed for the excess Gibbs energy of mixing. These are developed in this section. For simplicity, and because some of these relations are required only for the binary systems, we confine our considerations to the A–C binary. The equations for the B–C binary are analogous since we will assume the existence of species a, b, c, ac, and bc, where a corresponds to uncombined element A, b to uncombined B, c to uncombined C, ac to a diatomic molecule consisting of one atom of A and one of C, and bc to the B–C analog. Since only monatomic, uncombined, species occur in the A–B binary, the species and components are identical and the various quantities expressed in terms of species or components are identical. The equations obtained can be generalized in a straightforward way for a ternary system.

The total number of gram-atoms of the component A is N_1, that of component C is N_2, Then if n_1, n_2, and n_3 are the number of moles of a, c, and ac species, respectively, conservation of the number of atoms of each type gives

$$N_1 = n_1 + n_3, \tag{19}$$

$$N_2 = n_2 + n_3. \tag{20}$$

The Gibbs energy change upon mixing N_1 gram-at. of A and N_2 of C to form a homogeneous solution—the Gibbs energy of mixing—is given by

$$\Delta G_M = N_1 \bar{\mu}_1(\text{CO}) + N_2 \bar{\mu}_2(\text{CO}), \tag{21}$$

where CO indicates temporarily the chemical potential of a component and the overhead line indicates a relative chemical potential. Thus $\bar{\mu}_1(\text{CO})$ is the difference between the chemical potential of component 1 or A in the solution and that of the pure liquid element A, and $\bar{\mu}_2(\text{CO})$ is the difference between the chemical potential of component 2 or C in the solution and that of the pure liquid element C. Using Eqs. (19) and (20) to eliminate N_1 and N_2, Eq. (21) becomes

$$\Delta G_M = n_1 \bar{\mu}_1(\text{CO}) + n_2 \bar{\mu}_2(\text{CO}) + n_3 [\bar{\mu}_1(\text{CO}) + \bar{\mu}_2(\text{CO})]. \tag{22}$$

Since ΔG_M is a state function that is extensive in n_1, n_2, and n_3, i.e., a homogeneous function of the first degree in n_1, n_2, and n_3, Euler's theorem gives

$$\partial \Delta G_M(n_1, n_2, n_3, T, P)/\partial n_j = \bar{\mu}_j(\text{CO}), \qquad j = 1, 2, \tag{23}$$

$$\partial \Delta G_M/\partial n_3 = \bar{\mu}_1(\text{CO}) + \bar{\mu}_2(\text{CO}), \tag{24}$$

where the variables to be held constant in the partial differentiations can be inferred from the variables listed as the arguments of ΔG_M. On the other hand, the left-hand members of Eqs. (23) and (24) are the natural definitions for the relative chemical potentials of the species, so one can write

$$\bar{\mu}_j(\text{species}) = \bar{\mu}_j(\text{CO}), \qquad j = 1, 2, \tag{25}$$

$$\bar{\mu}_3(\text{species}) = \bar{\mu}_1(\text{CO}) + \bar{\mu}_2(\text{CO}). \tag{26}$$

Thus the relative chemical potential of each component equals that of the corresponding uncombined species and we shall no longer distinguish between them. We note then that we can rewrite Eq. (26) as

$$\bar{\mu}_3 = \bar{\mu}_1 + \bar{\mu}_2, \tag{27}$$

where the species designation is superfluous for $\bar{\mu}_3$ since there is no component 3. We also note that the relative chemical potential of the molecular species 3 as defined here is the difference between the chemical potential of species 3 and the sum of the chemical potentials of pure components 1 and 2. These are the first important relations of solution thermodynamics that we require.

Designating the atom fractions of the components as x_j and the mole fractions of the species as y_j, these are given by the usual definitions as

$$x_j = N_j/(N_1 + N_2), \qquad j = 1, 2, \tag{28}$$

$$y_j = n_j/(n_1 + n_2 + n_3), \qquad j = 1, 2, 3. \tag{29}$$

3. SOLUTION MODEL FOR Ga–In–Sb AND Hg–Cd–Te

Using Eqs. (19), (20), and (29) for $j = 3$ gives the ratio of the number of gram-atoms of components to the number of moles of species as

$$(N_1 + N_2)/(n_1 + n_2 + n_3) = 1 + y_3. \tag{30}$$

Rewriting the Gibbs energy of mixing by Eqs. (21) and (22) and using Eqs. (25) and (26) gives

$$\Delta G_M = N_1 \bar{\mu}_1 + N_2 \bar{\mu}_2, \tag{31}$$

$$\Delta G_M = n_1 \bar{\mu}_1 + n_2 \bar{\mu}_2 + n_3 \bar{\mu}_3. \tag{32}$$

Dividing Eq. (31) by $N_1 + N_2$ to convert N_1 and N_2 to the corresponding atom fractions with Eq. (28), multiplying Eq. (32) by $(n_1 + n_2 + n_3)/(N_1 + N_2)(n_1 + n_2 + n_3)$ and using Eqs. (29) and (30), and equating gives

$$x_1 \bar{\mu}_1 + x_2 \bar{\mu}_2 = (1 + y_3)^{-1}(y_1 \bar{\mu}_1 + y_2 \bar{\mu}_2 + y_3 \bar{\mu}_3) \tag{33}$$

or

$$\Delta G_M \text{ (gram-atom components)} = (1 + y_3)^{-1} \Delta G_M \text{ (mole species)}. \tag{34}$$

We now show that equations analogous to Eq. (34) follow for the enthalpy and entropy of mixing, ΔH_M and ΔS_M, but that, in contrast to the chemical potentials, the partial molar enthalpies and entropies for the components differ from those for the species. Finally we show that the equation for the constant pressure relative heat capacity is of a slightly more complicated form than Eq. (34). Equation (34) and its analogs for ΔH_M and ΔS_M are necessary for comparison of model predicted quantities with experiment.

From basic thermodynamic equations we have

$$\partial[\Delta G_M(N_1, N_2, T, P)/T]/\partial(1/T) = \Delta H_M, \tag{35}$$

$$\partial \Delta G_M/\partial T = -\Delta S_M. \tag{36}$$

Using Eq. (31) in Eq. (35) and using Euler theorem again, since ΔH_M is a homogeneous function of the first degree in N_1 and N_2, gives

$$\bar{h}_j(CO) \equiv \partial \Delta H_M(N_1, N_2, T, P)/\partial N_j = [\partial(\bar{\mu}_j/T)/\partial(1/T)]_{N_1, N_2, P}, \quad j = 1, 2, \tag{37}$$

so that

$$\Delta H_M = N_1 \bar{h}_1(CO) + N_2 \bar{h}_2(CO). \tag{38}$$

Equation (36) gives

$$\bar{s}_j(CO) = \partial \Delta S_M(N_1, N_2, T, P)/\partial N_j = -(\partial \bar{\mu}_j/\partial T)_{N_1, N_2, P}, \quad j = 1, 2. \tag{39}$$

However, using Eq. (32) for ΔG_M in Eq. (35) and viewing the chemical potentials as functions of the mole fractions of the species as well as of T and

P gives

$$\Delta H_M(n_1, n_2, n_3, T, P) = (1/T) \sum_{i=1}^{3} \bar{\mu}_i [\partial n_i/\partial(1/T)]_{N_1, N_2}$$

$$+ \sum_{i=1}^{3} \sum_{j=1}^{3} n_i \left[\frac{\partial(\bar{\mu}_i/T)}{\partial y_j} \right] \left[\frac{\partial y_j}{\partial(1/T)} \right]_{N_1, N_2}$$

$$+ \sum_{i=1}^{3} n_i \, \partial(\bar{\mu}_i/T)/\partial(1/T). \qquad (40)$$

The relative partial molar enthalpies of the species are defined as the partial derivatives, in the last term,

$$\bar{h}_j \text{ (species)} \equiv [\partial(\bar{\mu}_j/T)/\partial(1/T)]_{y_1, y_2, y_3, P}, \quad j = 1, 2, 3. \qquad (41)$$

These are different from the corresponding quantities for the components given in Eq. (37). The order of the double sum in the second term on the right can be reversed to give

$$\sum_{j=1}^{3} [\partial y_j/\partial(1/T)]_{N_1, N_2} (1/T) \left[\sum_{i=1}^{3} n_i (\partial \bar{\mu}_i/\partial y_j) \right],$$

where we have utilized the fact that $\mu_i = \mu_i(y_1, y_2, y_3, T, P)$ and so extracted the factor $1/T$ from within the partial derivative. Since the partial derivatives in the sum over i are at constant T and P, this sum is zero by the Gibbs–Duhem relation. Using Eqs. (19) and (20), the partial derivatives in the first sum in Eq. (40) are related by

$$\partial n_1/\partial(1/T) = \partial n_2/\partial(1/T) = -\partial n_3/\partial(1/T) \qquad (42)$$

so that this sum can be written as

$$(\bar{\mu}_3 - \bar{\mu}_1 - \bar{\mu}_2) \partial n_3/\partial(1/T).$$

But by Eq. (27) the chemical potential difference in parentheses is zero when the solution is in internal equilibrium. Therefore using Eq. (41), defining the relative partial molar enthalpies of the species, Eq. (40) reduces to

$$\Delta H_M = \sum_{i=1}^{3} n_i \bar{h}_i \text{ (species)}. \qquad (43)$$

Dividing Eq. (38) by $N_1 + N_2$ to obtain the enthalpy of mixing per gram-atom of components, multiplying Eq. (43) by $(n_1 + n_2 + n_3)/(N_1 + N_2) \times (n_1 + n_2 + n_3)$, and using Eq. (30), equating then gives [compare the derivation of Eq. (34)]

$$\Delta H_M \text{ (gram-atom components)} = (1 + y_3)^{-1} \Delta H_M \text{ (mole species)}. \qquad (44)$$

One can proceed in the same fashion starting with Eq. (36) for ΔS_M. If the relative partial molar entropies of the species are defined, analogously

3. SOLUTION MODEL FOR Ga–In–Sb AND Hg–Cd–Te

to those for the components, as

$$\bar{s}_j(\text{species}) = -(\partial \bar{\mu}_j/\partial T)_{y_1, y_2, y_3, P}, \quad j = 1, 2, 3, \quad (45)$$

then

$$\Delta S_M = \sum_{i=1}^{3} n_i \bar{s}_i \text{ (species)} \quad (46)$$

and

$$\Delta S_M(\text{gram-atom components}) = (1 + y_3)^{-1} \Delta S_M \text{ (mole species)}. \quad (47)$$

The relative constant pressure heat capacity is defined as

$$\Delta C_P = (\partial \Delta H_M/\partial T)_{N_1, N_2, P}. \quad (48)$$

This, of course, is the difference between the heat capacity of the solution and the sum of those of the unmixed liquid elements. Using Eq. (38) and defining relative partial molar heat capacities of the components as

$$\bar{c}(CO)_{P,j} = [\partial \bar{h}_j(CO)/\partial T]_{N_1, N_2}, \quad (49)$$

Eq. (48) can be written as

$$\Delta C_P = N_1 \bar{c}(CO)_{P,1} + N_2 \bar{c}(CO)_{P,2}. \quad (50)$$

On the other hand, if one starts with Eq. (48) but considers ΔH_M to be a function of the species mole fractions, one obtains

$$\Delta C_P = \partial \Delta H_M(n_1, n_2, n_3, T, P)/\partial T + \sum_{j=1}^{3} (\partial \Delta H_M/\partial n_j)(\partial n_j/\partial T)_{N_1, N_2}. \quad (51)$$

Defining the first term on the right as the heat capacity in terms of species, ΔC_p (species), and using Eq. (42) gives

$$\Delta C_p(\text{components}) = \Delta C_p(\text{species}) + [\bar{h}_3(\text{species}) - \bar{h}_1(\text{species}) - \bar{h}_2(\text{species})](\partial n_3/\partial T)_{N_1, N_2}. \quad (52)$$

Again dividing by $N_1 + N_2$ gives

$$\Delta C_p(\text{gram-atom components}) = (1 + y_3)^{-1} \Delta C_p(\text{mole species}) + y_3(1 + y_3)^{-2}(\bar{h}_3 - \bar{h}_1 - \bar{h}_2)(\partial \ln y_3/\partial T)_{x_2}, \quad (53)$$

where the definition of y_3 in Eq. (29) has been used to convert the last partial derivative of n_3 in Eq. (52) to one of y_3 and where for brevity we have dropped the notation, indicating that the relative partial molar enthalpies in the last term are for species and not for components.

The extension of Eqs. (25)–(27) to a ternary system states that the chemical potential of each monatomic, uncombined species equals that of the corresponding component element, whereas the chemical potential of each molecular species is equal to the sum of the chemical potentials of the component elements that comprise the species. The first of these results is immediately applicable to the liquidus equations given by Eqs. (12) and (13). The second is useful in obtaining the relationship between the atomic fractions of the components and the mole fractions of the species in the next section. Equations (44) and (47) are used in satisfying the auxiliary conditions, Eqs. (16) and (17), which apply at the melting points of AC(s) and BC(s). Equation (53) is useful in matching the calculated heat capacity with experimental values in the binary systems. For the associated solution model used here and described in the following sections, the composition-independent parameters or interation coefficients appearing in the excess Gibbs energy of mixing are assumed to be linear functions of T. Then ΔH_M (mole species) is independent of T and ΔC_p (mole species) is zero. However, by Eq. (53), ΔC_p (gram-atom components) is generally not zero except for the limiting cases of complete dissociation, $y_3 = 0$, or complete association, $y_3 = $ a constant, determined by the composition.

IV. Associated Solution Model for the Liquid Phase

In the following three sections the model of the liquid phase is defined and developed. Once the species are chosen it is possible to obtain the general relations, giving the mole fractions of the species in terms of the atom fractions of the component elements without specifying how the species interact, i.e., without giving a specific form of the excess Gibbs energy of mixing. These relations are obtained in Section 1 below while the definition of the model is completed by giving ΔG_M^x in 2.

1. RELATION BETWEEN SPECIES MOLE FRACTIONS AND COMPOSITION

The components are written as A, B, and C and numbered consecutively 1 through 3. The species are the uncombined, monatomic forms a, b, and c and the diatomic molecular forms ac and bc. These are numbered 1 through 5. Thus for the Ga–In–Sb system the species assumed in the liquid phase are Ga, In, Sb, GaSb, and InSb, and these are numbered in order. For the Hg–Cd–Te system the liquid species are Hg, Cd, Te, HgTe, and CdTe. The molecular species assumed correspond to compositions at which narrow-homogeneity-range solid compounds form. We note that there is of course no distinguishable diatomic unit in the zinc-blende structure of the solids. The number of nearest neighbors in the structure is 4 and the molecule is as large as

3. SOLUTION MODEL FOR Ga–In–Sb AND Hg–Cd–Te

the crystal. Nor are molecules corresponding to the assumed liquid molecules predominant in the vapor phase. The numbering is now different than in Part III. If N_i is the number of gram-atoms of component element i and n_j is the number of moles of species j, then the conservation of atom types requires that

$$N_1 = n_1 + n_4, \tag{54}$$

$$N_2 = n_2 + n_5, \tag{55}$$

$$N_3 = n_3 + n_4 + n_5. \tag{56}$$

Defining atom fractions of the components x_i and mole fractions of species y_i by a generalization of Eqs. (28) and (29), the mole fractions of the uncombined species are obtained from Eqs. (54)–(56) as

$$\begin{aligned} y_i = x_i(1 + y_4 + y_5) &+ [\delta(i, 2) - 1] y_4 \\ &+ [\delta(i, 1) - 1] y_5, \quad i = 1, 2, 3 \end{aligned} \tag{57}$$

$$\delta(i, j) = \begin{cases} 1 & \text{if } i = j, \\ 0 & \text{if } i \neq j. \end{cases} \tag{58}$$

Adding Eqs. (54)–(56), the ratio of the number of gram-atoms of components to the number of moles of species is

$$r = \sum_{i=1}^{3} N_i \bigg/ \sum_{j=1}^{5} n_j = 1 + y_4 + y_5, \tag{59}$$

which is an obvious generalization of Eq. (30) for a binary system. The relative chemical potentials of the species are written as

$$\bar{\mu}_p = RT \ln \gamma_p y_p - \delta(p, 4) \Delta G_4^\circ - \delta(p, 5) \Delta G_5^\circ, \quad p = 1, \ldots, 5, \tag{60}$$

where ΔG_4° is the standard Gibbs energy of dissociation of species 4, or ac, into the liquid elements A and C and ΔG_5° is the corresponding quantity for species 5 or bc. Thus $\gamma_4 = 1$ in the hypothetical liquid consisting entirely of ac species, $y_4 = 1$, while $\gamma_5 = 1$ in the hypothetical liquid consisting entirely of bc species, $y_5 = 1$. The activity coefficients γ_j are given a specific form in the next subsection. Generalizing Eq. (27) and remembering the numbering is now different, internal equilibrium in the liquid requires that

$$\bar{\mu}_4 = \bar{\mu}_1 + \bar{\mu}_3, \tag{61}$$

$$\bar{\mu}_5 = \bar{\mu}_2 + \bar{\mu}_3. \tag{62}$$

With Eq. (60) these become upon rearrangement

$$y_1 y_3 / y_4 = (\gamma_4 / \gamma_1 \gamma_3) \exp(-\Delta G_4^\circ / RT) = \kappa_4, \tag{63}$$

$$y_2 y_3 / y_5 = (\gamma_5 / \gamma_2 \gamma_3) \exp(-\Delta G_5^\circ / RT) = \kappa_5, \tag{64}$$

where κ_4 and κ_5 are effective equilibriun constants that in general depend on composition through the ratio of activity coefficients. With specific forms given for the activity coefficients, Eqs. (63) and (64) become two transcendental equations in y_4 and y_5 once y_1, y_2, and y_3 are eliminated with Eq. (57). These can be solved simultaneously by numerical methods. Following Szapiro (1976), we find it convenient to proceed differently. Equation (57) is used to eliminate y_1 from Eq. (63) and then $y_4 + y_5$ is eliminated using Eq. (57) with $i = 3$ to give

$$y_4 = x_1 y_3 (1 - y_3)/(1 - x_3)(y_3 + \kappa_4). \tag{65}$$

Starting with Eq. (64), y_2 and $y_4 + y_5$ are eliminated using Eq. (57) with $i = 2$ and $i = 3$ to give

$$y_5 = x_2 y_3 (1 - y_3)/(1 - x_3)(y_3 + \kappa_5). \tag{66}$$

With Eqs. (65) and (66), y_4 and y_5 can be eliminated from Eq. (57) with $i = 3$ to give

$$y_3 = x_3 - y_3(1 - y_3)[x_1/(y_3 + \kappa_4) + x_2/(y_3 + \kappa_5)]. \tag{67}$$

This equation is used in an iterative scheme to obtain the mole fraction of the species for a fixed T and composition. First κ_4 and κ_5 are approximated by setting the activity coefficients in Eqs. (63) and (64) equal to one. Equation (67) becomes a cubic in y_3, which is solved by Cardan's method. Equations (65) and (66) are then used to obtain y_4 and y_5 and Eq. (57) to obtain y_1 and y_2. In the second step the approximate values for the mole fractions are used to calculate approximate values for the activity coefficients and then for κ_4 and κ_5. The new values for κ_4 and κ_5 are inserted into Eq. (67), which is solved to give a second approximation for y_3. These steps are repeated until κ_4 and κ_5 change by less than 10^{-9} in successive steps. A generally more reliable criterion for termination would involve the fractional changes in κ_4 and κ_5 upon successive iterations.

2. Interaction Terms

Having chosen the species in the liquid phase, the thermodynamic characterization of the model is completed by assuming an equation for the excess Gibbs energy of mixing—of forming the solution from the liquid elements. The most general polynomial form complete through cubic terms in the species mole fractions is

$$\Delta G_M^x = \sum_{j=1}^{5} \sum_{i=1}^{5} (\alpha_{ij} + \beta_{ij} y_j) y_i y_j + \sum_{i=1}^{5} \sum_{j>i}^{5} \sum_{k>j}^{5} \gamma_{ijk} y_i y_j y_k - y_4 \Delta G_4^0 - y_5 \Delta G_5^0$$

where $\alpha_{ij} = \alpha_{ji}$, $\alpha_{jj} = 0$, $\beta_{ij} = -\beta_{ji}$. $\tag{68}$

The α_{ij}, β_{ij}, γ_{ijk} are composition-independent interaction coefficients, which are the adjustable parameters of the model along with the Gibbs energy of dissociation of the molecular species, ΔG_4^0 and ΔG_5^0. This equation is a more general form than given by Chang and Sharma (1979) who omitted the γ_{ijk} terms and the terms with ΔG_4^0 and ΔG_5^0 and so gave an excess Gibbs energy of mixing relative to the species rather than relative to the pure liquid elements. Equation (68) is also equivalent to the general model used by Kellogg and Larrain (1979), who, however, gave the relative excess chemical potentials and not ΔG_M^x itself. When the atom fraction of any one component is unity, then ΔG_M^x is zero as required by its definition. When all the β_{ij} and γ_{ijk} parameters are zero, then the species interact as a quasiregular or conformal solution. The γ_{ijk} parameters drop out in all of the binary systems. Equation (68) can be generalized and given a wider interpretation than here as discussed briefly by Liao et al. (1982). If N is an arbitrary total number of moles of species in the liquid, then the relative excess chemical potential of species j is

$$\bar{\mu}_j^x = (\partial(N \Delta G_M^x)/\partial n_j)_{T,P,n_i \neq j}. \tag{69}$$

With Eq. (68) for ΔG_M^x this becomes

$$\bar{\mu}_p^x = 2 \sum_{i=1}^{5} [\alpha_{ip} + (y_i/2 - y_p)\beta_{pi}]y_i - \sum_{i=1}^{5} \sum_{j=1}^{5} (\alpha_{ij} + 2\beta_{ij}y_j)y_i y_j$$

$$+ \sum_{i=1}^{5} \sum_{j>i}^{5} [\varepsilon(i,p)\gamma_{pij} + \varepsilon(p,i)\varepsilon(j,p)\gamma_{ipj} + \varepsilon(p,j)\gamma_{ijp}]y_i y_j$$

$$- 2 \sum_{i=1}^{5} \sum_{j>i}^{5} \sum_{k>j}^{5} \gamma_{ijk}y_i y_j y_k - \delta(p,4)\Delta G_4^0 - \delta(p,5)\Delta G_5^0, \tag{70}$$

where

$$\varepsilon(i,p) = \begin{cases} 0 & \text{if } i \leq p, \\ 1 & \text{if } i > p. \end{cases}$$

Thus

$$\varepsilon(p,i)\,\varepsilon(j,p) = \begin{cases} 1 & \text{only if } \quad i < p < j, \\ 0 & \text{otherwise.} \end{cases}$$

It has been possible to obtain good fits without using any of the ternary parameters, γ_{ijk}, and so from this point these are all set to zero. The simplest version of the general model that is capable of describing asymmetric behavior in the A–C or 1–3 binary can be established by examining the excess chemical potentials at infinite dilution. These are

$$\bar{\mu}_3^x(x_1 = 1) = 2\alpha_{13} - \beta_{13}, \tag{71}$$

$$\bar{\mu}_1^x(x_3 = 1) = 2\alpha_{13} + \beta_{13}, \tag{72}$$

$$\bar{\mu}_4^x(x_1 = 1) = 2\alpha_{14} + \beta_{41} - \Delta G_4^0, \tag{73}$$

$$\bar{\mu}_4^x(x_3 = 1) = 2\alpha_{43} + \beta_{43} - \Delta G_4^0. \tag{74}$$

Nonzero values of β_{13} are required if the excess chemical potentials of 1 and 3 are to be different at infinite dilution. Since α_{14} and α_{43} are, in general, different, nonzero values are not required for β_{41} and β_{43}. A similar consideration of the 2–3 or B–C binary shows that β_{23} must be different from zero if the chemical potentials of 2 and 3 are to be different at infinite dilution. This simplified version of Eqs. (68) and (70) in which all the γ_{ijk} are zero and all the β_{ij}, except $\beta_{13} = -\beta_{31}$ and $\beta_{23} = -\beta_{32}$, are zero proves adequate for the Ga–In–Sb system but not for the Hg–Cd–Te system. For the latter, more of the β_{ij} parameters, particularly for the Cd–Te binary, are required, although all of the γ_{ijk} parameters can be set at zero.

The relative chemical potentials can be obtained from the relative excess chemical potentials given by Eq. (70) by the definition

$$\bar{\mu}_j = \bar{\mu}_j^x + RT \ln y_j. \tag{75}$$

The relative partial molar enthalpies of the *species* are obtained by using Eqs. (70) and (75) in Eq. (41). When the interaction coefficients $\alpha_{ij}, \beta_{ij}, \gamma_{ijk}$ are linear functions of T as assumed here, these enthalpies can be written down directly from Eq. (70) since the partial derivatives defining them in Eq. (41) are all taken at constant values for the species mole fractions. Since the concept of excess quantities measures a quantity for a solution relative to its value in an ideal solution, all nonzero enthalpy quantities are excess. The total enthalpy of mixing is then the same as the excess enthalpy of mixing and a relative partial molar enthalpy is the same as the excess relative partial molar enthalpy. Therefore for brevity the adjective excess is not used here in connection with enthalpy quantities. By definition the relation between the relative partial molar entropy of species j, $\bar{s}_j$, and the excess relative partial molar entropy $\bar{s}_j^x$ is

$$\bar{s}_j = \bar{s}_j^x - R \ln y_j. \tag{76}$$

Thus Eq. (45) defining $\bar{s}_j$ can (after generalization to a ternary system) be rewritten for the corresponding excess quantity as

$$\bar{s}_j^x \text{ (species)} = -(\partial \bar{\mu}_j^x / \partial T)_{y_1, y_2, y_3, y_4, y_5, P}. \tag{77}$$

Again these can be written down directly using Eq. (70).

3. Development of the Liquid Model for the A–C Binary

In the notation used here, congruently melting, narrow-homogeneity-range compounds form in the A–C and B–C binaries of the A–B–C system. These are, of course, the Ga–Sb and In–Sb binaries for the Ga–In–Sb system and the Hg–Te and Cd–Te binaries for the Hg–Cd–Te system. For these binaries it is desired to apply the auxiliary conditions of Eqs. (16) and (17) as well as fit other experimental data before fitting the liquidus lines and then the ternary data. For this purpose it is necessary to carry the development of the model somewhat further. At the same time some insight into the behavior of the model can be attained. We show this development specifically for the A–C or 1–3 binary. The equations for the B–C or 2–3 binary are completely analogous and can be written directly once those for the A–C binary are given.

For the A–C binary, $x_2 = y_2 = y_5 = 0$ and only species 1, 3, and 4 occur. For the sake of a simplified notation, the atom fraction of component 3 or C, x_3, and the mole fraction of species 4 or ac, y_4, are written as

$$x_3 = x, \tag{78}$$

$$y_4 = z. \tag{79}$$

Eliminating y_1 and y_3 from the mass action law, Eq. (63), using Eq. (57) with $i = 1, 3$, gives

$$z = -1 + (1 + \kappa_4)(1 - A)/2x(1 - x), \tag{80}$$

where

$$A = [1 - 4x(1 - x)/(1 + \kappa_4)]^{1/2}. \tag{81}$$

The mathematically possible choice of a positive sign in front of A in Eq. (80) leads to values for the mole fraction z greater than one and so is rejected. In turn the effective equilibrium constant κ_4 can be written in terms of x and z. The starting point is the definition of κ_4 in terms of the activity coefficient ratio given by the rightmost equation of Eq. (63). Using Eq. (60) to eliminate each activity coefficient in favor of a relative chemical potential, Eq. (75) relating the relative and relative excess chemical potentials, and Eq. (70) for the relative excess chemical potentials for our model, gives κ_4 in terms of y_1, y_3, z, and x. Using Eq. (57) again to eliminate the first two gives

$$\begin{aligned}
RT \ln \kappa_4 &= -2(\Omega_S + \Omega_R) + (1 + z)^2[\Omega_R - (2x - 1)^2\Omega_S - (2x - 1)\Omega_A] \\
&\quad + (2x - 1)(1 + z)^2[2x(1 - x)(1 + z) - 1]\beta_{13} \\
&\quad - \beta_{14}[(1 - x)^2 - 2(1 - x)(1 + 2x)z - (1 - 2x - 5x^2)z^2 + 2x(1 + x)z^3] \\
&\quad - \beta_{34}[x^2 - 2x(3 - 2x)z + (6 - 12x + 5x^2)z^2 + 2(1 - x)(2 - x)z^3] \\
&\quad - \Delta G_4^0,
\end{aligned} \tag{82}$$

where we have set all of the pure ternary parameters γ_{ijk} to zero and where we have replaced the quadratic interaction parameters by convenient linear combinations given by

$$\alpha_{13} = 2\Omega_s, \tag{83}$$

$$\alpha_{14} = (\Omega_S - \Omega_R + \Omega_A)/2, \tag{84}$$

$$\alpha_{34} = (\Omega_S - \Omega_R - \Omega_A)/2. \tag{85}$$

It can be seen that the Ω_A and β_{13} terms in Eq. (82) are asymmetric about $x = \frac{1}{2}$ and the Ω_S terms are symmetric. On the other hand, the contribution of Ω_R is greater the greater z is and the greater the extent of the association reaction is.

For use below we now explicitly write out the assumed linear temperature dependence of the interaction parameters as

$$\Omega_i = \omega_i - v_i T, \quad i = S, R, A, \tag{86}$$

$$\beta_{ij} = B_{ij} - C_{ij} T. \tag{87}$$

Upon insertion of Eq. (82) into Eq. (81) one has an implicit transcendental equation for z that must in general be solved numerically for specified values of x, T, and the interaction coefficients. Then y_1 and y_3 can be obtained with Eq. (57) and various thermodynamic properties calculated. Analytical expressions can be obtained for the various properties in terms of x, T, and z. However, these are somewhat cumbersome and in general we shall only write them for $x = \frac{1}{2}$.

At $x = \frac{1}{2}$ the relative chemical potentials for the A–C binary can be obtained starting with Eq. (70) and using Eq. (83)–(85) to eliminate the α_{ij} parameters and Eq. (57) to eliminate y_1 and y_3. The result is

$$\bar{\mu}_1(\tfrac{1}{2}) = RT\ln[(1-z)/2] + \Omega_S + \Omega_A z - \Omega_R z^2 - (1-z)^2(\beta_{13}/4) \\ + z(3z^2 - 1)(\beta_{14}/2) - z(1-z)(3z-1)(\beta_{34}/2), \tag{88}$$

$$\bar{\mu}_3(\tfrac{1}{2}) = RT\ln[(1-z)/2] + \Omega_S - \Omega_A z - \Omega_R z^2 + (1-z)^2(\beta_{13}/4) \\ - z(1-z)(3z-1)(\beta_{14}/2) + z(3z^2 - 1)(\beta_{34}/2), \tag{89}$$

$$\bar{\mu}_4(\tfrac{1}{2}) = RT\ln z - (1-z)^2 \Omega_R + (1-z)^2(6z-1)(\beta_{14} + \beta_{34})/4 - \Delta G_4^0. \tag{90}$$

These can be combined to give the sum and difference

$$\bar{\mu}_1 + \bar{\mu}_3 = 2RT\ln[(1-z)/2] + 2\Omega_S \\ - 2\Omega_R z^2 + z^2(3z-2)(\beta_{14} + \beta_{34}), \tag{91}$$

$$\bar{\mu}_1 - \bar{\mu}_3 = 2\Omega_A z - (1-z)^2(\beta_{13}/2) + z(2z-1)(\beta_{14} - \beta_{34}). \tag{92}$$

Internal equilibrium in the liquid requires that $\bar{\mu}_1 + \bar{\mu}_3 = \bar{\mu}_4$ and therefore both Eq. (90) and (91) give the Gibbs energy of mixing at $x = \frac{1}{2}$ for 2 gram-at. of liquid. Inspection of Eq. (91) shows that as the model parameters are varied to make z approach zero, the Gibbs energy of mixing approaches $-RT\ln 2 + \Omega_S$ per gram-atom. On the other hand, as z approaches unity, Eq. (90) shows that the Gibbs energy of mixing approaches $-\Delta G_4^0/2$ per gram-atom. Setting $u = 0$, $\Gamma_{AC} = 1$ in Eq. (12) gives the equation for the liquidus of AC(s) as

$$\bar{\mu}_A^L + \bar{\mu}_C^L \equiv \bar{\mu}_1 + \bar{\mu}_3 = \Delta G_f^0[\text{AC(s)}]. \tag{93}$$

Thus at $x = \frac{1}{2}$ the Gibbs energy of formation of the solid compound must equal the right side of both Eq. (90) and Eq. (91). When $\Delta G_f^0[\text{AC(s)}]$ is a large negative number and the solid compound is very stable, then z must be near unity and the Gibbs energy of dissociation of the ac species ΔG_4^0 must be approximately equal to $-\Delta G_f^0[\text{AC(s)}]$ if the interaction coefficients, Ω_R and $\beta_{14} + \beta_{34}$, are to remain relatively small.

Equation (92) gives the negative of the slope of ΔG_M with respect to x at $x = \frac{1}{2}$ and constant T. The equation is valid for all degrees of dissociation and shows that the slope at $x = \frac{1}{2}$ depends on Ω_A, β_{13}, and $\beta_{14} - \beta_{34}$ and is zero if these are all zero. Approximate treatments for the case of complete association ($z = 1$) have sometimes been given that incorrectly yield a V shape to the ΔG_M isotherms and hence a discontinuity in $\bar{\mu}_1 - \bar{\mu}_3$ at $x = \frac{1}{2}$.

The relative partial enthalpies and entropies for the *species* at $x = \frac{1}{2}$ can be obtained directly from Eqs. (88)–(90) as discussed after Eq. (75). The enthalpy of mixing per mole of species is

$$\Delta H_M \text{ (mole species, } x = \tfrac{1}{2}) = (1 - z)\omega_S - z(1 - z)\omega_R$$
$$+ z(1 - z)(3z - 1)(B_{14} + B_{34})/4 - z\Delta H_4^0. \tag{94}$$

The enthalpy of mixing per gram-atom of components is obtained by multiplying the right side of Eq. (94) by $(1 + z)^{-1}$, as can be seen by changing the notation of Eq. (44) to correspond to that used here.

In the remainder of this section it is desired to obtain the relative, constant-pressure heat capacity of the liquid at $x = \frac{1}{2}$ and the concentration fluctuation factor for all compositions. Since the latter equation is complicated, it is not written out in full here. This has been done in Eqs. (37)–(45) of the paper by Liao et al. (1982) for the special case that $\beta_{14} = \beta_{34} = 0$ and β_{13} is the only nonzero cubic interaction term, i.e., the version of the model applied here to the Ga–Sb and In–Sb binaries. Bhatia and Hargrove (1974) have given equations for the composition fluctuation factor at zero wave number for the special cases of complete association or dissociation and only quadratic interaction coefficients.

Bhatia and co-workers have reexpressed the partial scattering factors for binary (1970) and multicomponent (1977) liquids in terms of wave-number-dependent fluctuation factors. At zero wave number these are related to the thermodynamic properties of the liquid, and thus a linkage has been provided between the thermodynamic properties and the structure as reflected by diffraction effects. In particular, the composition fluctuation factor for a binary system is given at zero wave number by

$$S_{cc}(0) = [(\partial^2(\Delta G_M/RT)/\partial x^2)_{T,P}]^{-1} = (1-x)RT/(\partial \bar{\mu}_3/\partial x)_{T,P}. \quad (95)$$

The relative chemical potentials of component 3 and species 3 are equal and are obtained from Eq. (70). Inserting the quadratic interaction coefficients Ω_i in place of α_{ij} with Eqs. (83)–(85) and eliminating y_1 and y_3 with Eq. (57) gives $\bar{\mu}_3$ as a function of x, T, and z:

$$\begin{aligned}\bar{\mu}_3 = {} & RT\ln[x(1+z)-z] + [4(1-x)(1+z) \\ & - 4x(1-x)(1+z)^2 + z^2]\Omega_S - z^2\Omega_R \\ & - z[1+(1-2x)(1+z)]\Omega_A + 2[1-x(1+z)][(1-x)(2x-1) \\ & \times (1+z)^2 - x(1+z)/2 + 1/2]\beta_{13} \\ & - z\{z - 2(1-x)(1+z)[2z - x(1+z)]\}\beta_{34} \\ & + 2z[(1-z) + x(1+z)(z-2) + x^2(1+z)^2]\beta_{14}. \end{aligned} \quad (96)$$

Thus the partial derivative required for $S_{cc}(0)$ can be obtained in terms of x, T, z and the corresponding derivative of z. The latter can be obtained in terms of x, T, and z also, as shown immediately below. Therefore for a given x and T and values of the interaction parameters, z is obtained numerically from Eq. (80) and the value of $S_{cc}(0)$ is then calculated. The required composition derivative of z can be obtained starting with the mass action law given by Eq. (63). Eliminating y_1 and y_3 with Eq. (57) and rearranging, this is

$$x(1-x)z^2 + [2x(1-x) - \kappa_4 - 1]z + x(1-x) = 0. \quad (97)$$

Taking the partial derivative with respect to x at constant T and P gives

$$(\partial z/\partial x)_{T,P} = [(2x-1)(1+z)^2 + z\kappa_x]/[2x(1-x)(1+z) - \kappa_4 - 1 - z\kappa_z], \quad (98)$$

where

$$\kappa_x \equiv \partial \kappa_4(x, T, z)/\partial x, \quad (99)$$

$$\kappa_z = \partial \kappa_4(x, T, z)/\partial z. \quad (100)$$

Equation (82) is used for κ_4 and the variables to be held constant in the partial derivatives are inferred from the argument of κ_4.

The corresponding temperature derivative of z occurs in the equation for the relative heat capacity. Starting again with Eq. (97) gives

$$(\partial z/\partial T)_{x,P} = z\kappa_T/[2x(1-x)(1+z) - \kappa_4 - 1 - z\kappa_z]. \quad (101)$$

Note that the denominators in Eqs. (98) and (101) are identical. Rewriting Eq. (53) in the notation of this section gives

$$\Delta C_p \text{ (gram-atom components)} = (1+z)^{-1} \Delta C_p \text{ (mole species)}$$
$$+ z(1+z)^{-2}(\bar{h}_4 - \bar{h}_1 - \bar{h}_3)(\partial \ln z/\partial T)_x, \quad (102)$$

where the relative partial molar enthalpies in the second term are for the species and where, since z is an intensive function, the partial derivative of $\ln z$ can be taken at constant x. Now the quantity ΔC_p (mole species) is defined as the first term on the right side of Eq. (51) as the temperature derivative of ΔH_M at constant number of moles for each species. But when the interaction coefficients are linear functions of T, as assumed in Eqs. (86) and (87), then the relative partial molar enthalpies of the species defined by Eq. (41) and consequently ΔH_M defined by Eq. (43) are not explicit functions of T, and ΔC_p (mole species) is zero. The difference in the relative partial enthalpies of the species occurring in the second term of Eq. (102) can be replaced by the temperature derivative of κ_4 starting with the thermodynamic equation

$$\bar{\mu}_j = \bar{h}_j - T\bar{s}_j, \quad j = 1, 3, 4. \quad (103)$$

The result is

$$\Delta C_p \text{ (gram-atom components)} = z(1+z)^{-2} R (\partial \ln \kappa_4 / \partial (1/T))_{x,z,P} (\partial \ln z / \partial T)_{x,P} \quad (104)$$

$$= -z(1+z)^{-2} RT^2 \kappa_T^2 / [2x(1-x)(1+z)$$
$$- \kappa_4 - 1 - z\kappa_z] \kappa_4, \quad (105)$$

where Eq. (101) has been used to get Eq. (105) from (104). At $x = \tfrac{1}{2}$ this becomes

$$\Delta C_p \text{ (gram-atom components, } x = \tfrac{1}{2}) = N/D,$$
$$N = z(1-z)[2(\omega_S + \omega_R) - (1+z)^2 \omega_R$$
$$+ (1/4 - 2z + 5z^2/4 + 3z^3/2)(B_{14} + B_{34}) + \Delta H_4^0]^2,$$
$$D = T(1+z)^2 \{(1+z)RT + z(1-z)[2(1+z)\Omega_R$$
$$+ (2 - 5z/2 - 9z^2/2)(\beta_{14} + \beta_{34})]\}. \quad (106)$$

The possibility of a zero denominator and an infinite heat capacity can readily be seen in the special case that $\beta_{14} = \beta_{34} = 0$. This occurs when

$$\Omega_R = -RT/2z(1-z), \quad x = \tfrac{1}{2}. \quad (107)$$

This behavior is linked to the temperature derivative of $\ln z$ in Eq. (104). The denominator in Eq. (101) for $(\partial z/\partial T)_{x,p}$ and that for $(\partial z/\partial x)_{T,p}$ in Eq. (98) is given at $x = \frac{1}{2}$ by

$$D_2(x = \tfrac{1}{2}) = (4zRT)^{-1}(z-1)\{(1+z)RT + z(1-z)$$
$$\times [2(1+z)\Omega_R + (2 - 5z/2 - 9z^2/2)(\beta_{14} + \beta_{34})]\}. \quad (108)$$

The factor in braces also appears in Eq. (106). Thus at $x = \frac{1}{2}$, both partial derivatives of z are infinite when Eq. (107) is satisfied and when $\beta_{14} = \beta_{34} = 0$. These points are discussed in more detail in Appendix B.

4. Constraints on Interaction Coefficients in the A–C and B–C Binaries

The constraints imposed upon the interaction coefficients in the A–C binary are given here. Those for the B–C binary can be obtained by analogy. The constraints generally serve to reduce the number of independent coefficients and at the same time insure that the liquid model will match selected experimental values exactly regardless of how well the liquidus points and other data can be fit.

Inserting Eq. (91) into Eq. (93) for the liquidus line of AC(s) gives the liquidus temperature at $x = \frac{1}{2}$. The experimental value T_{AC} is treated as special and it is required that the resulting equation be satisfied for $T = T_{AC}$. The equation then becomes a restriction on the interaction coefficients. We use it to provide a relation between ω_S and v_S in terms of the other coefficients and a new model parameter z^*, the value of z at $x = \frac{1}{2}$ and $T = T_{AC}$. The relation is

$$\omega_S - T_{AC}v_S = \Delta G_f^0/2 - RT_{AC}\ln[(1-z)/2] + \Omega_R z^2 - (\tfrac{1}{2})z^2(3z-2)(\beta_{14} + \beta_{34}), \quad (109)$$

where for simplicity of notation the asterisk superscript has been omitted but it is understood that z, ΔG_f^0, and the interaction coefficients are evaluated at $x = \frac{1}{2}$ and/or $T = T_{AC}$.

Secondly we apply the auxiliary constraint on ΔH_M at $x = \frac{1}{2}$, $T = T_{AC}$ given by Eq. (16). Converting ΔH_M (mole species) given by Eq. (94) to ΔH_M (gram-atom components) via Eq. (44) and inserting this into Eq. (16) gives an equation for the enthalpy of dissociation of ac, ΔH_4^0, and ω_S in terms of the other coefficients and known thermodynamic quantities:

$$\omega_S - z^*\Delta H_4^0/(1-z^*) = z^*\omega_R - z^*(3z^*-1)(B_{14}+B_{34})/4$$
$$+ (\tfrac{1}{2})(\Delta H_f^0 + \Delta H_{AC}^m)(1+z^*)/(1-z^*), \quad (110)$$

where ΔH_f^0 and ΔH_{AC}^m are both for 1 mol of AC(s).

If ΔC_p (gram-atom components) is known experimentally at $x = \frac{1}{2}$ and $T = T_{AC}$, then Eq. (106) can be used to provide a second equation for $\Delta H_4^0 + 2\omega_S$ in terms of z^* and the other coefficients so that ω_S and ΔH_4^0

can be obtained individually and then v_S from Eq. (109). For the Ga–Sb and In–Sb binaries, experimental values for $\Delta C_p (x = \frac{1}{2})$ are available and both ω_S and ΔH_4^0 are dependent interaction coefficients. The equation to match the experimental value of ΔC_p is given explicitly by Liao et al. (1982) as their Eq. (71) for the special case of the model used here in which the cubic interaction coefficients β_{14} and β_{34} are both zero. For the Hg–Te and Cd–Te binaries there are no experimental values for ΔC_p ($x = \frac{1}{2}$) and Eq. (110) is used to express ΔH_4^0 in terms of ω_S and the other parameters.

Equations (109) and (110) imply that Eq. (17) for the auxiliary condition on ΔS_M is satisfied. The next independent constraint applied is then the requirement that the relative chemical potential of the species ac must equal the Gibbs energy of formation of AC(s) at the congruent melting point. This allows one to express the entropy of dissociation of ac as

$$\Delta S_4^0 = (\Delta H_f^0 + \Delta H_4^0)/T_{AC} - \Delta S_f^0 - R \ln z$$
$$+ [(1-z)^2 \Omega_R - (1-z)^2 (6z-1)(\beta_{14} + \beta_{34})/4]/T_{AC}, \quad (111)$$

where all quantities including the enthalpy and entropy of formation of AC(s), ΔH_f^0, and ΔS_f^0, respectively, are understood to be evaluated at $x = \frac{1}{2}$ and/or $T = T_{AC}$. The enthalpy of dissociation ΔH_4^0 is of course obtained from either Eq. (110) or from Eq. (110) and the equation to match the experimental value of ΔC_p.

Finally in the case of the Ga–Sb and In–Sb binaries the relative chemical potentials of the Group III element in the liquid phase have been determined experimentally. The experimental value at $x = \frac{1}{2}$ and some temperature $T = T^\dagger$ can be matched exactly using Eq. (88). The left-hand side is the experimental value so that the equation can be used to express one asymmetric interaction coefficient, say Ω_A, in terms of the other, β_{13}.

For the Ga–Sb and In–Sb binaries $\beta_{14} = \beta_{34} = 0$ and Ω_A and β_{13} are taken as constants, i.e., $v_A = C_{13} = 0$. The independent interaction coefficients that can be adjusted to obtain an optimum fit to the liquidus points are then ω_R, v_R, ω_A, and z^*. For the Hg–Te and Cd–Te binaries the independent coefficients are $\omega_R, v_R, \omega_S, \omega_A, v_A, B_{13}, C_{13}, B_{14}, C_{14}, B_{34}, C_{34}$, and z^*.

V. Ga–In–Sb Ternary

5. Ga–In Binary

The 156.6°C melting point of In is the highest melting point in this system. No attempt is made to fit the phase diagram but the thermodynamic properties of the melt are fit. The only species assumed for the liquid phase are uncombined Ga and In and these are numbered 1 and 2, respectively. Remembering that the mole fractions of uncombined Sb, of GaSb, and of InSb are y_3, y_4, and y_5, respectively, that these are all zero and so $y_1 = x_1$, $y_2 = x_2$, and setting $\beta_{12} = 0$, then the relative excess chemical potentials are obtained from Eq. (70) as

$$\bar{\mu}_1^x = 2\alpha_{12} x_2^2, \qquad \bar{\mu}_2^x = 2\alpha_{12} x_1^2. \tag{112}$$

Assuming a linear temperature dependence for the interaction coefficient,

$$\alpha_{12} = \omega_{12} - v_{12} T, \tag{113}$$

the enthalpy of mixing and excess entropy of mixing are obtained using basic thermodynamic equations as

$$\Delta H_M = 2\omega_{12} x_1 x_2, \tag{114}$$

$$\Delta S_M^x = 2v_{12} x_1 x_2. \tag{115}$$

The above equations agree well with the collected data of Hultgren et al. (1973a) if $\omega_{12} = 530$ cal/gram-at. and $v_{12} = -0.5746$ cal/°K gram-at., i.e., the liquid phase is a quasiregular solution.

6. In–Sb Binary

The thermodynamic properties of InSb(s) and the relative constant pressure heat capacity of the stoichiometric liquid used as known input data are given in Table I along with analogous quantities for GaSb(s) (Hultgren et al., 1973a; Brebrick, 1977). In both cases the Gibbs energy of formation of the solid compound from its pure liquid elements is a linear function of T within experimental error and may be expressed in terms of a constant effective enthalpy of formation, $\overline{\Delta H_f}$, and a constant effective entropy of formation, $\overline{\Delta S_f}$, as

$$\Delta G_f^0 = \overline{\Delta H_f} - T \overline{\Delta S_f}. \tag{116}$$

TABLE I

Thermodynamic Properties of InSb and GaSb[a]

AC	T_{mp} (°C)	ΔH_{AC}^m	$\Delta H_f^0(T_{mp})$	$\Delta S_f^0(T_{mp})$	$\overline{\Delta H_f}$	$\overline{\Delta S_f}$	$\overline{\Delta C_p}(\tfrac{1}{2}, T_{mp})$
InSb	525	11.41	−13.27	−11.53	−13.02	−11.22	1.00
GaSb	709.2	15.80	−16.30	−13.10	−16.23	−13.03	0.575

[a] Enthalpies in kcal/mol, and entropies and relative heat capacities in cal/degree mol. ΔC_p is for 2 gram-at. of the liquid phase at 50 at. % and the melting point T_{mp}.

TABLE II

Best-Fit Parameters, Measure of Fit to the Liquidus Line, and Calculated InSb–Sb Eutectic for the In–Sb System[a]

z^*	ω_R	v_R	ω_S	v_S	Ω_A	β_{23}
0.6092	1902.99 cal	1.5858 cal/°K	1720.62 cal	1.1627 cal/°K	−460.68 cal	1321.78 cal
ΔH_5^0	ΔS_5^0	σ_T (°C)	σ_V (°C)	T (eut., °C)	x_{sb} (eut.)	
2817 cal	−0.45723 cal/°K	8.43	1.99	495.2	0.691	

[a] In is species 2, Sb is species 3, and InSb is species 5 in the liquid. ΔH_5^0 and ΔS_5^0 are the standard enthalpy and entropy of dissociation of the InSb species into the pure liquid elements.

The independent interaction coefficients, z^*, ω_R, v_R, and ω_A, were varied to find a minimum value for σ_T where

$$\sigma_T^2 = \sum_{j=1}^{M} (T_{j,\,cal} - T_{j,\,obs})^2/M \qquad (117)$$

and where $T_{j,\,cal}$ and $T_{j,\,obs}$ are the calculated and observed liquidus temperatures for the jth observed composition. Omission of a possible factor, $T_{j,\,obs}^{-2}$, in the definition of σ_T^2 is considered to be unimportant since the extremes in the observed temperatures differ by less than an order of magnitude. Moreover, the significance of a value for σ_T, with the dimensions of temperature, is immediately apparent.

For the liquidus of InSb(s) there are 13 points from thermal analysis, which include a eutectic point at 494 ± 0.5°C and $x_{Sb} = 0.69$ (Liu and Peretti, 1952), as well as three points from dissolution experiments by Hall (cited in Shunk, 1969). The best-fit interaction coefficients, the fit to the liquidus line of InSb(s), σ_T, the fit to 12 Sb liquidus points (Liu and Peretti, 1952), σ_V, and the calculated eutectic temperature and composition are given in Table II. It is emphasized that the values of the interaction coefficients are such that Eqs. (16) and (17) are satisfied with the experimental values for the enthalpy and entropy of formation of InSb(s) at 525°C and the enthalpy of fusion given in Table I. Moreover the value for ΔC_p (2 gram at. components) obtained by doubling the value calculated with Eq. (106) also equals the experimental value given in Table I. The measures of fit given in Table II are comparable to what we judge to be the experimental accuracy. The calculated liquidus line and experimental points are shown in Fig. 1.

Various high temperature thermochemical properties of the liquid phase were then calculated for comparison with experiment using the parameter set

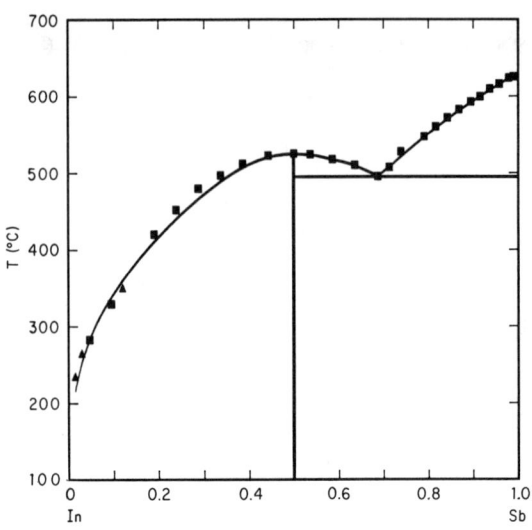

FIG. 1. Calculated liquidus line for In–Sb and experimental points. Squares are from Liu and Peretti (1952), triangles are from Shunk (1969).

TABLE III

FITS TO THE THERMOCHEMICAL DATA IN THE In–Sb BINARY[a]

$\sigma(a_{In})$ (%)			$\sigma(\Delta H_M)$ (%)							$\sigma(\bar{h}_{In})$ (%)	$\sigma(\bar{h}_{Sb})$ (%)
627°C	700	800	627	680	684	713	778	860	911	627	627
4.2	10.6	5.0	5.0	4.2	3.7	3.2	4.0	4.1	7.2	15.2	15.3

[a] The values of σ are multiplied by 100 to convert to a percentage standard deviation.

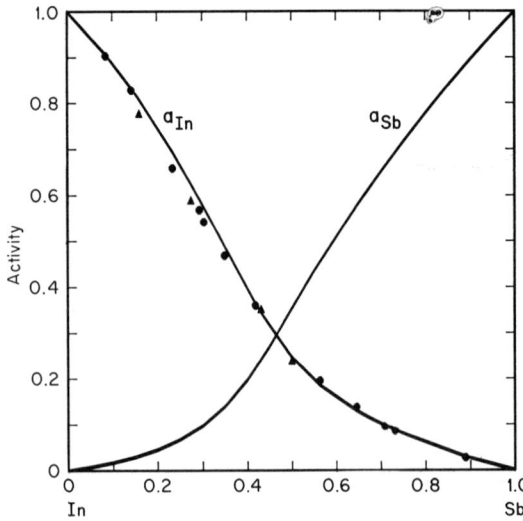

FIG. 2. Calculated activities of In and Sb in In–Sb melt at 627°C and experimental data. Circles are from Hoshino et al. (1965), triangles are from Chatterji and Smith (1973).

in Table II. The measure of fit is defined as $\sigma(Z)$ where

$$[\sigma(Z)]^2 = \sum_{j=1}^{M} \left(\frac{Z_{j,\,\text{obs}} - Z_{j,\,\text{cal}}}{Z_{j,\,\text{obs}}}\right)^2 \Big/ M \tag{118}$$

and where Z stands for activity, enthalpy of mixing, or partial molar enthalpy. Thus $\sigma(Z)$ can be described as the standard deviation of the fractional difference between the observed and calculated values.

The activity of In has been determined electrochemically at 627, 700, and 800°C (Hoshino et al., 1965; Chatterji and Smith, 1973; Anderson, 1978). The fits are listed in the first three columns of Table III and the experimental points and calculated curves for 627°C are shown in Fig. 2. The enthalpies of mixing in In–Sb liquid have been determined for various temperatures between 627 (Hultgren et al., 1973b) and 911°C. The data of Predel and Oehme (1976) at 680°C are tabulated. For the other temperatures we have scaled the data (Predel, 1979) from a graph. Figures 3 and 4 show the calculated curves and

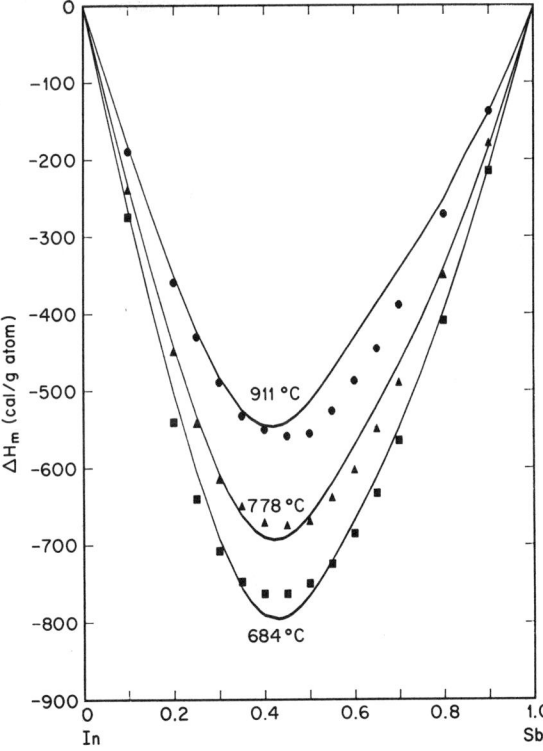

FIG. 3. Calculated enthalpy of mixing of In–Sb melt at various temperatures relative to pure liquid In and Sb and experimental points from Predel and Oehme (1976) and Predel (1979).

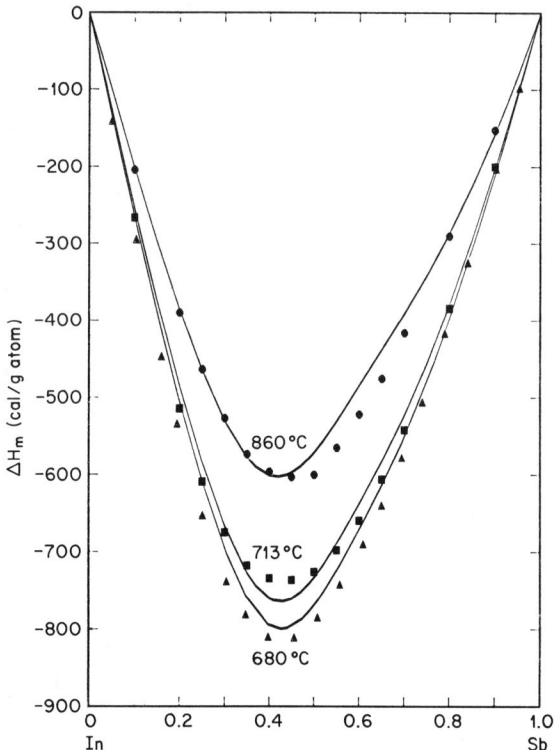

FIG. 4. Calculated enthalpy of mixing of In–Sb melt at additional temperatures and experimental points.

experimental points at various temperatures and the quantitative measures of fit are given in Table III. In contrast to the good agreement obtained here, the quasiregular model and parameters used by Gratton and Wooley (1980) imply a positive, temperature-independent value for ΔH_M which is 292 cal/gram-at. at 50 at. %. The calculated relative partial molar enthalpies of In and Sb are shown as the curves in Fig. 5 and the selected values of Hultgren *et al.* (1973a) are shown as symbols. The measures of fit are given in the last two columns of Table III.

The composition fluctuation factor of Eq. (95) is shown in Fig. 6 as a function of the atomic fraction of Sb in the liquid at 525°C. The minimum is due primarily to the assumed presence of a InSb species whose mole fraction is 0.61 at 50 at. %. The minimum is shifted to about 41 at. % Sb by the interactions among the species, which must be assumed to fit the phase diagram and thermochemical data.

3. SOLUTION MODEL FOR Ga–In–Sb AND Hg–Cd–Te

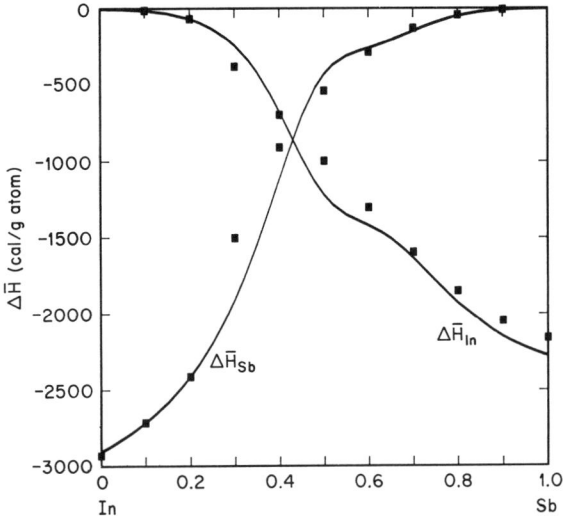

FIG. 5. Calculated relative partial molar enthalpies of In and Sb in In–Sb melt at 627°C and experimental points (■) from Hultgren et al. (1973b).

We have calculated quantitative measures of fit using the parameters given by Kaufman et al. (1981). The fit to the InSb(s) liquidus is 28°C, with a calculated InSb–Sb eutectic temperature of 488.5°C compared to the experimental value of 494 ± 0.5°C. Moreover the enthalpy of mixing for the

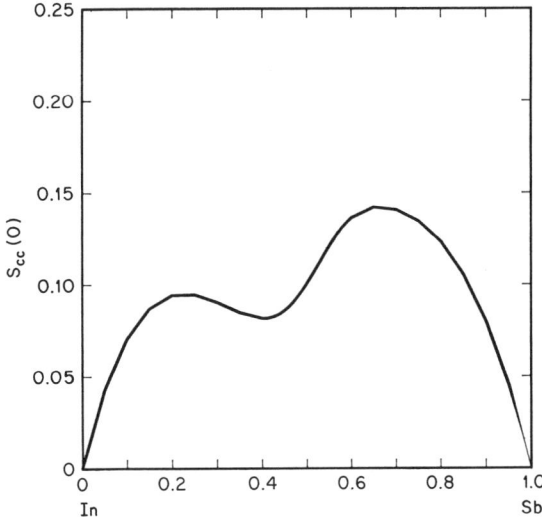

FIG. 6. Composition fluctuation factor at zero wave number for In–Sb melt at 525°C.

TABLE IV

Parameters and Fits in the Ga–Sb System[a]

z^*	ω_R cal.	ν_R (cal/°K)	Ω_A cal.	σ_T (°C)	σ_V (°C)	$\sigma(\Delta H_M)$ cal/gram-at.		T (eut.°C)	x_{Sb} (eut.)
						722°C	750		
0.6531	−579.3	−1.915	3.641	7.6(1.6)	0.91	65.6	103.5	588.6	0.883
0.3404	−2838	−5.248	209.3	16.1(7.7)	2.4	9.1	26.9	578.7	0.8742
0.2231	−243.3	3.800	−228.9	9.5(7.4)	0.85	19.6	48.5	590.0	0.889

[a] σ_T is the fit to the GaSb liquidus and σ_V is that to the Sb liquidus. The calculated GaSb–Sb eutectic is given in the last two columns.

liquid phase calculated from the model is temperature independent in contrast to the experimental data shown in Figs. 3 and 4. Therefore the fits are significantly poorer than those obtained here.

7. Ga–Sb Binary

As with the In–Sb binary, the interaction coefficients are constrained so that Eqs. (16) and (17) are satisfied with the experimental values of ΔH_f^0 and ΔS_f^0 shown in Table I. Also, the experimental value of ΔC_p for the stoichiometric liquid is matched. The best fit to the liquidus points does not provide a sufficiently good fit to ΔH_M in our opinion and conversely. Therefore both types of data were fit simultaneously to obtain the best overall fit shown as the last row in Table IV. The experimental liquidus points consist of 22 tabulated values determined using differential thermal analysis by Maglione and Potier (1968) and 6 points we have scaled from a graph representing dissolution experiments by Hall (1963). The value under σ_T enclosed in parentheses gives the fit if these last 6 points are omitted. The experimental GaSb–Sb eutectic is at 589 ± 0.5°C and 88 at. % Sb. There are 26 values of ΔH_M at 722°C from Ansara et al. (1976) and Gambino and Bros (1975) and 11 values at 750°C from Predel and Stein (1971). The first row of Table IV shows the best fit to the liquidus points, the second row the best fit to the enthalpy of mixing, and the third row the compromise fit that we adopt. The latter are slightly different than those we chose earlier (Liao et al., 1982), as discussed in Appendix B. Note that the interaction coefficients giving a best fit to the enthalpy of mixing also yield an incorrect eutectic temperature. Here as in the case of In–Sb the liquidus line of the Sb-rich solid solution was calculated using the melting point and enthalpy of fusion for Sb given by Hultgren et al. (1973b) and assuming that the chemical potential of Sb in the solid solution is equal to that of pure Sb(s). The values for the standard enthalpy and entropy of dissociation of the GaSb liquid species corresponding to row 3, ΔH_4^0 and ΔS_4^0 are 1821.5 cal/mol and -1.1015 cal/°K mol. Figure 7 shows the experimental liquidus points, the curve calculated here and the curve calculated using the sub-regular model of Eq. (1) for the liquid. The parameters of Eq. (1) are changed from those given by Brebrick (1977) in order to obtain a better fit to the ΔH_M data. The new values are $W = -1000$ cal, $a = 1.1498$, $V = 0.46024$ cal/°K, and $b = -3.500$. These parameters give a much better fit than those we obtain with the parameters given by Kaufman et al. (1981). Kaufman's parameters give a 25°C fit to the GaSb(s) liquidus and a calculated GaSb–Sb eutectic temperature of 577°C compared to the experimental value of 589 ± 0.5°C. Moreover the calculated enthalpy of mixing for the liquid is indepen-

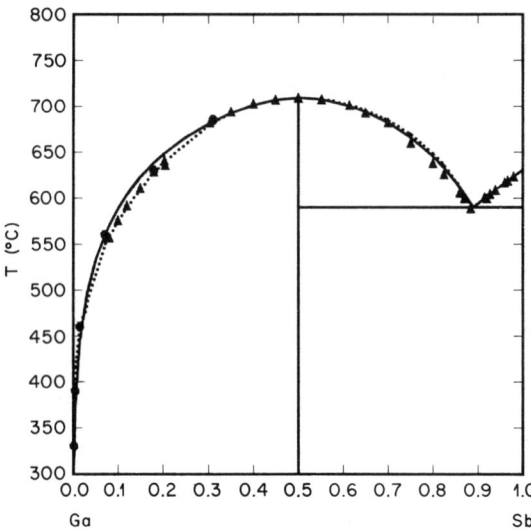

FIG. 7. Calculated liquidus for Ga–Sb and experimental points. Circles are from Hall (1963), triangles from Maglione and Potier (1968). Dotted line is calculated using subregular model and parameters given in text.

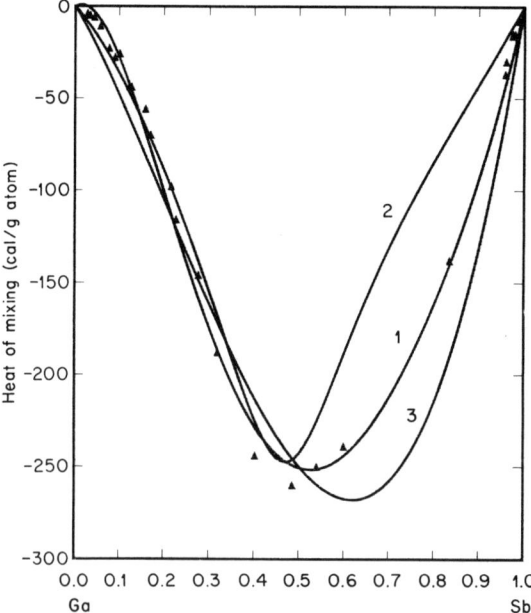

FIG. 8. Calculated enthalpy of mixing of Ga–Sb melt at 721.9°C and experimental points from Gambino and Bros (1975). Curve 1 is fifth-power polynomial fit from Ansara et al. (1976). Curve 2 is our calculation with associated solution parameters giving the best compromise fit to the enthalpy of mixing and the liquidus points. Curve 3 is calculated using the subregular solution model and parameters given in the text.

dent of T, changes sign with composition at 722°C, and fits the data at 722°C with a standard deviation of 113 cal. We therefore conclude that the parameters given by Kaufman et al. (1981) are not satisfactory. Our subregular model parameters given above of course also predict an enthalpy of mixing that is independent of T. However, they fit the GaSb liquidus to within 13.5°C and give a eutectic in agreement with experiment. Figure 8 shows the experimental points for ΔH_M at 995°K, our calculated curve, the curve calculated by Ansara et al. (1976) using a fifth-power polynomial in atomic fraction to fit the ΔH_M data alone, and the calculated curve using the subregular model with the parameters cited above. The model and parameters used by Gratton and Wooley (1980) again imply a positive, temperature-independent ΔH_M that is symmetric about $x = \frac{1}{2}$ and that is 1423.5 cal/gram-at. at $x = \frac{1}{2}$. This is in contrast to the negative experimental value of about -250 cal/gram-at. Figure 9 shows the calculated activities of Ga and Sb in the liquid at 727°C and experimental points from Bergman et al. (1974), Chu et al. (1966), and Anderson (1978). Other experimental results in poor argeement with the above are shown by Gambino and Bros (1975).

The calculated composition fluctuation factor at zero wave number for the liquid at 709.2°C is shown in Fig. 10. For comparison the uppermost curve shows the same quantity for an ideal solution of species Ga and Sb,

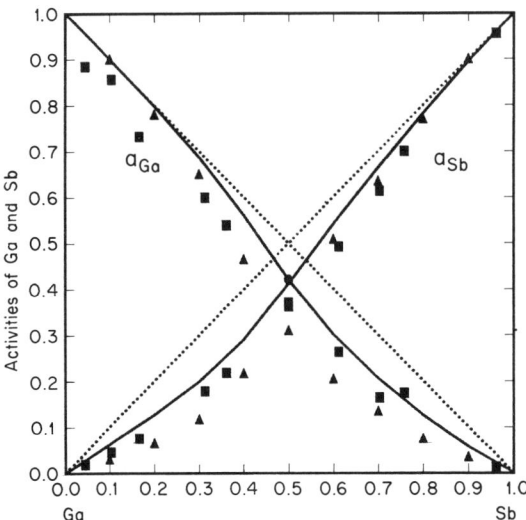

FIG. 9. Solid curves give our calculated values for the activities in Ga–Sb melt at 726.9°C. Squares are from Anderson (1978) for 730°C, circles from Chu et al. (1966), and triangles from Bergman (1974).

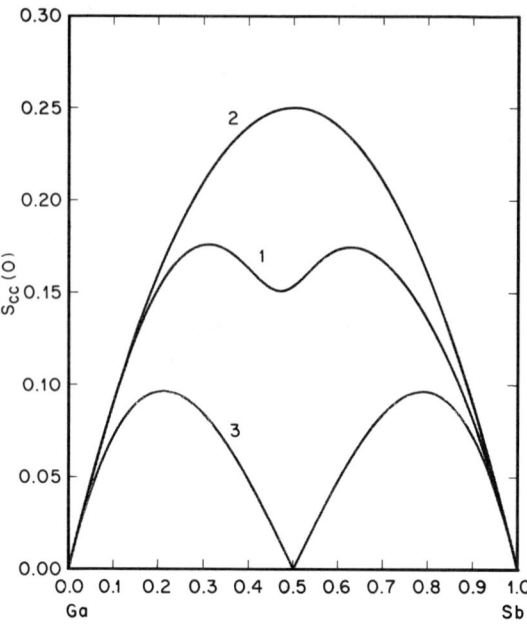

FIG. 10. Composition fluctuation factor at zero wave number for Ga–Sb melt at 709.2°C. Curve 1 is for the parameters established here, whereas curve 2 is for a completely dissociated, ideal solution and curve 3 for a completely associated ideal solution.

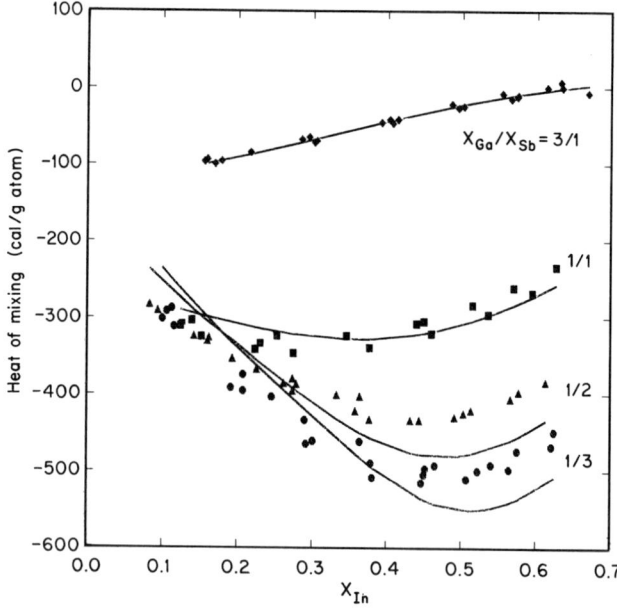

FIG. 11. Calculated enthalpy of mixing for the ternary Ga–In–Sb liquid at 722°C and experimental points from Ansara *et al.* (1976).

whereas the lowest curve is for a completely associated, ideal solution of Ga, Sb, and GaSb species.

8. Ga–In–Sb TERNARY

The three coefficients entering only for ternary compositions are α_{15}, α_{24} and α_{45} and these describe the interactions of Ga–InSb, In–GaSb, and GaSb–InSb, respectively. They are assumed here to be independent of T. (The corresponding β_{ij} and the pure ternary coefficients γ_{ijk} have already been set equal to zero.) The α-coefficients above were determined by fitting 90 experimental values of the enthalpy of mixing in the ternary liquid phase (Ansara et al., 1976). A standard deviation between experiment and calculation of 15.6% is obtained. The results are shown in Fig. 11 for various x_{Ga}/x_{Sb} ratios. The fit can be seen to be very good for the larger Ga contents. The liquidus and solidus points of the GaSb–InSb pseudobinary section as well as the ternary liquidus points were then fit by varying W_s and V_s, the parameters characterizing the quasiregular solid solution in Eqs. (14) and (15). The calculated lines in the pseudobinary section and the experimental points are shown in Fig. 12, and the best fit values of the ternary interaction coefficients and measures of fit are listed in Table V. Figure 13 shows calculated liquidus isotherms and

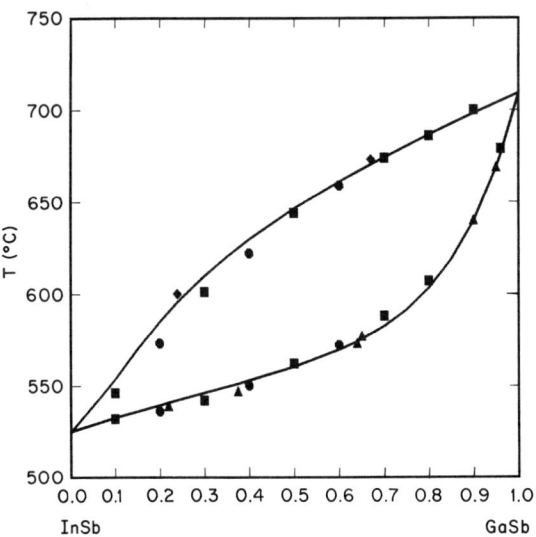

FIG. 12. Calculated liquidus and solidus in the InSb–GaSb pseudobinary section and experimental points. Squares from differential thermal analysis, circles from thermal analysis, and triangles from x-ray diffraction, all from Wooley and Lees (1959). Diamonds are from Blom and Plaskett (1971).

TABLE V

VALUES OF TERNARY PARAMETERS AND MEASURES OF FIT[a]

α_{15}	α_{24}	α_{45}	W_s	V_s	σ_L	σ_s	$\sigma_L(\text{ter})$	$\sigma(\overline{\Delta H_M})$
570.5	−185.0	1158.4	563.7	−1.500	5.812	2.9	11.5	15.6%

[a] σ_L and σ_s in degrees Celsius are the fits to the GaSb–InSb pseudobinary liquidus and solidus. $\sigma_L(\text{ter})$ is the fit to ternary liquidus in degrees Celsius. $\sigma(\overline{\Delta H_M})$ is the fit to 90 enthalpies of mixing for the ternary liquid. The α interaction parameters and W_s are in calories, V_s in calories per degrees kelvin.

experimental points. Figures 14 and 15 describe tie-line data at 400 and 500°C and again the agreement between calculation and experiment is very good. The calculations of Gratton and Wooley (1978) do not agree well with experiment although their calculated liquidus isotherms in Fig. 13 do. Finally, Fig. 16 shows the calculated liquidus isotherms and solid isoconcentration lines across the entire ternary. The dashed line across the top of the Gibbs triangle is the calculated eutectic valley. This was obtained using values of 631°C and 4750 cal/gram-at. for the melting point and enthalpy of fusion of Sb

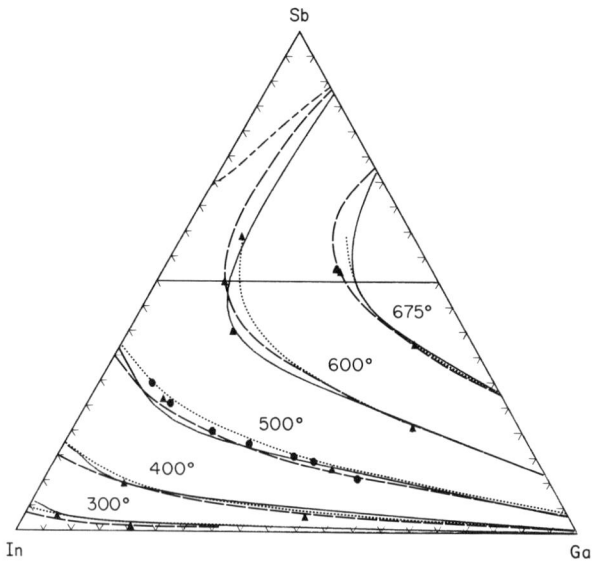

FIG. 13. Calculated ternary liquidus isotherms and experimental points. Triangles are from Blom and Plaskett (1971) and circles from Antypas (1972). Dotted curves are calculated by Gratton and Wooley (1978), dashed curves are calculated by Blom and Plaskett.

3. SOLUTION MODEL FOR Ga–In–Sb AND Hg–Cd–Te

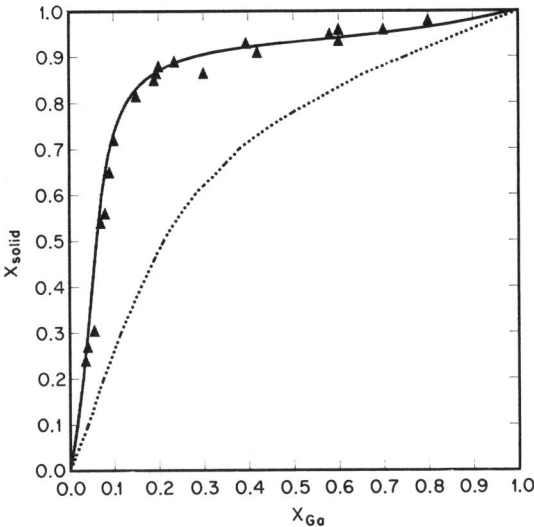

FIG. 14. Mole fraction of GaSb in the GaSb–InSb solid solution as a function of atom fraction Ga in the liquid at 400°C. Solid line is our calculated result, dotted line is calculated by Gratton and Wooley (1978). Experimental points are from Miki *et al.* (1975).

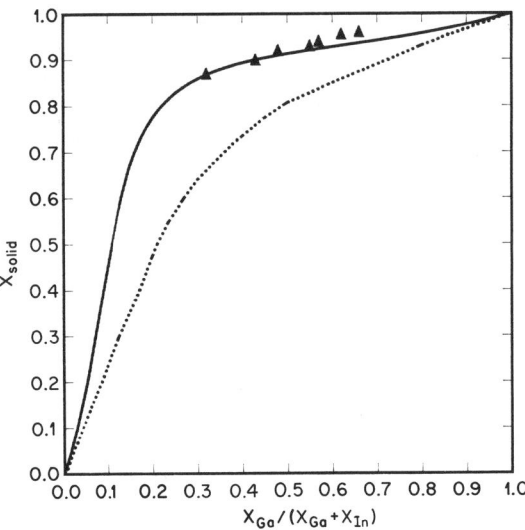

FIG. 15. Mole fraction of GaSb in the GaSb–InSb solid solution as a function of liquid composition at 500°C. Experimental points are from Antypas (1972). Dashed lines calculated by Gratton and Wooley (1978).

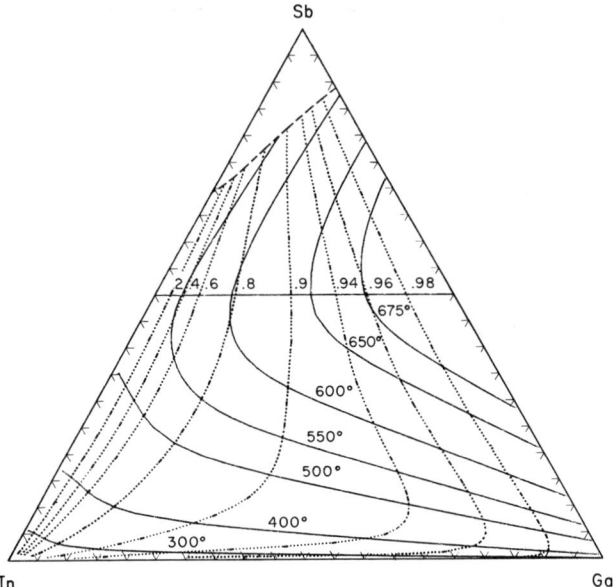

FIG. 16. Calculated liquidus isotherms (solid lines) and solid isoconcentration lines (dotted lines). The labels on the isotherms are in degrees Celsius. The numbers along the InSb–GaSb section label the solid isoconcentration lines and are the mole fraction GaSb in the GaSb–InSb solid solution. The dashed line across the upper part of the figure is the calculated eutectic valley.

and the assumption that the solubility of Ga and In in the Sb terminal solid solution is small enough that the chemical potential of Sb is essentially that of the pure element. The abrupt swing of the solid isoconcentration lines toward the In corner at low temperatures is particularly noteworthy. The experimental data (Gratton and Wooley, 1978) show this trend for 0.90 mole fraction GaSb but not for higher mole fractions. However, these data are not completely experimental, involving the intersection of experimentally determined tie-line directions with liquidus isotherms calculated with a quasi-regular model [see Eq. (1) and discussion] for the liquid phase.

We have not included the calculations for the Ga–In rich half of the phase diagram made by Szapiro (1980) using Jordan's simplification of the regular associated solution model. These have been criticized by Tung *et al.* (1981a) as using a starting liquidus equation that is in principle incorrect although the actual error incurred was not determined.

Finally the interaction coefficients used for the Ga–In–Sb system are collected in Table VI where the Ω_j coefficients of Eq. (83)–(85) have been replaced by the original α_{ij} coefficients of Eq. (68). Using Eq. (109), which is the liquidus equation at the maximum melting point of GaSb(s), and its analog for

3. SOLUTION MODEL FOR Ga–In–Sb AND Hg–Cd–Te

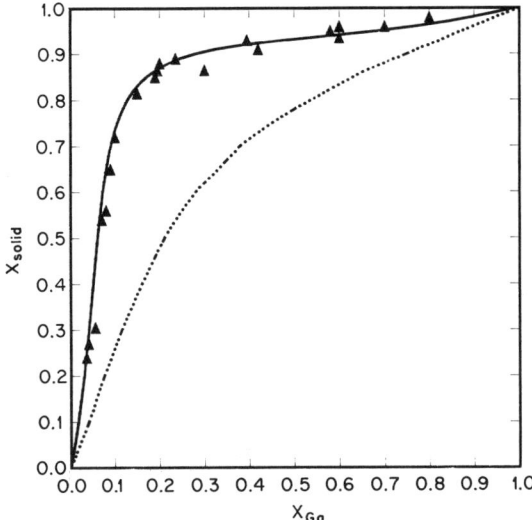

FIG. 14. Mole fraction of GaSb in the GaSb–InSb solid solution as a function of atom fraction Ga in the liquid at 400°C. Solid line is our calculated result, dotted line is calculated by Gratton and Wooley (1978). Experimental points are from Miki et al. (1975).

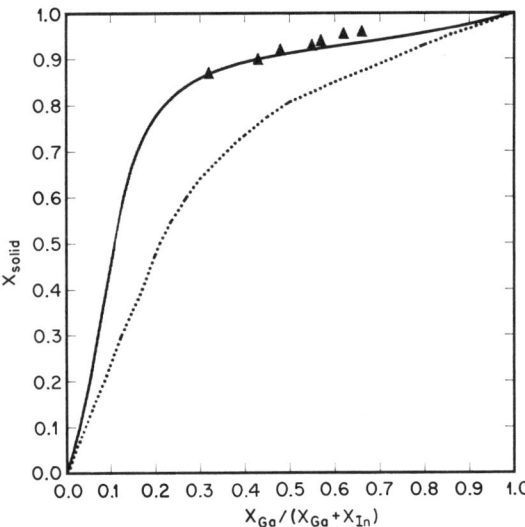

FIG. 15. Mole fraction of GaSb in the GaSb–InSb solid solution as a function of liquid composition at 500°C. Experimental points are from Antypas (1972). Dashed lines calculated by Gratton and Wooley (1978).

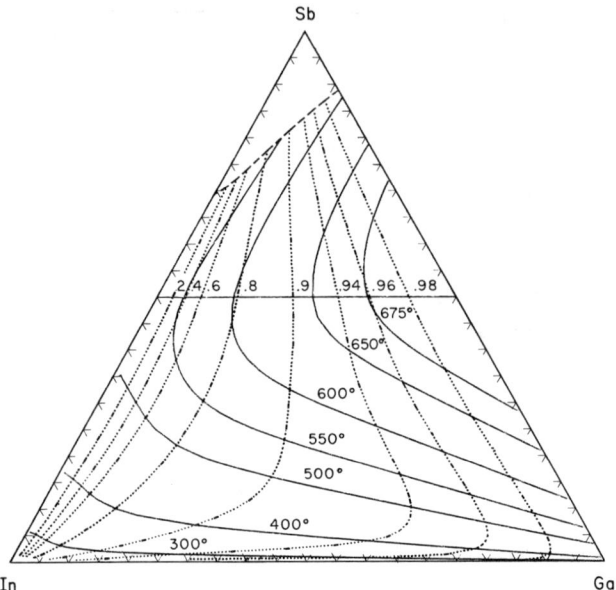

FIG. 16. Calculated liquidus isotherms (solid lines) and solid isoconcentration lines (dotted lines). The labels on the isotherms are in degrees Celsius. The numbers along the InSb–GaSb section label the solid isoconcentration lines and are the mole fraction GaSb in the GaSb–InSb solid solution. The dashed line across the upper part of the figure is the calculated eutectic valley.

and the assumption that the solubility of Ga and In in the Sb terminal solid solution is small enough that the chemical potential of Sb is essentially that of the pure element. The abrupt swing of the solid isoconcentration lines toward the In corner at low temperatures is particularly noteworthy. The experimental data (Gratton and Wooley, 1978) show this trend for 0.90 mole fraction GaSb but not for higher mole fractions. However, these data are not completely experimental, involving the intersection of experimentally determined tie-line directions with liquidus isotherms calculated with a quasi-regular model [see Eq. (1) and discussion] for the liquid phase.

We have not included the calculations for the Ga–In rich half of the phase diagram made by Szapiro (1980) using Jordan's simplification of the regular associated solution model. These have been criticized by Tung et al. (1981a) as using a starting liquidus equation that is in principle incorrect although the actual error incurred was not determined.

Finally the interaction coefficients used for the Ga–In–Sb system are collected in Table VI where the Ω_j coefficients of Eq. (83)–(85) have been replaced by the original α_{ij} coefficients of Eq. (68). Using Eq. (109), which is the liquidus equation at the maximum melting point of GaSb(s), and its analog for

TABLE VI
Best-Fit Interaction Coefficients for the Ga–In–Sb System in Calories

Ga–In	$\alpha_{12} = 530 + 0.5746T$	$\beta_{12} = 0$		
Ga–Sb	$\alpha_{13} = -150.4 - .2918T$	$\alpha_{14} = 44.8 + 1.827T$	$\alpha_{34} = 273.7 + 1.827T$	$\Delta H_4^0 = 1821.5$ cal
	$\beta_{13} = -441.0$	$\beta_{14} = \beta_{34} = 0$	$z^* = 0.2231$	$\Delta S_4^0 = -1.1015$ cal/°K
In–Sb	$\alpha_{23} = 3441.2 - 2.3255T$	$\alpha_{25} = -321.49 + 0.2115T$	$\alpha_{35} = 139.13 + 0.2116T$	$\Delta H_5^0 = 2817.0$
	$\beta_{23} = 1321.78$	$\beta_{25} = \beta_{35} = 0$	$z^* = 0.609$	$\Delta S_5^0 = -0.4572$
Ternary	$\alpha_{15} = 570.5$	$\alpha_{24} = -185.02$	$\alpha_{45} = 1158.4$	
	$\beta_{15} = 0$	$\beta_{24} = 0$	$\beta_{45} = 0$	

Solid solution	$W_s - V_s T = 563.7 + 1.5T$				
Liquid species	Ga	In	Sb	GaSb	InSb
Number	1	2	3	4	5

InSb(s), one can set all of the interaction coefficients to zero to calculate the mole fraction of the associated species at $x = \frac{1}{2}$ and $T = T_{AC}$ for an ideal associated solution. For GaSb the value is 0.415, while for InSb it is 0.28. These are to be compared with the best fit values of z^* in Table VI.

VI. Hg–Cd–Te Ternary

A significant portion of the thermodynamic data for this system consists of the partial pressures of Hg, Cd, and Te$_2$; P_{Hg}, P_{Cd}, and P_2, respectively, and quantities derived from these. The equations used to represent the vapor pressures of the elements are

$$\log_{10} P^\circ_{Hg} \text{ (atm)} = -3099.0/T + 4.920, \quad P^\circ_{Hg} > 0.1 \text{ atm}, \tag{119}$$

$$\log_{10} P^\circ_{Hg} \text{ (atm)} = -3157.0/T + 5.028, \quad 10^{-3} \leq P^\circ_{Hg} \leq 0.1 \text{ atm}, \tag{120}$$

$$\log_{10} P^\circ_{Cd} \text{ (atm)} = -5808.0/T + 5.956, \quad T < 594°K, \tag{121}$$

$$\log_{10} P^\circ_{Cd} \text{ (atm)} = -5317.0/T + 5.119, \quad T \geq 594°K, \tag{122}$$

$$\log_{10} P^\circ_2 \text{ (atm)} = -5960.2/T + 4.7191, \quad T > 723°K. \tag{123}$$

These analytical equations used by Schwartz et al. (1981) and Tung et al. (1981a) incorporate corrections and extensions to the tabulated, selected values of Hultgren et al. (1973a). Equation (119) reproduces the tabulated selected values to within 1% between 0.5 and 10 atm. It also agrees to within 1% with measured values to 30 atm (Sugawara and Sato, 1962), with optical absorbance measurements of the vapor over pure Hg to 20 atm (Brebrick and Strauss, 1965), and with one point near 30 atm obtained by detecting the complete vaporization of a known weight of Hg in a known volume by optical absorbance of the vapor (Schwartz et al., 1981). Between 10^{-3} and 10^{-1} atm, Eq. (120) matches the selected tabulated values to within 1%. Selected, tabulated values (Hultgren et al., 1973a) for the vapor pressure of Cd between 10^{-8} and 1 atm are reproduced to within 2.5% by Eqs. (121) and (122). The saturated vapor over Te(l) is essentially all Te$_2$(g). Equation (123) agrees to within 1% with the selected, tabulated values of Hultgren et al. (1973a) at 600°C and higher. However, whereas Eq. (123) gives a straight line on a $\log_{10} P^\circ_2$ versus $1/T$ plot down to the 723°K melting point of Te, in agreement with optical absorbance measurements of Brebrick (1968), the selected values give the data of Machol and Westrum (1958) the most weight and show the vapor pressure of Te falling increasingly below those calculated with Eq. (123). At the 449.5°C melting point Eq. (123) gives a vapor pressure about 20% higher. We choose Eq. (123) because of the agreement with optical absorbance measurements cited and because the data of Machol and Westrum are relatively few in number between 450 and 500°C and the values are near the experimental limits of precision.

3. SOLUTION MODEL FOR Ga–In–Sb AND Hg–Cd–Te

At high pressure Hg(g) is nonideal. Using the equation of state given by Sugawara and Sato (1962), the chemical potential or fugacity f are given by

$$\mu(\text{Hg}(g)) \equiv RT \ln f$$
$$= 1.987 T \ln P_{\text{Hg}}$$
$$+ 4857.6 P_{\text{Hg}}(-1.793/T + 251/T^2 + 0.001071), \quad (124)$$

where P_{Hg} is in atmospheres and μ in calories per gram-atom. For the data considered here the fugacity differs significantly from P_{Hg} only for the HgTe–CdTe pseudobinary melt.

The Gibbs energy of formation of CdTe(s) according to the reaction

$$\text{Cd}(l) + \text{Te}(l) \rightarrow \text{CdTe}(s) \quad (125)$$

is selected following Tung et al. (1981a) as

$$\Delta G_f^\circ \text{ (cal.)} = -30{,}024 + 10.346 T. \quad (126)$$

In the absence of heat capacity data the enthalpy of formation at the 1092°C melting point is taken as $-30{,}024$ cal/mol and the entropy of formation as -10.346 cal/mol °K. Recent partial pressure measurements by Su et al. (1981) establish the enthalpy of fusion of HgTe(s) as 8727 cal/mol and give

$$\Delta G_f^0 \text{ (cal.)} = -12{,}829 + 8.339 T - 4.0403(10^8)$$
$$\times (-1.793/T + 251/T^2 + 0.001071)$$
$$\times \exp(-7135.68/T) \quad (127)$$

for the reaction

$$\text{Hg}(l) + \text{Te}(l) \rightarrow \text{HgTe}(s). \quad (128)$$

The enthalpy and entropy of formation at the melting point are calculated from Eq. (127) using standard thermodynamic formulas.

In addition to Eqs. (126) and (127), the other necessary thermodynamic properties of HgTe(s) and CdTe(s) are given in Table VII. The enthalpy of fusion for CdTe(s) is based upon a single determination by Kulwicki (1963). The Te-rich eutectics listed are from Kulwicki for CdTe and from Strauss (1967) for HgTe.

TABLE VII

THERMODYNAMIC PROPERTIES FOR HgTe(s) AND CdTe(s) AND Te-RICH EUTECTIC

	Melting point (°C)	Enthalpy of fusion (cal/mol)	$-\Delta H_f^0$ (mp) (cal/mol)	$-\Delta S_f^0$ (mp) (cal/°K mol)	T (eut.) (°C)	x_{Te} (eut.)
HgTe	670	8,727.0	13,302.0	8.96	413.3	0.84
CdTe	1092	12,012	30,024.0	10.346	449.9	Degenerate

TABLE VIII

Liquid Phase Interaction Coefficients in the Hg–Cd–Te System

Hg–Cd	$\alpha_{12} = -1254$	$\beta_{12} = 0$			
Hg–Te	$\alpha_{13} = -498.0 - 0.6557T$	$\alpha_{14} = 1532.0 - 0.2898T$	$\alpha_{34} = 162 + 0.029T$	$\Delta H_4^0 = 8535.0$ cal
	$\beta_{13} = 628.0$	$\beta_{14} = -717.0$	$\beta_{34} = -689$	$\Delta S_4^0 = 5.11$ cal/°K
Cd–Te	$\alpha_{23} = 5832.0 - 5.86T$	$\alpha_{25} = 5938.0 - 3.272T$	$\alpha_{35} = 734 - 0.2346T$	$\Delta H_5^0 = 19{,}169.0$
	$\beta_{23} = -6783.0 + 0.392T$	$\beta_{25} = -513 + 0.1488T$	$\beta_{35} = 2.0 - 0.5577T$	$\Delta S_5^0 = 2.53$
Ternary	$\alpha_{15} = 5621.0 - 2.891T$	$\alpha_{24} = 0$	$\alpha_{45} = -3429.0 + 1.866T$	
	$\beta_{15} = 0$	$\beta_{24} = 0$	$\beta_{45} = 0$	
Solid solution	$W_s - V_s T = 1384 - 0.8452T$			

All interaction coefficients are in calories

Liquid species	Hg	Cd	Te	HgTe	CdTe
Number	1	2	3	4	5

3. SOLUTION MODEL FOR Ga–In–Sb AND Hg–Cd–Te

9. Hg–Cd BINARY

Similar to the Ga–In system no molecular species are assumed so that the enthalpy and entropy of mixing are given in our model by Eqs. (114) and (115). The selected values of Hultgren *et al.* (1973b) are fit well taking $\alpha_{12} = -1254$ cal, with ν_{12} and all the cubic interaction coefficients as zero. No attempt is made here to fit the phase diagram in which the 321°C melting point of Cd is the highest temperature.

10. Hg–Te AND Cd–Te BINARIES

After satisfying the constraints given by Eqs. (109)–(111) the remaining interaction coefficients were varied to find an acceptable fit to the liquidus points, eutectic temperature, and partial pressures along the three-phase curve of HgTe(s). These parameters along with those remaining to characterize the entire ternary are given in Table VIII. A similar procedure was followed for the Cd–Te binary. In neither case was it possible to obtain satisfactory fits with all of the β_{ij} parameters set to zero. Figure 17 shows the calculated liquidus lines

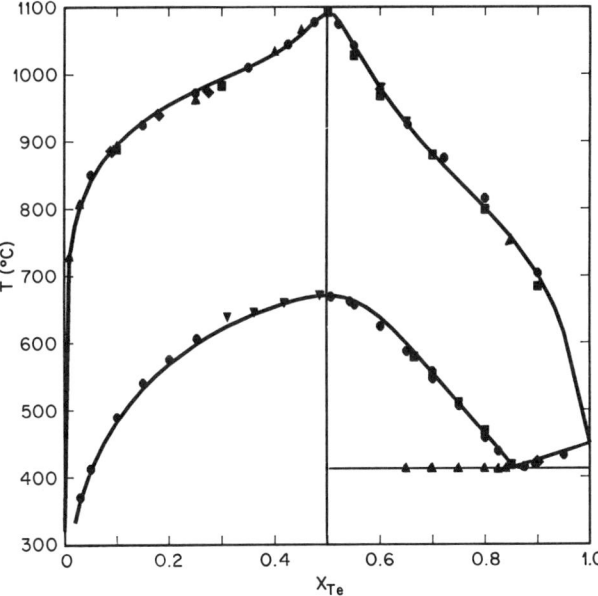

FIG. 17. Liquidus lines for CdTe (upper) and HgTe (lower). For CdTe squares are from Steininger *et al.* (1970), circles from Lorenz (1962), triangles from Kulwicki (1963), inverted triangles from Brebrick (1971b), and diamonds are selected data from de Nobel (1959). For HgTe the circles and triangles are from Strauss (1967), inverted triangles from Brebrick and Strauss (1965), and squares from Harman (1980).

TABLE IX

CALCULATED EUTECTIC TEMPERATURES T_e IN DEGREES CELSIUS AND
COMPOSITIONS IN ATOMIC FRACTION Te, x_e

	Hg–HgTe	HgTe–Te	Cd–CdTe	CdTe–Te
T_e	−38.9	413.5	321	447.2
x_e	$2(10^{-7})$	0.854	$4.2(10^{-6})$	0.991

in both binaries and the experimental points. The fit to the Cd–Te liquidus is 9.8°C and that to the HgTe liquidus is 6.3°C. The calculated eutectic points, which agree with experiment, are listed in Table IX. The composition fluctuation factor at zero wave number [Eqs. (95), (96), (98)] is shown for both binary liquids at the melting point of the corresponding solid compound in Fig. 18. The strong minimum near 50 at. % for Cd–Te is due to the strong association of the CdTe species, $z^* = 0.93$. The maximum at $x = 0.275$ exceeds the value for an ideal solution, which is $x(1 - x)$, and is indicative of clustering of like atoms and incipient phase separation. This trend is also indicated by the flat shape of the Cd-rich liquidus line in Fig. 17. For Hg–Te the minimum in $S_{cc}(0)$ is not so strong and occurs near 55 at. % Te. The mole fraction of the

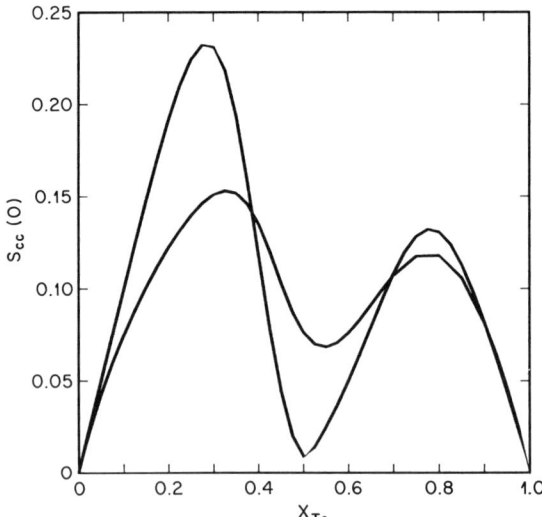

FIG. 18. Calculated composition fluctuation factor at zero wave number in the liquid phase versus atom fraction Te. The highest curve at 0.28 atom fraction is for Cd–Te at 1092°C. The other curve is for Hg–Te at 670°C.

3. SOLUTION MODEL FOR Ga–In–Sb AND Hg–Cd–Te

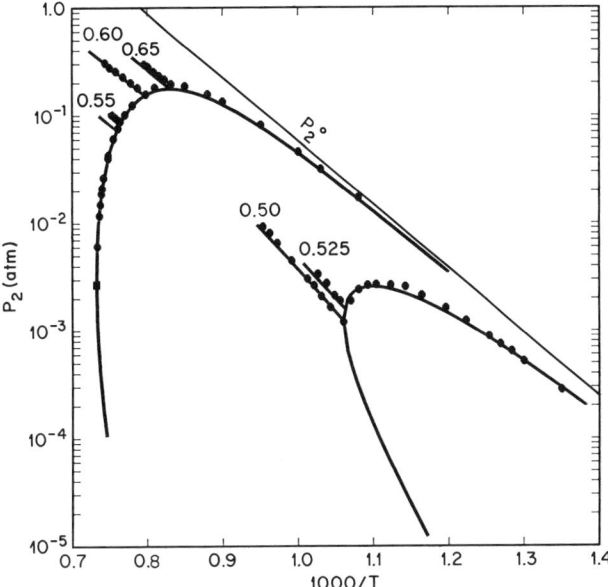

FIG. 19. Partial pressure of $Te_2(g)$ along the CdTe(s) three-phase curve, upper left, and along the HgTe(s) three-phase curve plotted on a log scale versus $1000/T$. Uppermost line labeled P_2^0 is the vapor pressure of Te(l). Short line segments are for liquids with labeled atom fraction Te. Experimental points for CdTe are from Brebrick (1971b), those for HgTe are from Su et al. (1981) and Brebrick and Strauss (1965).

HgTe liquid species at $x = \frac{1}{2}$ and 670°C is $z^* = 0.415$. Setting Eq. (91) equal to the Gibbs energy of formation of the solid compound at its melting point, the values of z^* for an ideal associated solution are obtained as 0.45 for Hg–Te and 0.89 for Cd–Te, close to the best-fit values obtained.

Figure 19 shows the calculated partial pressure of Te_2, P_2, along the three-phase curves for CdTe(s) and HgTe(s) along with calculated pressures for some melts, shown as short line segments. Again the experimental data are shown as points. The fit is 7% for HgTe, 4.6% along the three-phase curve of CdTe, and 12.6% for the three Cd–Te melts. For the 55-at. % Te melt the calculated line is distinctly below the experimental points. The calculated partial pressures at the maximum melting point are given in Table X. The partial pressure of Hg along the three-phase curve of HgTe(s) and for two melts is shown in Fig. 20. The fit is 4.3% along the three-phase curve and 2.9% for the 50- and 52.5-at. % melts. The fit to the cadmium partial pressure along the CdTe(s) three-phase curve shown in Fig. 21 is 6.9% if all of the triangle points and the scattered circle points below 0.15 atm are excluded.

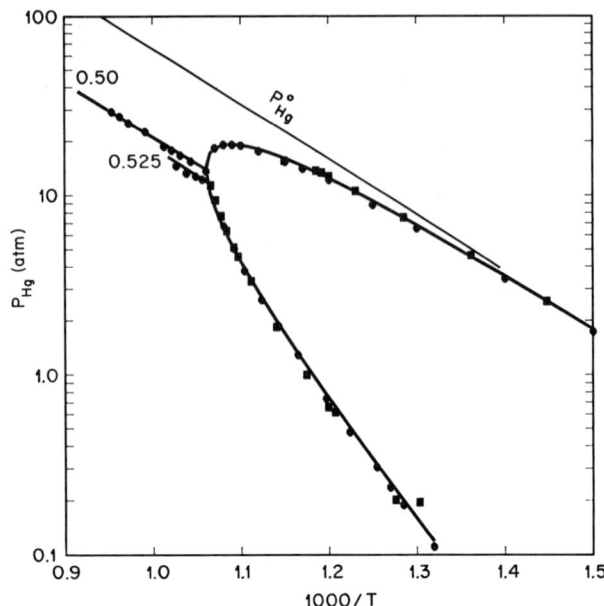

FIG. 20. Partial pressure of Hg along the three-phase curve for HgTe(s). Vapor pressure of Hg(l) given by line labeled P^0_{Hg}, short line segments are for liquids. Circles are from Su *et al.* (1981) and Brebrick and Strauss (1965). Squares are from Levitskaya *et al.* (1970).

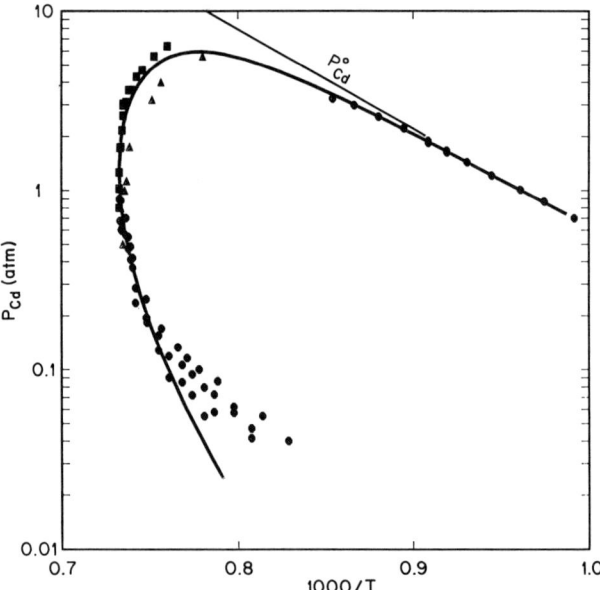

FIG. 21. Partial pressure of Cd along the three-phase curve for CdTe(s). Vapor pressure of Cd(l) is line labeled P^0_{Cd}. Squares are from Lorenz (1962), circles from Brebrick (1971b), and triangles from de Nobel (1959).

11. Hg–Cd–Te Ternary

The remaining ternary liquid phase interaction coefficients and the solid solution parameters, W_s and V_s, from Eq. (14, 15) were established by requiring a good fit to 7 solidus and 7 liquidus points in HgTe–CdTe pseudobinary plus 23 Te-rich liquidus points determined by visual observation by Harman (1980). Several sets of parameters giving almost equally good fits were found. For all of these $\alpha_{24}, \beta_{24}, \beta_{15}$, and β_{45} are zero. We choose the set given in Table VIII to get a good fit in the Hg-rich corner (to be discussed below) but at the sacrifice of a slightly poorer fit to some Te-rich liquidus points determined (Lawley, 1980) by differential thermal analysis. The calculated liquidus and solidus along with the experimental points in the pseudobinary are shown in Fig. 22. The fits are 7.0 and 3.3°C, respectively. The experimental solidus point at 90 mol % CdTe from Szofran and Lehoczky (1981) is obviously low as is their 1082°C melting point for CdTe(s). Three solidus points at 10, 20, and 41 mol % CdTe were obtained from the discontinuities in the slopes of log (optical absorbance) versus $1/T$ plots (Schwartz et al., 1981; Tung et al., 1981b). Calculated temperatures along the pseudobinary solidus and calculated partial pressures are listed in Table X. The values of P_{Hg} in the pseudobinary melt shown in Fig. 23 are fit to 11% although slightly different lines are calculated for different mole percent CdTe in contrast to experiment (Steininger, 1976; Su et al., 1981).

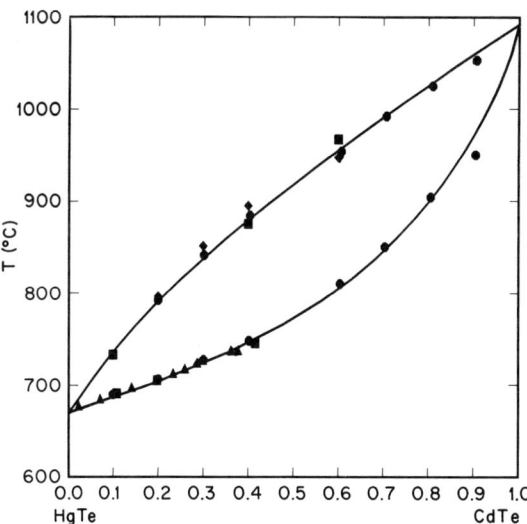

FIG. 22. Liquidus and solidus lines in the HgTe–CdTe pseudobinary section. Circles are from Szofran and Lehoczky (1981), squares on the solidus from Tung et al. (1981b), squares on the liquidus from Blair and Newnham (1961), diamonds from Steininger (1976), and triangles from Harman (1967).

TABLE X

CALCULATED PARTIAL PRESSURES IN ATMOSPHERES ALONG
THE SOLIDUS LINE OF $Hg_{1-x}Cd_xTe(s)$

x	T (mp, °C)	$P_2 = P_{Te_2}$	P_{Hg}	P_{Cd}
0	670	$1.24(10^{-3})$	13.95	0
0.10	686.2	$1.58(10^{-3})$	16.1	$5.71(10^{-6})$
0.20	703.6	$2.03(10^{-3})$	18.7	$1.80(10^{-5})$
0.40	745.5	$3.53(10^{-3})$	26.0	$1.05(10^{-4})$
0.60	803.8	$6.65(10^{-3})$	39.2	$6.70(10^{-4})$
0.80	897.5	$1.35(10^{-2})$	65.1	$7.73(10^{-3})$
0.90	969.4	$1.68(10^{-2})$	79.9	$4.21(10^{-2})$
0.925	991.8	$1.63(10^{-2})$	80.8	$7.14(10^{-2})$
0.95	1018.2	$1.48(10^{-2})$	77.9	0.133
1.00	1092	$2.17(10^{-3})$	0	1.38

Figures 24–26 show the partial pressures of Hg, Te$_2$, and Cd along the three-phase curves for various solid solutions, $Hg_{1-x}Cd_xTe(s)$. (We used the letter u in Parts II and IV to denote the composition of the solid solution and x for the atom fractions of the components in the liquid. In these figures and

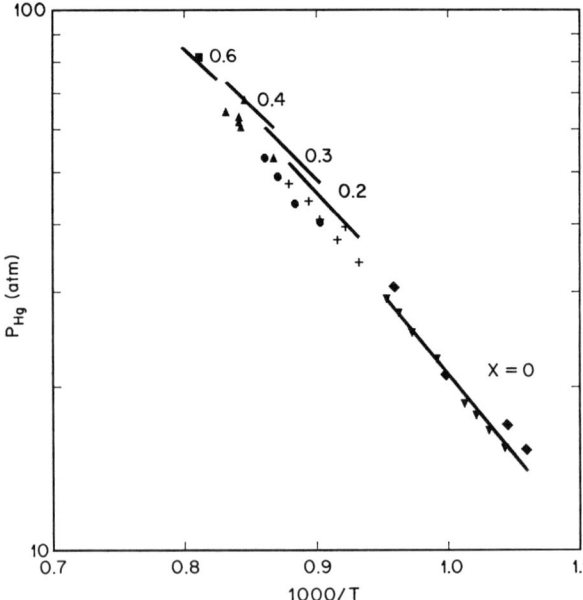

FIG. 23. The partial pressure of Hg for various pseudobinary melts. Inverted triangles are from Su *et al.* (1981), other experimental points are from Steininger (1976).

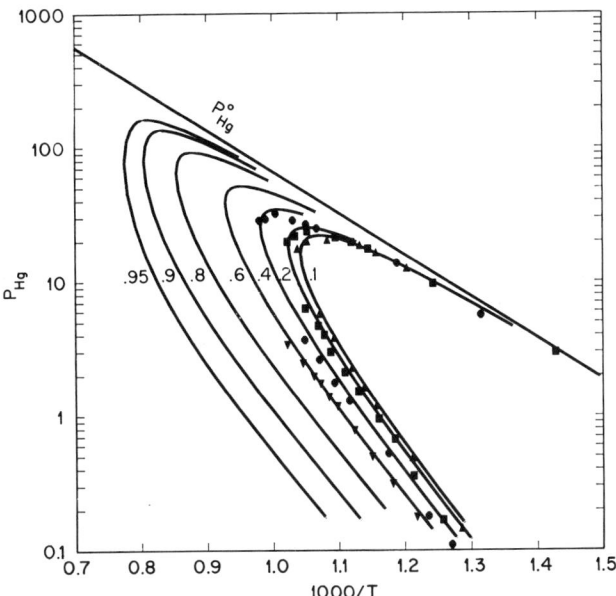

FIG. 24. Partial pressure of Hg along the three-phase curves for various solid solutions. The labels are the value of x in the formula, $Hg_{1-x}Cd_xTe(s)$. Experimental points from Schwartz et al. (1981) and Tung et al. (1981b, 1982).

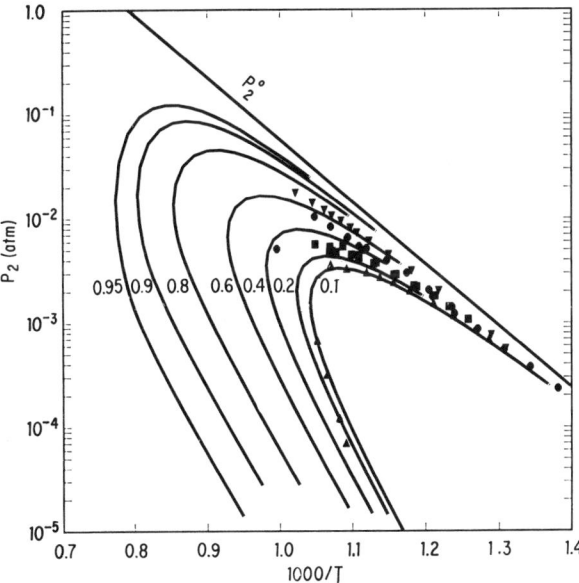

FIG. 25. Partial pressure of Te_2 along the three-phase curves for various solid solutions. Same credit source as for Fig. 24.

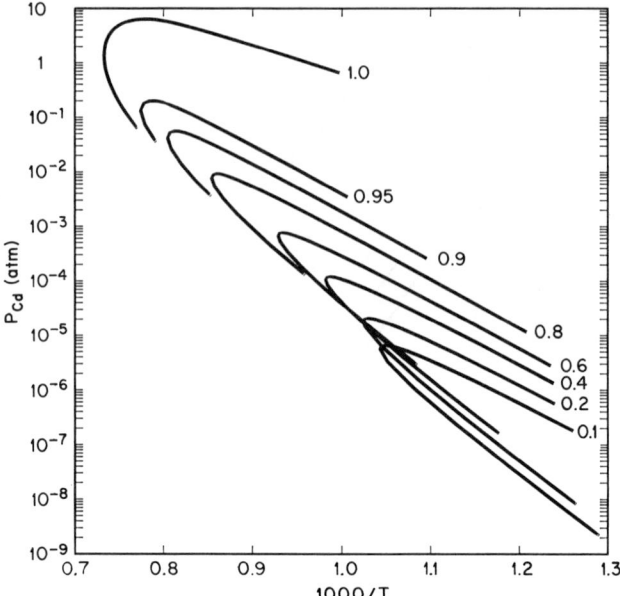

FIG. 26. Partial pressure of Cd along the three-phase curves for various solid solutions.

those that follow we revert to common usage and use x for the solid-solution composition, and x_A, $A =$ Hg, Cd, Te, for the atom fraction in the liquid.) The data shown for $x = 0.2$ and 0.4 are a redetermination and extension of our earlier results (Tung et al., 1982). Calculated curves are also shown for x values for which there are no experimental data as yet. The calculated maximum value of P_{Hg} along the pseudobinary solidus (or liquidus), which is the largest value of P_{Hg} at maximum temperature for any x value in Fig. 24, is 81 atm for the 92.5-mole % CdTe solid solution. The mercury pressures in Fig. 24 are fit to 18%. The Te$_2$ pressures in Fig. 25 are fit to 12%. The calculated curves for Te saturation at low temperature are omitted for clarity in Fig. 25. However, the calculated values for different x values remain distinct and larger at a given T the larger x (but still below the vapor pressure line for Te), whereas the experimental values for different x values tend to run together at the lowest temperatures. This is undoubtedly due in part to the fact that these lowest pressures were about at the level of experimental detection. The Cd pressures in Fig. 26 fit the experimental points (which are omitted here but shown in Tung et al., 1981b) to 48%. This is the largest discrepancy observed between calculated and experimental partial pressures.

Figure 27 shows calculated lines and experimental points for the Te-rich liquidus. The squares are optical absorbance measurements in which the liquidus temperature is taken as the temperature where $\log P_{Hg}$ versus $10^3/T$

3. SOLUTION MODEL FOR Ga–In–Sb AND Hg–Cd–Te   225

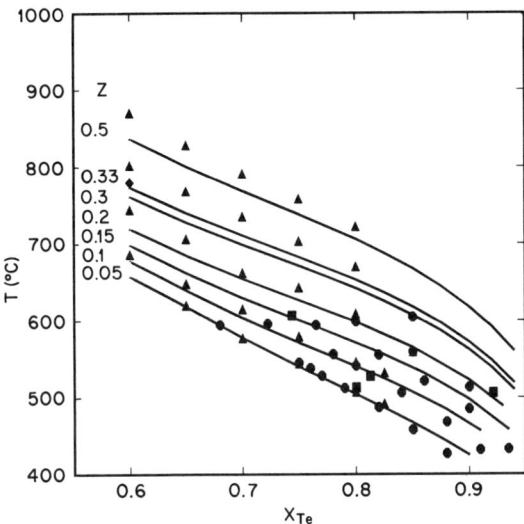

FIG. 27. Te-rich liquidus of $Hg_{1-x}Cd_xTe(s)$ shown by plotting temperature versus atom fraction Te for various labeled values of the atom fraction ratio in the liquid, $Z = x_{Cd}/(x_{Cd} + x_{Hg})$. Circles are from Harman (1980), triangles from Lawley (1980), squares and diamond from Tung et al. (1982).

shows a discontinuity in slope. Some of the earlier experimental results in poor agreement with those shown here have been plotted or cited elsewhere (Tung et al., 1981a). The overall fit to the data shown here is 14°C. The fit to the circle points of Harman (1980) is very good at the higher temperatures but discrepancies appear at low T and high atom fraction of Te. The calculated curves are consistently below Lawley's (1980) results from differential thermal analysis heating curves. The experimental partial pressures for four Te-rich liquids from which the square points of Fig. 27 are obtained are not shown here but are generally fit well. The results range from 3 to 12% for Hg, 5% to 9% for Te_2, and 17 to 80% for Cd.

The liquidus isotherms and solid-solution isoconcentration lines over the entire Gibbs composition triangle are shown in Fig. 28. A point to be elaborated upon further below is that in the Hg–Cd rich half the solid-solution isoconcentration lines turn to the Hg corner even for very high x values. An expanded plot near the Te-rich corner is shown in Fig. 29. The liquidus isotherms match Harman's (1980) experimental values for $x_{Cd} < 0.005$ and are close to his for 550 and 600°C. However, his 450 and 500°C isotherms are displaced from the calculated lines towards higher x_{Cd} by about 0.003 in x_{Cd}. The calculated isoconcentration lines agree with the composition analysis of films grown epitaxially upon CdTe substrates by Harman (1980) for $x = 0.2$ and 550°C, $x = 0.3$ and 575°C, $x = 0.4$ and 580°C, and $x = 0.5$ and 580°C.

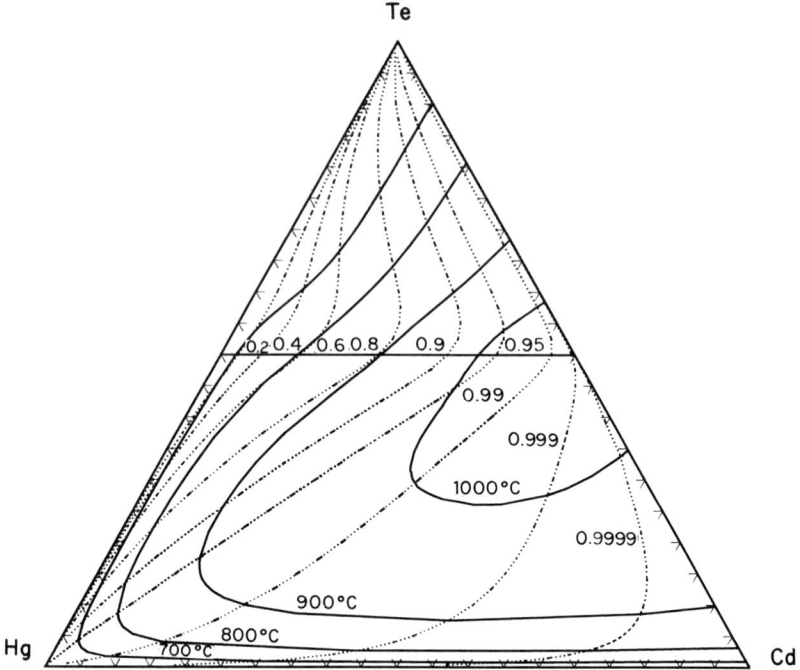

FIG. 28. Liquid isotherms (solid lines) at high temperatures and solid-solution isoconcentration lines (dashed lines) plotted across the entire Gibbs composition triangle.

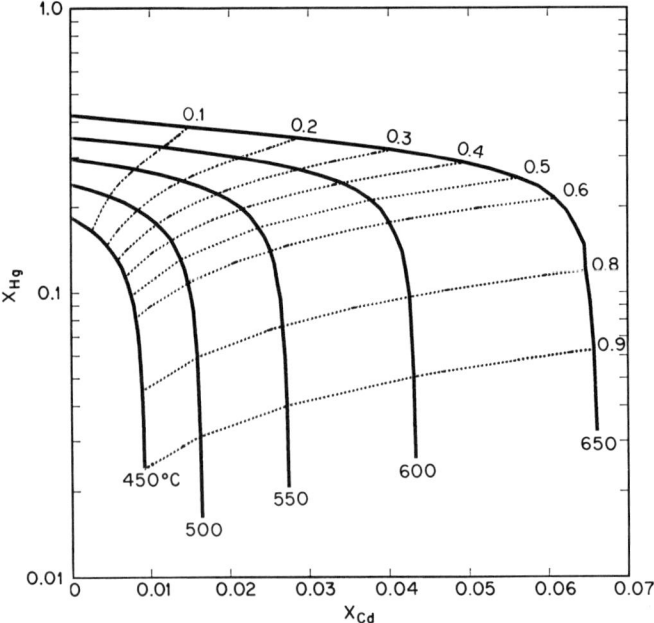

FIG. 29. Liquidus isotherms (solid) and solid-solution isoconcentration lines (dashed) for Te-rich liquids. The numbers adjacent to the isoconcentration lines are the values of x in $Hg_{1-x}Cd_xTe(s)$. The ordinate is the atom fraction of Hg in the liquid phase on a log scale, the abscissa is the atom fraction of Cd in the liquid on a linear scale.

3. SOLUTION MODEL FOR Ga–In–Sb AND Hg–Cd–Te

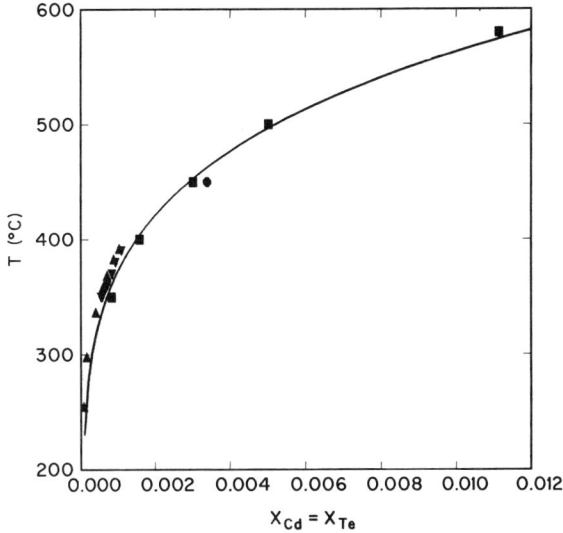

FIG. 30. Calculated curve and experimental points for the solution of CdTe(s) in Hg. Squares and circles are from Vanyukov et al. (1977) and inverted triangles from Wong (1980).

However, discrepancies occur at lower temperatures. At $x_{Cd} = 0.01$ the calculated isoconcentration lines for x between 0.20 and 0.60 are at too high values of x_{Hg}. The discrepancy is largest for high x, our line for $x = 0.6$ lying at higher values of x_{Hg} than his line for $x = 0.4$. Tie-lines obtained by Schmit near 500°C (tabulated in Tung et al., 1981a) fall between the calculated values and Harman's results for $x = 0.20$ and 0.29 but are somewhat closer to the calculated values for $x = 0.37$ and 0.40.

The solubility of CdTe in Hg is shown in Fig. 30. The overall fit is 17°C. If only the squares and circle are included the fit is 7°C. An expanded view of the Gibbs composition triangle near the Hg corner showing the low temperature liquidus isotherms and the solid-solution isoconcentration lines is shown in Fig. 31. Note the hundredfold expansion of the Cd scale. It can be seen that the tie-lines at 360°C converge toward the Hg corner for x values as high as 0.95. Another type of plot showing the same calculated results as well as dotted lines representing the smoothed experimental data (Wong, 1980; Riley, 1980) is shown in Fig. 32. The agreement can be seen to be very good. Another set of experimental isotherms from Vanyukov et al. (1978) are very different, giving a pseudobinary section between Hg and CdTe. These are ignored.

Finally, Table XI summarizes the calculated requirements for low temperature metal saturation of $Hg_{1-x}Cd_xTe(s)$ following Su et al. (1982). For each x value between 0.1 and 0.9 and each crystal temperature of 250, 275, or 300°C the table gives two entries. The top one is the calculated value of

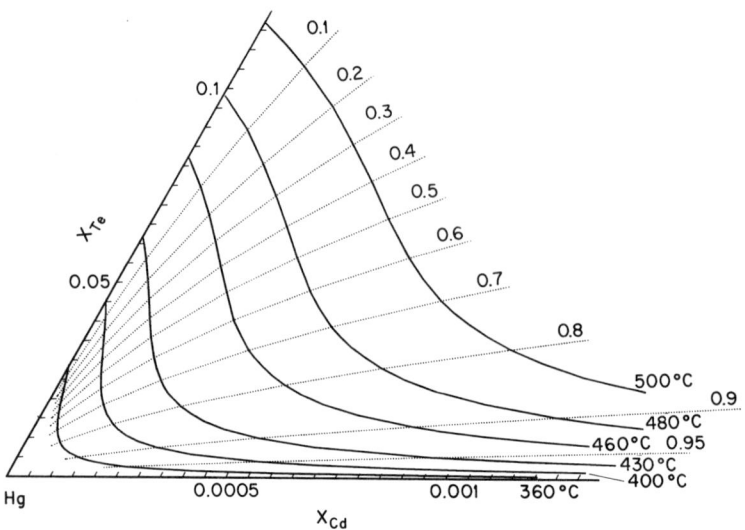

FIG. 31. Liquid isotherms and solid-solution isoconcentration lines in the Hg-rich corner of the Gibbs triangle.

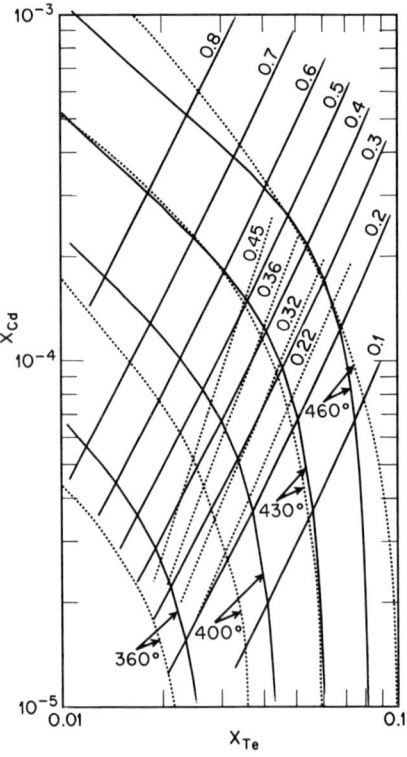

FIG. 32. Liquidus isotherms and solid isoconcentration lines for Hg-rich liquids. Axes are atom fractions in the liquid. Nearly straight lines slanting up towards the right are the solid isoconcentration lines for labeled values of x in the formula $Hg_{1-x}Cd_xTe(s)$—solid lines are calculated, dotted lines are the average of narrow bands established experimentally (Riley, 1980). Isotherms with labeled values of temperature in degrees Celsius are curves bending downwards left to right. Solid curves are calculated, dashed curves are smoothed curves through experimental points.

TABLE XI

Values for P_{Hg}/P_{Hg}^0, First Entry, and $T_C - T_R$ in Degrees Celsius, Second Entry, for Various Mole Fractions of CdTe in $Hg_{1-x}Cd_xTe(s)$ and Crystal Temperatures, T_C[a]

T_C (°C)	x	0.1	0.2	0.3	0.4	0.5	0.6	0.7	0.8	0.9
250	P_{Hg}/P_{Hg}^0	0.99493	0.99537	0.99577	0.99613	0.99651	0.99691	0.99739	0.99801	0.99883
	$T_C - T_R$	0.193	0.176	0.161	0.147	0.133	0.118	0.099	0.076	0.044
275	P_{Hg}/P_{Hg}^0	0.99225	0.99294	0.99356	0.99414	0.99473	0.99538	0.99613	0.99707	0.99830
	$T_C - T_R$	0.325	0.296	0.270	0.245	0.220	0.193	0.162	0.122	0.071
300	P_{Hg}/P_{Hg}^0	0.98853	0.98957	0.99051	0.99141	0.99231	0.99329	0.99442	0.99581	0.99758
	$T_C - T_R$	0.526	0.478	0.435	0.394	0.352	0.307	0.255	0.192	0.110

[a] T_R is the temperature of a pure Hg reservoir with vapor pressure equal to P_{Hg}, the partial pressure over the metal saturated crystal at T_C.

P_{Hg}/P_{Hg}^0, the ratio of the partial pressure of Hg over the metal-saturated crystal to the vapor pressure of Hg at that temperature. The second entry is the difference between the crystal temperature T_C and the temperature of a pure Hg reservoir T_R, which would have a vapor pressure equal to the partial pressure of Hg required for metal saturation of the crystal at T_C. The calculation of this temperature difference requires the value of P_{Hg}/P_{Hg}^0 and the value of A in the vapor pressure equation

$$\log_{10} P_{Hg}^0 \text{ (atm)} = -A/T + B. \qquad (129)$$

A least-squares fit of the tabulated values of the vapor pressure given by Hultgren et al. (1973a) at 500, 550, and 600°K to an equation of this form gives $A = -3125.0$. It can be seen that the differences in crystal and reservoir temperatures are all small and reach a maximum of 0.52°C for $x = 0.1$ and 300°C. Moreover, the difference decreases with increasing x value up to the largest x value of 0.9 shown. This trend must of course reverse somewhere between $x = 0.9$ and 1.0. This somewhat surprising trend is consistent with the phase-diagram sections shown in Figs. 31 and 32. For example, in Fig. 31, at 360°C a solid solution with $x = 0.95$ is in equilibrium with a liquid phase that is closer to pure Hg than is the case for solid solutions with lower x values. Although the accuracy of the calculation is expected to be good, it may not be good enough in view of the small temperature differences obtained. Therefore a conservative procedure would be to use a Hg reservoir temperature 1 or 2°C lower than given in Table XI. One would thereby give up the possibility of metal saturation for near-metal saturation but would, on the other hand, minimize the possibility of annealing under a mercury pressure higher than that on the metal-saturated leg of the three-phase curve for the particular x value of the crystal and the composition changes that would be driven to follow.

VII. Summary

We have obtained quantitatively good fits to the extensive phase diagram and thermodynamic data in the Ga–In–Sb and Hg–Cd–Te systems using a single associated solution model and a single set of interaction coefficients in each system. These fits are the most comprehensive and accurate ones achieved to date, as far as we are aware. They are useful in smoothing and extrapolating the experimental data upon which they are based. Although the values of the interaction coefficients used lead to the fits shown, and we believe these fits are essentially the best that can be obtained within the general framework of the liquid model used and the values selected for the thermodynamic properties of the binary solid compounds, the interaction coefficients are not claimed to be unique. Correlated but significant changes in the coefficients presented could lead to essentially the same fits, but this has not been investigated. The discussion of the Ga–Sb and In–Sb binaries suggests that a generalization of the subregular model of Eq. (1) to allow a temperature-dependent enthalpy of mixing, such as used by Brebrick (1977), might provide fits comparable to those obtained here with the associated solution model, but this remains to be demonstrated.

In the development of the associated solution model used a number of general features have been obtained, which hopefully should prove useful. In particular it is shown in Appendix B that the model behavior includes the occurrence of miscibility gaps in the liquid that do not require interaction coefficients that are large in magnitude and that are associated with an uncommon form for the Gibbs energy of mixing isotherms.

Appendix A

Analytical equations are derived for the liquidus surface of the solid solution $A_{1-u}B_uC(s)$, which form a complete mathematical solution in the special case of an ideal associated solution but that do not in the general case, when the equations contain the activity coefficients of the species. Nevertheless these equations are useful in achieving some understanding of the behavior of the liquidus surface when the liquid consists almost entirely of one element. Then the activity coefficients of the species can be approximated by setting the mole fractions of all species equal to zero, except that of the uncombined species corresponding to the predominant element, and explicit equations then obtained. If the liquidus surface approaches one of the pure elements in composition at all, it will generally do so at relatively low temperatures. This happens in the In corner of the Ga–In–Sb system and in the Hg and Te corners of the Hg–Cd–Te system. The equations obtained are useful in identifying those factors responsible for the low temperature convergence of the tie-lines toward the In corner and towards the Hg corner in these two systems.

As in Part IV the species are taken to be a, b, c, ac, and bc and numbered consecutively 1 through 5. The mole fraction of the molecular species bc is replaced by another variable, C, according to

$$y_5 = (C - 1)y_4, \qquad (A1)$$

and the mole fraction of the molecular species ac is rewritten for brevity as

$$y_4 = z. \qquad (A2)$$

Starting with Eqs. (63) and (64), describing the equilibria among the liquid species, the mole fractions of species 1, 2, and 3 are eliminated with Eq. (57), and y_5 is eliminated with Eq. (A1), to give

$$(1 + Cz)x_1^2 + (1 + Cz)x_1x_2 - (1 + z)x_1 - zx_2$$
$$= -z(1 + \kappa_4)/(1 + Cz), \tag{A3}$$

$$(1 + Cz)x_2^2 + (1 + Cz)x_1x_2 - z(C - 1)x_1 - [1 + (C - 1)z]x_2$$
$$= -z(C - 1)(1 + \kappa_5)/(1 + Cz), \tag{A4}$$

where again x_1 and x_2 are the atomic fractions of components A and B and κ_4 and κ_5 are the effective equilibrium constants that depend on the species activity coefficients. Equations (A3) and (A4) are added and an equation obtained for $x_1 + x_2$ in terms of z, C, κ_4, and κ_5. Then Eq. (A4) is subtracted from (A3) to obtain an equation for $x_1 - x_2$ in terms of $x_1 + x_2$, z, C, κ_4, and κ_5. Using these, x_1 and x_2 are obtained as

$$x_1 = \{z(C - 1)\kappa_5 + [1 + (2 - C)z](\kappa_4/2) \pm \kappa_4 F\}/D, \tag{A5}$$

$$x_2 = \{z(C - 1)\kappa_4 + [1 + (C - 2)z](C - 1)(\kappa_5/2) \pm (C - 1)\kappa_5 F\}/D, \tag{A6}$$

where $\quad F = \{(1 - Cz)^2/4 - z[(C - 1)\kappa_5 + \kappa_4]\}^{1/2}, \tag{A7}$

$$D = (1 + Cz)[(C - 1)\kappa_5 + \kappa_4]. \tag{A8}$$

Equations (A5) and (A6) describe the interspecies equilibrium in the liquid phase for some composition x_1, x_2. If this liquid is on the liquidus surface of $A_{1-u}B_uC(s)$, then the liquidus equations given by Eqs. (12) and (13) must also be satisfied. Using the general relation between relative chemical potential and activity coefficient given by Eq. (60), the liquidus equations are

$$y_1 y_3 = (1 - u)(\gamma_1 \gamma_3)^{-1} \exp(\{\Omega u^2 + \Delta G_f^0[AC(s)]\}/RT) \equiv Q_1, \tag{A9}$$

$$y_2 y_3 = u(\gamma_2 \gamma_3)^{-1} \exp(\{\Omega(1 - u)^2 + \Delta G_f^0[BC(s)]\}/RT) \equiv Q_2, \tag{A10}$$

where Ω is related to the solid-solution parameters in Eqs. (14) and (15) by

$$\Omega = W_s - V_s T \tag{A11}$$

and where the Gibbs energies of formation ΔG_f^0 are for the solid binary compounds. With Eqs. (63) and (64) for interspecies equilibrium the left side of Eq.(A9) can be replaced by $z\kappa_4$ and that of Eq. (A10) by $z(C - 1)\kappa_5$. This yields

$$z = Q_1/\kappa_4, \tag{A12}$$

$$C = 1 + Q_2\kappa_4/Q_1\kappa_5. \tag{A13}$$

If z and C are eliminated from Eqs. (A5) and (A6) using the above, one has two liquidus equations:

$$x_1 = [Q_2/\kappa_4 + \tfrac{1}{2}(1 + Q_1/\kappa_4 - Q_2/\kappa_5) \pm F_2]/(1 + Q_2/Q_1)D_2, \tag{A14}$$

$$x_2 = [Q_1/\kappa_5 + \tfrac{1}{2}(1 - Q_1/\kappa_4 + Q_2/\kappa_5) \pm F_2]/(1 + Q_1/Q_2)D_2, \tag{A15}$$

3. SOLUTION MODEL FOR Ga–In–Sb AND Hg–Cd–Te

where

$$F_2 = [(1 - Q_1/\kappa_4 - Q_2/\kappa_5)^2/4 - Q_1 - Q_2]^{1/2}, \quad \text{(A16)}$$

$$D_2 = 1 + Q_1/\kappa_4 + Q_2/\kappa_5. \quad \text{(A17)}$$

The presence of the $\pm$ sign in Eqs. (A14) and (A15) gives two liquid compositions in equilibrium with a solid solution of given u value, as expected. If the activity coefficients that appear in Q_1, Q_2, κ_4, and κ_5 can be established independently, or if the associated solution is assumed to be ideal so that they are all unity, then Eqs. (A14) and (A15) are two equations in four variables, x_1, x_2, u, and T. There are therefore two degrees of freedom in agreement with the Gibbs phase rule. One could for instance specify u and T and calculate the liquid composition, x_1, x_2. From the definitions of Q_1 and Q_2 given by the rightmost members of Eqs. (A9) and (A10), it can be seen that Q_2/Q_1 ranges from zero when $u = 0$ to infinity when $u = 1$. Thus over the same range in u, x_1 goes to zero as u goes to unity and x_2 goes to zero as u goes to zero because of the factors $(1 + Q_2/Q_1)$ and $(1 + Q_1/Q_2)$ in the denominators of, respectively, Eqs. (A14) and (A15). The liquid in equilibrium with a solid solution with $u = \frac{1}{2}$ will tend to be near pure element 1 if Q_1/Q_2 is large compared to unity and near pure element 2 if Q_1/Q_2 is a fraction small compared to unity.

Equations (A14) and (A15) are used to study the Hg–Cd rich liquidus surface of $Hg_{1-u}Cd_uTe(s)$ at 673°K. The activity coefficients are approximated by calculation using Eqs. (60), (70), and (75) with $y_1 = 1$. This is a reasonable approximation if it leads to a value of x_1 near unity. The activity coefficients for species 1 through 5 are, respectively, 1.0, 0.153, 0.153, 12.62 and 244. Because the Gibbs energy of formation of CdTe(s) is $-23,061$ cal/mol compared to -7227 cal/mol for HgTe(s), Q_1/Q_2 is 390 at $u = 0.99$, even though it must vanish at $u = 1$. Because of this and the fact that Q_1/κ_4, Q_2/κ_5, and Q_1 are all small compared to one, Eq. (A15) simplifies to

$$x_2 = Q_2/\kappa_5 + Q_2/Q_1. \quad \text{(A18)}$$

When $u = 0.90$ this gives $x_2 = x_{Cd} = 3.1(10^{-4})$ while Eq. (A14) gives $x_1 = x_{Hg} = 0.9922$. This is in agreement with the more exact computer calculations whose results are shown in Fig. 31. The activity coefficients in the metal-rich liquid in equilibrium with a solid solution with $u = 0.9$ and at 673°K obtained from a computer calculation also agree to within 3% with the approximate values listed above. Thus the relative stability of CdTe(s) compared to HgTe(s) is a major factor in the tie-lines converging toward the Hg corner. The smallness of Q_2/κ_5, which is determined in part by the large value of 244 for the activity coefficient of the CdTe liquid species, also enters and is less transparent.

The Ga–In–Sb liquidus is examined in a similar way at 573°K. We assume $x_2 = x_{In}$ to be near unity and approximate the activity coefficients calculated using Eqs. (60), (70), and (75) with $y_2 = 1$. Then the activity coefficients of Ga, In, Sb, GaSb, and InSb (species numbered 1 through 5) are, respectively, 4.52, 1.0, 12.7, 0.772, and 0.703. For $u = \frac{1}{2}$ in the solid solution formula, $Ga_{1-u}In_uSb(s)$, the ratio of Q_1/Q_2 is 0.030, primarily because ΔG_f^0 is more negative for GaSb(s) than for InSb(s). As a result, $x_1 = x_{Ga}$ is small, 0.029, and $x_2 = x_{In}$ is large, 0.92. These results are in good agreement with Fig. 16, showing that the approximation made in evaluating the activity coefficients is a good one. Again the tie-lines converge toward one corner, although not so strongly as for the Hg–Cd–Te system. Again the primary reason is the difference in the Gibbs energies of formation of the binary solid compounds.

Appendix B

In the discussion of the associated model for the A–C binary, the possibility of an infinite heat capacity at 50 at. % was discussed in connection with Eq. (106). This was shown to arise because the temperature derivative of the mole fraction of the molecular ac species becomes infinite for certain values of the interaction coefficients. This behavior is discussed here only for a simplified version of the model in which the interaction coefficients β_{14} and β_{34} of Eq. (68) for ΔG_M^x are equal and opposite. (These parameters then cancel out of κ_4 at $x = \frac{1}{2}$). The discussion below therefore covers the version of the model applied to Ga–In–Sb.

We find that in certain ranges of the interaction coefficients that the equation for equilibrium among the species is satisfied by more than one value of the mole fraction of ac. At $x = \frac{1}{2}$ and at the melting point of AC(s), where most of the parameter constraints are applied, it is possible to delineate those areas in a $\Omega_R/RT_{AC} - z^*$ plane in which the multiple roots occur and the curve along which the heat capacity is infinite. Thus as a consequence it is possible in advance to avoid these regions in varying the parameters to seek an optimum fit to the liquidus line. Our present and earlier (Liao, 1982) fit to the Ga–Sb binary are discussed in terms of these results. The results are then illustrated concretely with a calculation for Hg–Te. The origin of an infinite heat capacity is identified and the formation of a liquid miscibility gap arising from a butterfly shape in the ΔG_M versus x isotherms is shown, which results in the infinite heat capacity not being observable.

Consider the mass action law given by Eq. (63). In the A–C binary at atom fraction $\frac{1}{2}$, the effective equilibrium constant can be obtained from Eq. (82) with $\beta_{14} = -\beta_{34}$ to write the mass action law as

$$M = A, \qquad (A19)$$

where

$$M = \ln[(1-z)^2/4z], \qquad (A20)$$

$$A = [-2(\Omega_S + \Omega_R) + (1+z)^2\Omega_R - \Delta G_4^0]/RT, \qquad x = \tfrac{1}{2}. \qquad (A21)$$

The solution of Eq. (A19) can be depicted graphically by plotting M and A as a function of $(1 + z)^2$ in the range 1–4 as shown in Fig. 33. Then A is a straight line of slope Ω_R/RT and intercept $-(2\Omega_S + \Omega_R + \Delta G_4^0)/RT$ on the vertical axis at $(1 + z)^2 = 1$, whereas M is a curve going from plus infinity to minus infinity. We wish to characterize those cases in which A and M intersect three times, corresponding to three values of z that satisfy Eq. (A19). The curvature of M plotted against $(1 + z)^2$ changes sign at $z = \frac{1}{2}$ where the slope of M is a maximum of -2. The tangent to M at $z = \frac{1}{2}$ intersects the left vertical axis at a value of 0.42056. It can immediately be concluded that $\Omega_R > 0$ or $-(2\Omega_S + \Omega_R + \Delta G_4^0)/RT \leq 0.42056$ are sufficient conditions that only one value of z between zero and unity satisfies Eq. (A19). The equation for the intercept on the left vertical axis, I, of a line that is tangent to M at z_0 is

$$I = \ln[(1 - z_0)^2/4z_0] + (2 + z_0)/2(1 - z_0). \quad (A22)$$

For each $I > 0.42056$ this equation is satisfied by two values of z_0. Each of the two tangents so defined must intersect M in a second position. Moreover these

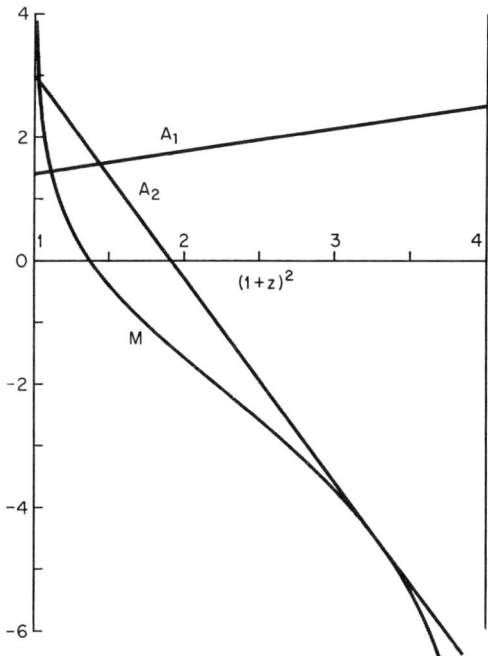

FIG. 33. Graphical solution of the equilibrium between species a, c, and ac in the A–C system at 50 at.% A. M and A are defined in Eqs. (A20) and (A21), z is the mole fraction of species ac. Two different A curves, originating from two different sets of interaction parameters, are shown.

two tangents enclose an infinite number of lines that have the same intercept, I, but intersect M three times. Table XII lists the solution of Eq. (A22) for a few values of I and gives the slope of the tangent at z_0:

$$dM/d(1+z)^2 = -1/2z(1-z), \quad z = z_0. \quad (A23)$$

A general solution for $x = \frac{1}{2}$ can now be stated. Equation (A19) for species equilibrium at $x = \frac{1}{2}$ is satisfied by more than one value of z whenever $\Omega_R/RT < 0$ and $A(z=0) \equiv -(2\Omega_S + \Omega_R + \Delta G_4^0)/RT > 0.42056$. For a given value of $A(z=0) > 0.42056$ three values of z satisfy Eq. (A19) for values of Ω_R/RT between the two slopes given in a table like Table XII for $I = A(z=0)$. Thus for $A(z=0) = 1$ three values of z are obtained whenever $-2.806 < \Omega_R/RT < -2.348$. When Ω_R/RT equals either of these bounds, then two values of z satisfy Eq. (A19).

The general solution given above can be developed further by fixing the temperature as the melting point of AC(s), T_{AC}, and applying the constraints given by Eqs. (109) and (111) to eliminate Ω_S and ΔG_4^0 from Eq. (A21) for A in favor of Ω_R and z^*. Here z^* is a desired value for the mole fraction of the molecular species at $x = \frac{1}{2}$ and $T = T_{AC}$. One obtains

$$A = [(1+z)^2 - (1+z^*)^2](\Omega_R^*/RT_{AC}) + \ln[(1-z^*)^2/4z^*], \quad (A24)$$

where Ω_R is evaluated at T_{AC} as indicated by the asterisk. Using this equation it can be seen that A is still a linear function of $(1+z)^2$ and that it intersects, or is tangent to, the curve M at a point corresponding to $z = z^*$, i.e., with A given by Eq. (A24), $z = z^*$ satisfies Eq. (A19). The possibility of other intersections or tangencies for given value for z^* and Ω_R^*/RT_{AC} can now be determined. At $z = 0$, Eq. (A24) is

$$A(z=0, T_{AC}) = [1 - (1+z^*)^2](\Omega_R^*/RT_{AC}) + \ln[(1-z^*)^2/4z^*]. \quad (A25)$$

TABLE XII

Solutions of Eq. (A22) for a Few Values of I and Slope from Eq. (A23)[a]

I	z_0	$dM/d(1+z)^2\|_{z_0}$	z_0	$dM/d(1+z)^2\|_{z_0}$
0.42506	0.500	−2.0	—	—
0.6	0.619	−2.120	0.354	−2.186
1	0.6925	−2.348	0.232	−2.806
2	0.7723	−2.843	0.0884	−6.204
3	0.8133	−3.293	0.0333	−15.53
4	0.840	−3.720	0.01237	−40.93
8.61	0.900	−5.555	$1.24(10^{-4})$	−4032.0

[a] Only one value of z_0 satisfies Eq. (A20) for $I = 0.42506$.

3. SOLUTION MODEL FOR Ga–In–Sb AND Hg–Cd–Te

An infinite number of pairs of values for z^* and Ω_R^*/RT_{AC} satisfy Eq. (A25) for any constant value k of $A(z = 0)$. Each pair in turn generates a straight line, A versus $(1 + z)^2$, through Eq. (A24) with a common intercept k on a vertical axis at $(1 + z)^2 = 1$. Those values of z^* and Ω_R^*/RT_{AC} for which z^* lies between the two roots z_0 of Eq. (A22) with $I = k$ lead to triple intersections of M and A. In this case two values of z in addition to the chosen value of z^* satisfy the mass action law at $x = \frac{1}{2}$ given by Eq. (A19). If k is less than 0.42056, no real values for z_0 between 0 and 1 satisfy Eq. (A22) and there are only single intersections of M and A. The result is shown in Fig. 34. All values for Ω_R^*/RT_{AC} and z^* falling on or below the uppermost, inverted U curve give more than one value of z that satisfies Eq. (A19), one of which is given by the value of z^*. The innermost curve corresponds to $\Omega_R^* = -RT_{AC}/2z^*(1 - z^*)$ and values falling along it give an infinite value for $\partial z/\partial T$ and ΔC_P at $x = \frac{1}{2}$, T_{AC}. Both curves have a common point at $z^* = \frac{1}{2}$ where $\Omega_R^*/RT_{AC} = -2$. For a point on the multiple-root boundary, such as point 1 or 2 in Fig. 34, there are two solutions to Eq. (A19), point 1 and point 1' or point 2 and point 2'. Conversely, if the selected value of z^* is such that the point falls on the inner curve at 1', then a second value

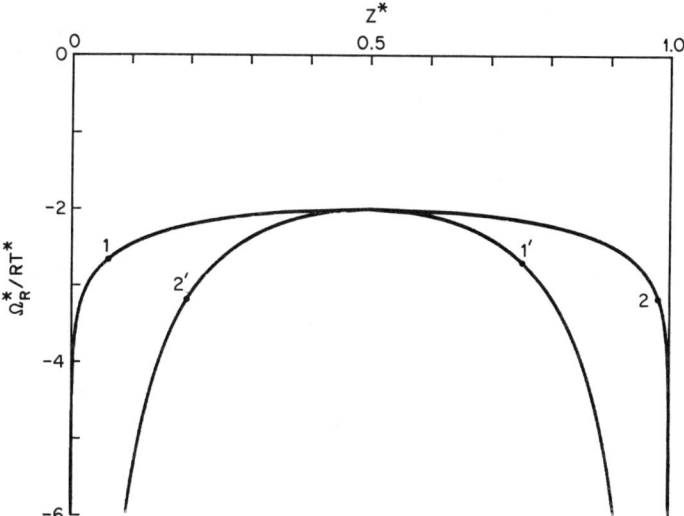

FIG. 34. Ω_R^* is the value of the interaction coefficient Ω_R evaluated at the melting point of AC(s), T_{AC}. z^* is a selected value for the mole fraction of the molecular species ac at T_{AC} and equal atom fractions of the *components* of the binary A–C system. Values of Ω_R^*/RT_{AC} and z^* that fall under the upper curve are associated with two other values of z that also satisfy Eq. (A19) for species equilibrium. Values of Ω_R^*/RT_{AC} and z^* falling on the inner curve give an infinite heat capacity.

of z that satisfies Eq. (A19) with the same value of Ω_R^*/RT_{AC} is given by point 1.

Other correlations between the value of z^* and the other two values of z that satisfy Eq. (A19) with the same value of Ω_R^*/RT_{AC} are the following:

(a) If the point is within the inner curve, then z^* lies between the other two values.

(b) If the point lies between the left sides of the inner and outer curves, then z^* is smaller than either of the other two values.

(c) If the point lies between the right sides of the inner and outer curves, then z^* is larger than either of the other two values.

The above results can be readily generalized to arbitrary temperature. Using the assumed linear temperature dependence of the interaction coefficients given by Eq. (86), one has

$$\Omega_j/RT = \Omega_j^*/RT_{AC} + (1/T - 1/T_{AC})(\omega_j/R), \quad j = R, S, A. \quad (A26)$$

Again starting with Eq. (A21) for A and eliminating Ω_S^*/T_{AC} and $\Delta G_4^0(T = T_{AC})$ with Eqs. (109) and (111), one obtains

$$A(z = 0, T) = A(z = 0, T_{AC}) - (1/T - 1/T_{AC})(2\omega_S + \omega_R + \Delta H_4^0)/R. \quad (A27)$$

Eliminating the reciprocal temperature difference between this equation and

$$\Omega_R/RT = \Omega_R^*/RT_{AC} + (1/T - 1/T_{AC})(\omega_R/R) \quad (A28)$$

gives

$$A(z = 0, T) = A(z = 0, T_{AC}) - (2\omega_S + \omega_R + \Delta H_4^0)$$
$$\times (\Omega_R/RT - \Omega_R^*/RT_{AC})/\Omega_R. \quad (A29)$$

Thus for a given set of interaction coefficients, $A(z = 0, T)$ is a straight line when plotted against Ω_R/RT. The temperature associated with any point on this line can be calculated from the value of Ω_R/RT through Eq. (A28). On the other hand, the boundaries of the multiple-root region in this plane can be fixed by calculating I from Eq. (A22) and $dM/d(1 + z)^2$ from Eq. (A23) for various values of z_0. Figure 35 results when the I values are plotted along the $A(z = 0, T)$ axis and the values of $dM/d(1 + z)^2$ along the Ω_R/RT axis. The wedge-shaped region terminating at $A(z = 0, T) = 0.42056$, $\Omega_R/RT = -2.0$ encloses the multiple-root region. Our earlier compromise fit for the Ga–Sb binary (row 3, Table IV; Liao et al., 1982) was made with $z^* = 0.2233$ and $\Omega_R^*/RT_{AC} = -2.2090$. From Fig. 34 this is just inside the multiple-root region. The variation of $A(z = 0, T)$ with Ω_R/RT is shown as straight line 1 in Fig. 35. The 709.2°C melting point is indicated by an x. Using Fig. 35 one concludes that triple roots for z at $x = \frac{1}{2}$ will occur between about 627 and 777°C. Computer calculations have been made to

3. SOLUTION MODEL FOR Ga–In–Sb AND Hg–Cd–Te

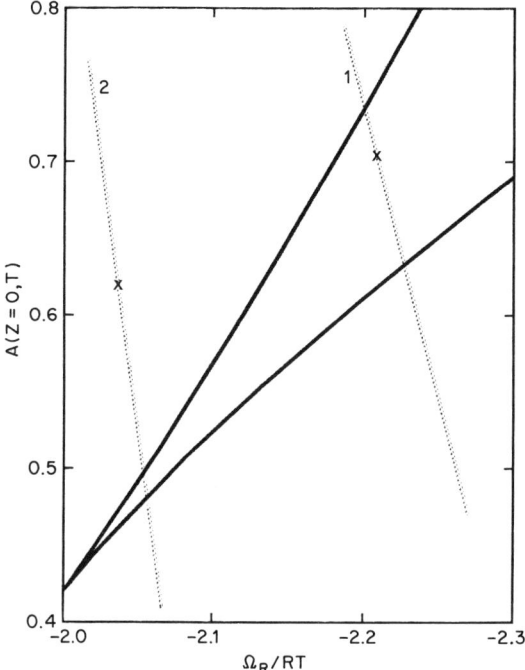

FIG. 35. Three values of the mole fraction of the species ac satisfy the mass action law at $x = \frac{1}{2}$ for points between the two solid lines. Ordinate defined by Eq. (A27); abscissa by Eq. (A26).

determine these roots at various compositions and temperatures and to then obtain the Gibbs energy of mixing isotherms. The triple-root region extends from about 47 to 51 at.% Sb at 730°C and widens slightly with decreasing temperature to about 45 to 53 at.% Sb at 660°C. At temperatures of about 705°C or above, the Gibbs energy of mixing consists of a small closed curve 20–30 cal above a continuous curve extending over the entire composition range. Below 705°C the Gibbs energy of mixing isotherm is a simple continuous, but triple-valued, curve (similar to Fig. 37 for HgTe). However, the most stable equilibrium is the desired one, i.e., a homogeneous liquid above 709.2°C, a GaSb(s)–liquid equilibrium below.

Nevertheless, a second parameter set was sought that would give equally good fits but for which the triple-root region would be less extensive. The result is the set in row 3 of Table IV, which gives a point just outside the multiple-root region in Fig. 34 for 50 at.% Sb and 709.2°C. Changing temperature generates line 2 in Fig. 35. The melting point indicated by an x is outside the triple-root region, which occurs only in the narrow temperature interval between 577 and 590°C. The values of the Gibbs energy of mixing of the liquid phase for the three values of z satisfying the mass action law for species equilibrium only differ by a few calories. On the other hand, the Gibbs energy of GaSb(s) is about 1 kcal more negative than the Gibbs energy of mixing of the liquid at its lowest point near 50 at.% As a consequence the most stable equilibrium is the desired one between GaSb(s) and the liquid.

Other features of this behavior are illustrated concretely using the Hg–Te system. The calculations were made using values for the thermodynamic properties that are slightly different than those shown in Table VII and used in the calculations described in the main text. Thus the calculations are not quite correct for Hg–Te but are still valid for illustrative purposes. The enthalpy of fusion is taken as 8,680 cal/mol and the enthalpy and entropy of formation of HgTe(s) at 943.1°K as $-13,933$ cal/mol and -9.538 cal/°K mol, respectively. After applying the constraints of Eqs. (109)–(111) and setting $\beta_{13} = \beta_{14} = \beta_{34} = 0$, the independent model parameters are z^*, Ω_R, and Ω_A. The former

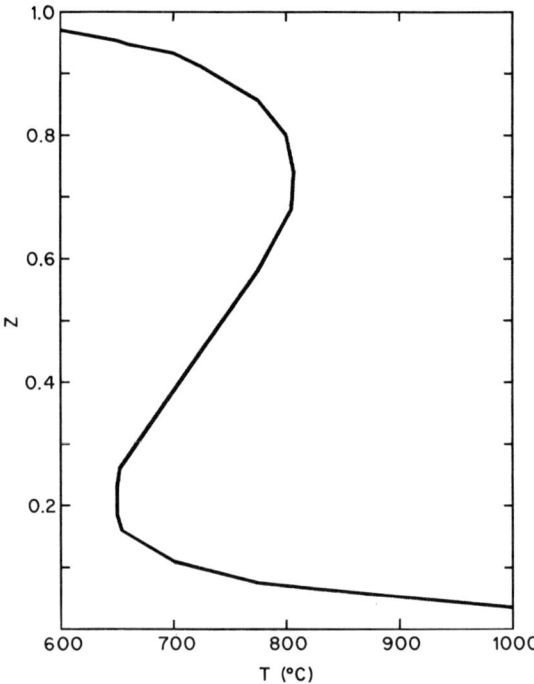

FIG. 36. The mole fraction of the liquid species HgTe z as a function of temperature for a 50-at.% Te–50-at.% Hg melt. Calculated assuming Hg, Te, and HgTe species and using values for the interaction coefficients given in Appendix B.

3. SOLUTION MODEL FOR Ga–In–Sb AND Hg–Cd–Te

is taken as 0.3 and the latter as temperature-independent constants equal to −5621.8 cal and 3123.0 cal, respectively. Since $\Omega_R/(1.987)(943.1) = -3.0$, the corresponding point in Fig. 34 is just within the inner curve. The value of z as a function of T is shown in Fig. 36 for the 50-at. % melt. At 670°C the chosen value of 0.3 for z^* is between the other two values of about 0.14 and 0.94 for z that also satisfy the mass action law. The triple-root region is seen to extend from about 650 to 810°C and infinite heat capacities would be predicted for a homogeneous liquid phase at 650 and 810°C. However, these infinities are not observable since a homogeneous liquid phase is metastable. This is shown in Fig. 37 where a portion of the Gibbs energy of mixing at 670°C is shown as a function of the atom fraction of Te. The butterfly shape arises because of the three values of z that satisfy the mass action law between about 43 and 52 at. % Te. Since a simultaneous tangent can be drawn to ΔG_M between about 30.0 and 50.2 at. % Te, the equilibrium state at 670°C consists of two liquid phases of these two compositions and does not include HgTe(s) at all. The liquid phase parameters chosen to illustrate the multiple-root behavior of the liquid species equilibrium obviously do not reproduce the observed phase diagram.

The miscibility gap behavior contained in the associated solution model would appear to be complex and has only been touched upon here. Miscibility

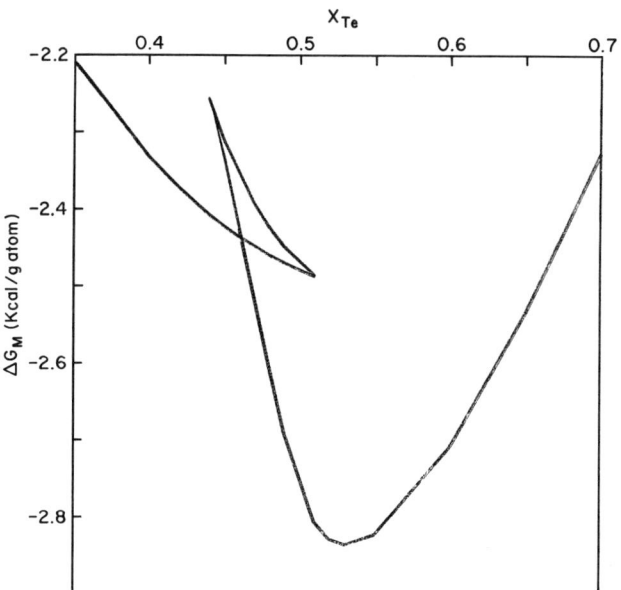

FIG. 37. A portion of the Gibbs energy of mixing isotherm for a Hg–Te melt at 670°C as a function of the atom fraction of the Te *component*. Calculated with the same interaction parameters as used for Fig. 36.

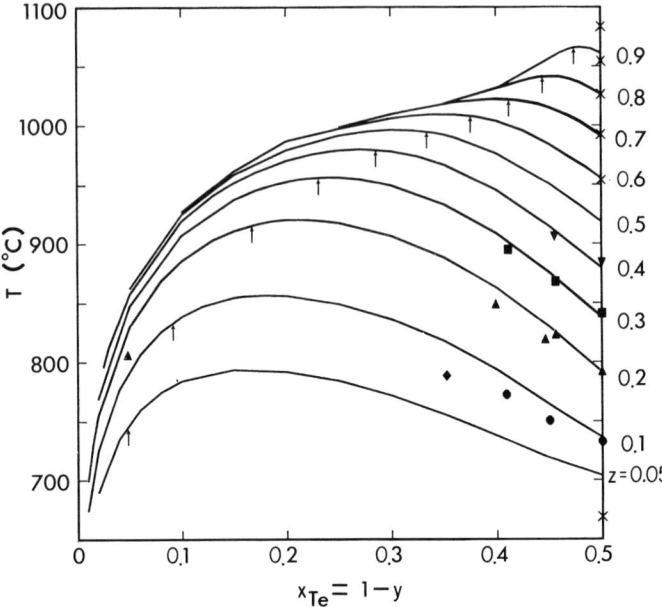

FIG. 38. Liquidus surface in the Hg–Cd–Te system on the Hg–Cd–rich side of the HgTe–CdTe pseudobinary section. The liquidus temperature in degrees Celsius is plotted against the atom fraction of Te for fixed values of z that are shown along the right vertical axis. The composition of the liquid phase is written as $(Hg_{1-z}Cd_z)_y Te_{1-y}$. Experimental points shown as symbols. The single diamond point is for $z = 0.09$. Arrows below each liquidus lines indicate where the atom fractions of Cd and Te are equal.

gaps arise from a behavior of the Gibbs energy isotherm that is quite different from that usually invoked, in which ΔG_M is a single-valued function of composition that has a negative curvature in some composition range. However, the butterfly shape for ΔG_M shown in Fig. 37 has a parallel (Gaskell, 1981) in the Gibbs energy of a one-component van der Waals gas as a function of T.

Appendix C*

The calculated liquidus isotherms for the Hg–Cd–Te system in Fig. 28 show that the liquidus temperature is not always a maximum in the HgTe–CdTe pseudobinary section. Figure 38 shows the liquidus surface between this pseudobinary and the Hg–Cd axis in more detail for comparison with ten experimental ternary liquidus points obtained by Szofran and Lehoczky

* With Tse Tung, Martin Marietta Laboratory, Baltimore, Maryland, as co-author.

(1983), using differential thermal analysis and with previously discussed pseudobinary points. The calculated liquidus temperatures are shown by solid curves as a function of the atom fraction of Te for the various fixed values of z shown along the right side of the figure. Here z is defined by the compositional formula $(Hg_{1-z}Cd_z)_y Te_{1-y}$ for the liquid phase. The experimental results are shown by various symbols. The agreement between calculation and experiment is very good and corresponds to a standard deviation of 13°C for the liquidus points off the pseudobinary section. For clarity only a short portion of the liquidii for $z = 0.8$ and 0.9 are shown. More extensive portions for these z values as well as for $z = 1$ are shown in Fig. 39. The liquidus curves for fixed z show a maximum removed from the pseudobinary, where $y = \frac{1}{2}$, for z values as low as 0.05 and as high as 0.9. For $z = 0$ or 1 the maxima occurs at $y = \frac{1}{2}$, as shown in Fig. 17. None of the maxima are higher than the 1092°C melting point of CdTe(s), and they tend to occur near the compositions indicated by vertical arrows where the atom fractions of Te and Cd are equal, i.e., where $1 - y = zy$.

Figure 40 shows the calculated partial pressure of Hg on the liquidus surface for fixed values of z as a function of 10^3 times the reciprocal absolute temperature, i.e., P_{Hg} along the three-phase curves for liquids of fixed z. The z value is placed along the Te-rich leg of the associated three-phase curve. The temperature where $y = \frac{1}{2}$ is indicated by a circle on each curve. For a given T in the low temperature range, P_{Hg} on the Te-rich leg is smaller the larger z is. On the metal-rich, upper leg, P_{Hg} at a given T is also smaller the larger z is and P_{Hg} is almost parallel to the vapor pressure of pure Hg, P_{Hg}^0. However, at the higher temperatures and for intermediate z values, $0.4 \le z \le 0.7$ in Fig. 40, the two legs of the three-phase curves cross and for a short temperature interval P_{Hg} is higher at a given temperature for a certain liquidus composition than it is for a liquidus composition richer in Hg. The maximum value of P_{Hg} is 169 atm, which is reached for $z = 0.4$ at $10^3/T = 0.80$. The partial pressure of Hg along the three-phase curve for a solid solution, $Hg_{1-x}Cd_xTe(s)$, of given x value is displaced to lower pressures and temperatures than that for a liquid for which z has the same numerical value as x. This can be seen by comparison with Fig. 24. Moreover, the three-phase curves for the liquid with a given z and for the solid solution with a value of x numerically equal to z are generally unrelated, e.g., for $z = 0.3$, the composition of the coexisting solid solution is not $x = 0.3$ at any temperature. Figure 41 shows P_{Hg} along a curve on the liquidus surface defined by the intersection of the liquidus surface and a plane perpendicular to the Gibbs composition triangle and through the pure Hg and CdTe composition points, i.e., P_{Hg} on the liquidus surface if one follows a path such that the atom fractions of Cd and Te on the liquidus surface are equal. The circles indicate the values of P_{Hg} and $10^3/T$ for the labeled values of the atom fraction of Hg in the liquid phase.

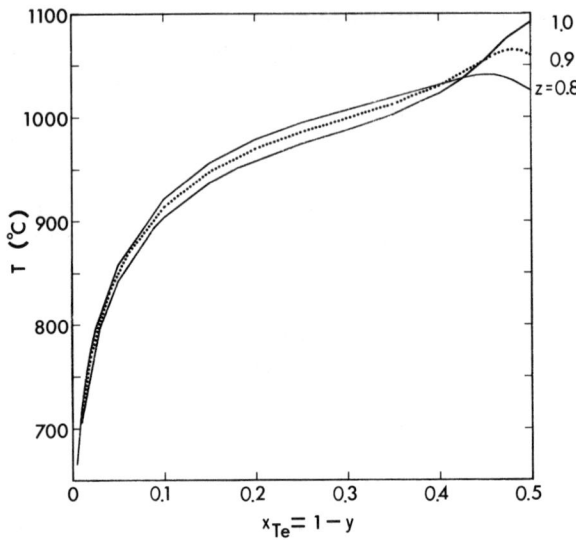

FIG. 39. Liquidus surface as in Fig. 38 but showing the crossing of the liquidus lines for high z values.

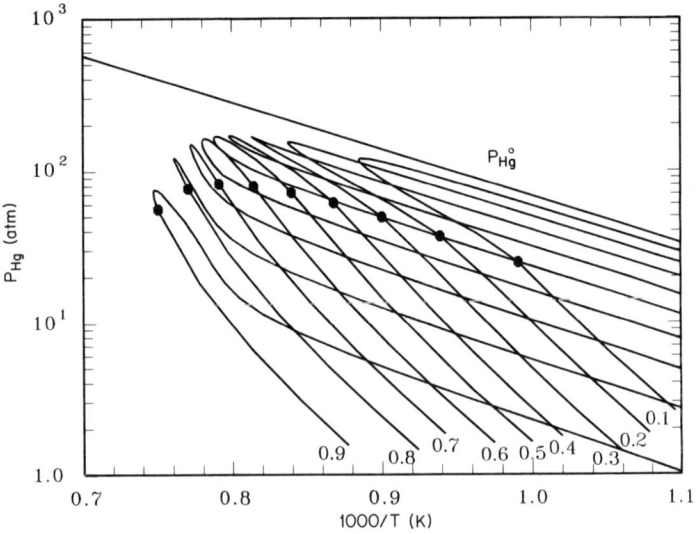

FIG. 40. Partial pressure of mercury in atmospheres as a function of 10^3 times the reciprocal absolute temperature along the three-phase curves for the liquids, $(Hg_{1-z}Cd_z)_yTe_{1-y}$. The value of z is shown near the bottom, Te-rich leg of each curve. The circles mark the pressure and temperature where $y = \frac{1}{2}$.

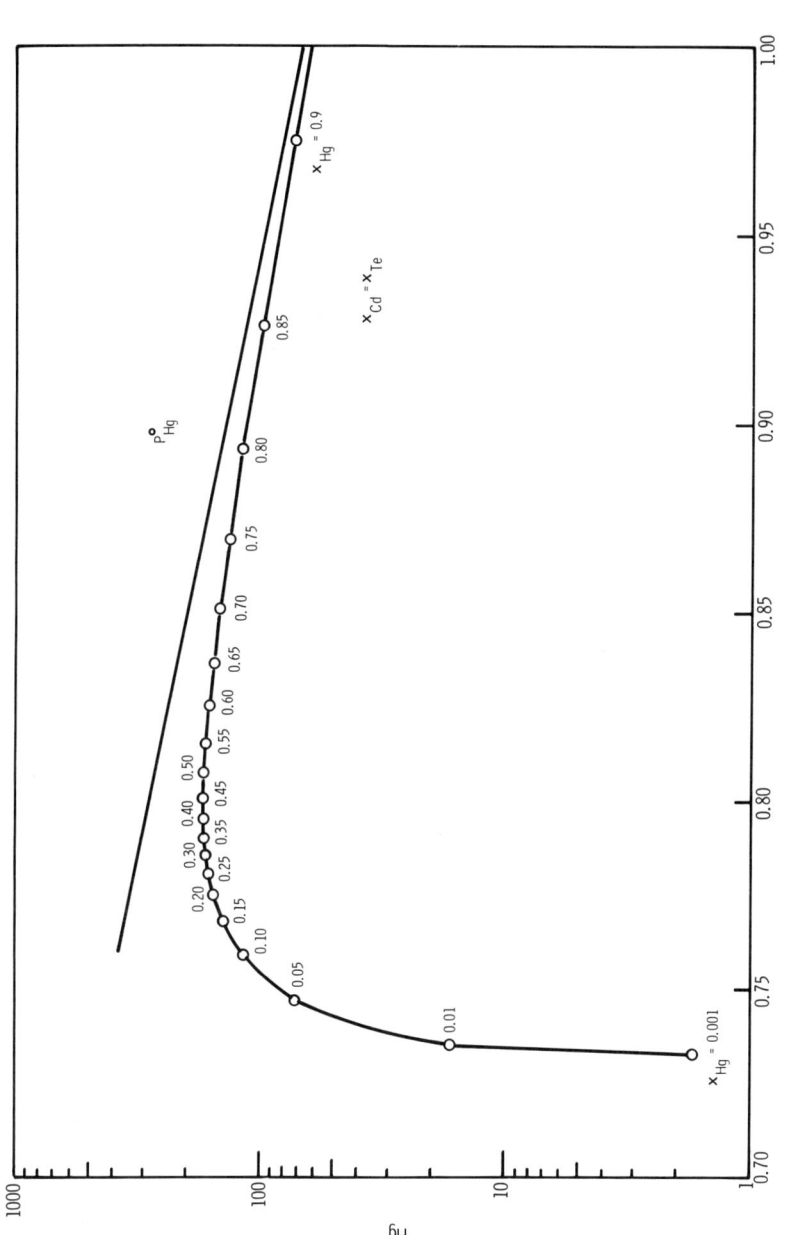

FIG. 41. The partial pressure of mercury in atmospheres plotted against 10^3 times the reciprocal absolute temperature for a traverse across the liquidus surface such that the atom fractions Cd and Te are always equal. The number near each circle give the atom fraction of mercury in the liquid phase at the pressure and temperature specified by the circle.

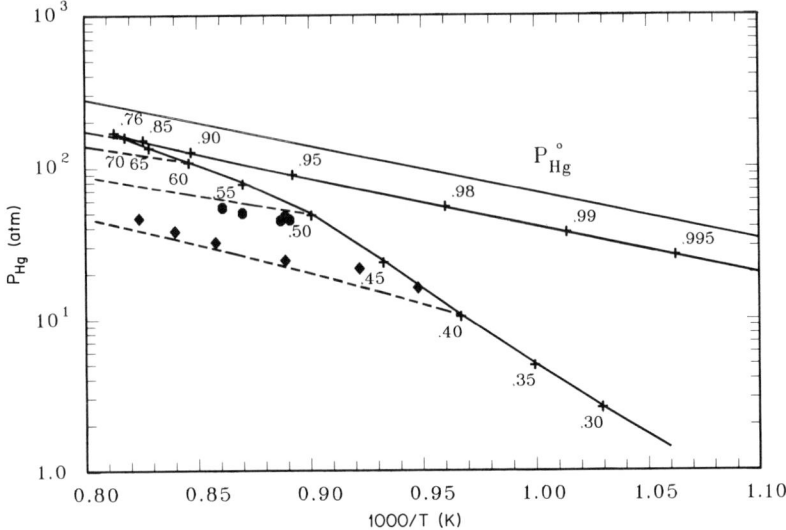

FIG. 42. Partial pressure of mercury in atm along the three-phase curve for the liquid $(Hg_{0.7}Cd_{0.3})_y Te_{1-y}$ is shown as a solid curve. The uppermost line gives the vapor pressure of pure mercury. Each cross along the three-phase curve marks the pressure and temperature where y is equal to the value given near that cross. The dashed lines are calculated results for the liquids $(Hg_{0.7}Cd_{0.3})_y Te_{1-y}$, $y = 0.4, 0.5, 0.6$, and 0.7, for temperatures above the liquidus temperature. The solid symbols are experimental values (Steininger, 1976). Solid circles are for $y = 0.50$; diamonds are for $y = 0.40$.

Figure 42 shows the calculated values of P_{Hg} along the three-phase curve for the liquid $(Hg_{0.7}Cd_{0.3})_y Te_{1-y}$ as a solid curve. The numbers near the crosses on this curve give the corresponding value of y. The dashed curves are calculated for liquids $(Hg_{0.7}Cd_{0.3})_y Te_{1-y}$ for $y = 0.40, 0.50, 0.60$, and 0.70 at temperatures for which these compositions are above the liquidus surface. The experimental points of Steininger (1976) are shown for $y = 0.40$ and 0.50 and are in good agreement with the calculations. Steininger used these points and others for $y = 0.60$ and 0.70 to construct a three-phase curve that has a maximum temperature at $y = \frac{1}{2}$. This is very different from the calculated result that shows a maximum P_{Hg} at $y = 0.76$, consistent with a maximum liquidus temperature of $957°C$ at $y = 0.76$ for $z = 0.30$. This discrepancy is due to the fact that Steininger obtained mercury pressures for the compositions $(Hg_{0.7}Cd_{0.3})_{0.6}Te_{0.4}$ and $(Hg_{0.7}Cd_{0.3})_{0.7}Te_{0.3}$ over a temperature range in which he assumed, based upon the observation of thermal arrests at lower temperatures, that these compositions were completely liquid. These experimental points are shown as the triangles and squares along the upper two curves of Fig. 43. They are well below the calculated liquidus temperatures

3. SOLUTION MODEL FOR Ga–In–Sb AND Hg–Cd–Te

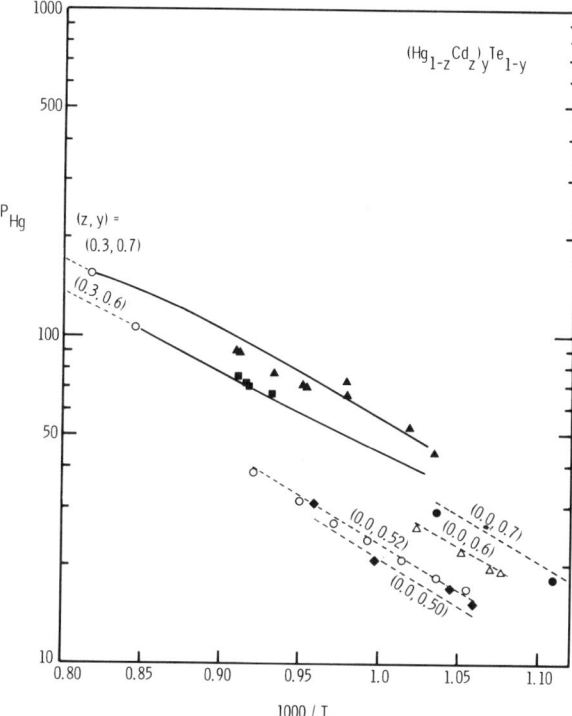

FIG. 43. Partial pressure of mercury in atmospheres plotted against $10^3/T$ for various compositions. Uppermost curve is the calculated result for $(Hg_{0.7}Cd_{0.3})_{0.7}Te_{0.3}$ with the liquidus temperature indicated by the open circle. Triangles are experimental results for the same composition. Second highest curve, with the liquidus temperature again indicated by an open circle, is for the composition $(Hg_{0.7}Cd_{0.3})_{0.6}Te_{0.4}$ and the squares are the experimental values. Lower four lines are the calculated results for various Hg_yTe_{1-y} melts along with the experimental points shown by symbols.

indicated by the open circles on each curve. The solid curves were obtained assuming the compositions are overall compositions for a mixture consisting of a liquid and a $Hg_{1-x}Cd_xTe(s)$ solid solution with some fixed value of x. For each of the two overall compositions and for each of a number of temperatures the compositions of the coexisting phases and value of P_{Hg} were calculated. Since the curves calculated are close to the experimental points, it seems plausible to speculate that Steininger misidentified thermal arrests for $(Hg_{0.7}Cd_{0.3})_{0.6}Te_{0.4}$ and $(Hg_{0.7}Cd_{0.3})_{0.7}Te_{0.3}$. These were accepted as reasonable since it was expected that the maximum liquidus temperature for any z value would occur in the pseudobinary section where $y = \frac{1}{2}$. The lower four sets of points in Fig. 43 are Steininger's (1976) measured values for Hg_yTe_{1-y} liquids and the dashed lines, in close agreement, are our calculated values. Figure 20 shows a similar plot and good agreement with other experimental data discussed previously.

Finally, Table XIII gives a tabulation of liquid composition, liquidus temperature, x value of the coexisting solid, and mercury pressure at the liquidus temperature. The table is arranged in sections for different z values in the liquidus composition $(Hg_{1-z}Cd_z)_y Te_{1-y}$. Since the calculated liquidus surface for Te-rich compositions and high temperatures is given only for z values up to 0.5 in Fig. 27, entries with y less than $\frac{1}{2}$ are also included for the high z values.

TABLE XIII

Liquid Phase Composition Written as $(Hg_{1-z}Cd_z)_y Te_{1-y}$, Liquidus Temperature in Degrees Celsius, Composition of Solid-Solution $Hg_{1-x}Cd_x Te$ in Equilibrium with the Liquid Phase at the Liquidus Temperature, and Mercury Pressure in Atmospheres at the Liquidus Temperature

$100y$	T	$1000x$	P_{Hg}	$100y$	T	$1000x$	P_{Hg}
$z = 0.10$				76.5	920	934	147.4
				78	921	943	150.4
50	736	360	24.3	80	921	953	152.7
55	764	491	37.3	81.5	919.6	960	153.1
60	794	612	52.1	83	917	965	152.1
65	818	706	68.1	85	911	972	149.1
70	837	781	84.4	87	904	977	143.9
75	849	842	99.8	90	886	984	131.5
80	857	892	112.5	92	869	987	120.1
82	857	909	115.9	95	831	992	96.5
85	855	931	118.8	98	754	996	59.8
88	848	949	117.4	98.5	731	996.8	50.8
90	839	960	113.6	99	699	997.6	40.3
92	826	969	106.8	99.5	648	998.5	26.8
94	807	977	96.7				
96	777	984	80.7	$z = 0.30$			
98	725	991	57.1				
99	674	994	39.0	50	838	685	48.3
99.5	626	996	26.2	55	876	774	77.5
				60	909	845	106.5
$z = 0.20$				65	934	898	133.6
				70	950	937	155.2
50	792	564	36.2	71	952	943	158.4
55	830	678	57.8	73	955	954	163.3
60	863	767	80.7	74	956	959	164.8
65	889	833	104.1	75	957	963	166.0
70	907	885	126.0	76	957	967	166.4
75	918	924	143.6	77	956.6	970	166.1

3. SOLUTION MODEL FOR Ga–In–Sb AND Hg–Cd–Te

TABLE XIII (continued)

$100y$	T	$1000x$	P_{Hg}	$100y$	T	$1000x$	P_{Hg}
78	955.8	973	165.4	65	992	971	167.4
79	955	976	164.4	66	994	975	167.6
80	953	979	162.6	68	996	981	165.2
81	951	981	160.4	70	996	986	160.2
83	945	984	155.0	72	995	989	153.5
85	938	987	147.8	75	991	992	142.4
90	907	992	124.0	80	979	995	123.0
95	848	996	88.0	85	958	997	103.6
98	768	998	53.6	90	924	997.8	82.6
99	711	998.8	35.8	95	862	998.8	56.3
99.3	683	999.1	29.1	97.5	798	999.3	37.9
99.6	642	999.4	20.8	98	779	999.4	33.3
				99	721	999.7	22.1
		$z = 0.40$		99.6	651	999.8	12.7
50	879	768	60.0			$z = 0.60$	
55	915	839	97.1				
60	945	839	130.7	10	639	935	.18
65	967	941	157.1	15	691	920	.50
67.5	974	957	165.4	20	730	907	1.06
70	978	969	169.1	25	764	894	1.99
72.5	980	977	168.5	30	796	881	3.60
73	980	978	167.9	35	831	868	6.77
75	979	983	164.3	40	870	856	14.13
76	978	985	161.8	45	914	855	33.7
77	976.5	986	159.0	50	955.6	882	78.0
77.5	975.8	987	157.6	51.5	965	896	94.5
80	970	990	149.3	55	984	930	130.2
83	960	992	137.8	57.5	995	952	149.6
85	951	994	129.6	59	1000	963	157.1
87	940	994.6	120.6	61	1005	974	161.7
90	919	996	105.5	63	1008	982	160.1
95	857	998	73.2	65	1009	987	153.7
97	811.1	998.6	55.3	68	1008	992	140.5
97.5	795	998.7	49.8	70	1005	994	131.1
98.75	736	999.3	33.4	75	997	996	110.0
99.5	664	999.6	19.4	80	986	997.5	93.6
				85	961	998	76.0
		$z = 0.50$		90	926	999	59.6
				95	862	999.4	39.6
50	918	832	70.4	97	816	999.6	29.6
52	932	855	89.0	98	780	999.7	23.3
55	951	890	115.6	99	722	999.8	15.4
60	976	939	150.9	99.5	668	999.9	10.1
64	990	966	166.0				

(Continued)

TABLE XIII (continued)

100y	T	1000x	P_{Hg}	100y	T	1000x	P_{Hg}
\multicolumn{4}{c}{z = 0.70}	56	1040	989	122.0
				58	1036	994	113.7
10	657	955	.17	60	1031	996	101.4
15	711	945	.46	65	1017	998	74.0
20	752	936	.96	70	1006	999	56.0
25	788	928	1.82	75	993	999.2	44.0
30	822	919	3.31	80	977	999.5	35.4
35	859	910	6.28	85	955	999.6	28.2
40	900	902	13.3	90	920	999.7	21.5
45	948	901.7	33.0	95	856	999.9	14.01
50	991	923.3	81.0	97.5	794	999.9	9.17
52	1003	939	105.7	98	775	999.9	7.99
53	1008	947	117.4	98.5	750	999.9	6.68
55	1015	963	136.6	99	717	1000.0	5.21
57	1020	976	147.3	99.5	664	1000.0	3.43
59	1022	984	147.9				
60	1022	988	145.0	\multicolumn{4}{c}{z = 0.90}
64	1018	994	123.7				
68	1012	996.6	102.4	10	690	987	.08
70	1008	997.2	93.5	15	745	984	.21
75	997	998.2	75.6	20	789	982	.44
80	982	998.8	61.8	25	828	979	.83
85	960	999.1	50.2	30	866	977	1.54
90	924	999.4	38.7	35	908	974	2.99
95	861	999.7	25.6	40	957	972	6.65
98	778	999.9	14.76	45	1013	972.5	18.19
99	721	999.9	9.69	50	1059	983	55.3
99.5	667	1000.0	6.36	51	1064	987	65.0
				52	1065	990	72.3
\multicolumn{4}{c}{z = 0.80}	53	1064	993	75.4
				54	1060	995	74.5
10	674	972	.13	55	1055	997	71.0
15	729	966	.36	60	1030	999	48.0
20	772	961	.76	65	1011	999.4	32.8
25	809	955	1.43	70	998	999.6	24.0
30	845	950	2.62	75	985	999.8	18.43
35	884	945	5.06	80	969	999.8	14.56
40	929	940	10.9	85	946	999.9	11.48
45	980	940	28.2	90	912	999.9	8.65
50	1025	957	75.7	94	866.2	999.9	6.23
52	1035	969	100.5	96	829	1000.0	4.82
54	1041	981	118.5	98	769	1000.0	3.12
55	1041	985	122.1	99	712	1000.0	2.02

Acknowledgment

The authors wish to gratefully acknowledge support from the Air Force Office of Scientific Research under grant AFOSR-78-3611.

References

Anderson, T. J. (1978). Ph.D. Dissertation, Univ. of California, Berkeley.
Ansara, I., Gambino, M., and Bros, J. P. (1976). *J. Cryst. Growth* **32**, 101.
Antypas, G. A. (1972). *J. Cryst. Growth* **16**, 181.
Bergman, C., Laffite, M., and Muggianu, Y. M. (1974). *High Temp. High Pressures* **6**, 53.
Bhatia, A. B., and Hargrove, W. H. (1974). *Phys. Rev.* **B10**, 3186.
Bhatia, A. B., and Ratti, V. K. (1977). *Phys. Chem. Liq.* **6**, 201.
Bhatia, A. B., and Thornton, D. E. (1970). *Phys. Rev.* **B2**, 3004.
Blair, J., and Newnham, R. (1961). *In* "Metallurgy of Elemental and Compound Semiconductors" (R. O. Grubel, ed.), p. 393. Wiley (Interscience), New York.
Blom, G. M., and Plaskett, T. S. (1971). *J. Electrochem. Soc.* **118**, 1831.
Brebrick, R. F. (1968). *J. Phys. Chem.* **72**, 1032.
Brebrick, R. F. (1971a). *Metall. Trans. AIME* **2**, 1657, 3377.
Brebrick, R. F. (1971b). *J. Electrochem. Soc.* 118, 2014.
Brebrick, R. F. (1977). *Metall. Trans. A* **8A**, 403.
Brebrick, R. F. (1979). *J. Phys. Chem. Solids* **40**, 177.
Brebrick, R. F., and Strauss, A. J. (1965). *J. Phys. Chem. Solids* **26**, 989.
Brebrick, R. F., Tung, T., Su, C.-H., and Liao, P.-K. (1981). *J. Electrochem. Soc.* **128**, 1595.
Chang, Y. A., and Sharma, R. C. (1979). *In* "Calculations of Phase Diagrams and Thermochemistry of Alloy Phases" (Y. A. Chang and J. F. Smith, eds.), p. 145. Metallurgical Soc. of AIME, Warrendale, Pennsylvania.
Chang, Y. A., and Smith, J. F., eds. (1979). "Calculations of Phase Diagrams and Thermochemistry of Alloy Phases." Metallurgical Soc. of AIME, Warrendale, Pennsylvania.
Chatterji, D., and Smith, J. V. (1973). *J. Electrochem. Soc.* **120**, 770.
Chu, H. H., Chao, P.-N., and Mo, C.-C. (1966). *Acta Metall. Sin.* **9**, 113.
Cutler, M. (1977). "Liquid Semiconductors," Chaps. 2 and 3. Academic Press, New York.
de Nobel, D. (1959). *Philips Res. Rept.* **14**, 361.
Gambino, M., and Bros, J.-P. (1975). *J. Chem. Thermodyn.* **7**, 443.
Gaskell, D. R. (1981). "Introduction to Metallurgical Thermodynamics," p. 200. Hemisphere Publishing Corp., Washington, D. C.
Gratton, M. F., and Wooley, J. C. (1978). *J. Electrochem. Soc.* **125**, 657.
Gratton, M. F., and Wooley, J. C. (1980). *J. Electrochem. Soc.* **127**, 55.
Guggenheim, E. A. (1952). "Mixtures." Oxford Univ. Press, London and New York.
Hall, R. N. (1963). *J. Electrochem. Soc.* **110**, 385.
Harman, T. C. (1967). *In* "Physics and Chemistry of II–VI Compound" (M. Aven and J. S. Prener, eds.), p. 769. North-Holland Publ., Amsterdam.
Harman, T. C. (1980). *J. Electron. Mater.* **9**, 945.
Hoshino, H., Nakamura, Y., Schimoji, M., and Niwa, K. (1965). *Ber. Bunsenges. Phys. Chem.* **69**, 114.
Hultgren, R., Desai, P., Hawkins, D., Gleiser, M., and Kelley, K. K. (1973a). "Selected Values of the Thermodynamic Properties of the Elements." Am. Soc. Metals, Metals Park, Ohio.
Hultgren R., Desai, P., Hawkins, D., Gleiser, M., and Kelley, K. K. (1973b). "Selected Values of the Thermodynamic Properties of Binary Alloys." Am. Soc. Metals, Metals Park, Ohio.

Jordan, A. S. (1970). *Metall. Trans.* **1**, 139.
Jordan, A. S. (1979). *In* "Calculations of Phase Diagrams and Thermochemistry of Alloy Phases" (Y. A. Chang and J. F. Smith, eds.), p. 100. Matallurgical Soc. of AIME, Warrendale, Pennsylvania.
Kaufman, L., Nell, J., Taylor, K., and Hayes, F. (1981). CALPHAD: *Comput. Coupling Phase Diagrams Thermochem.* **5**, 185.
Kellogg, H. H., and Larrain, J. M. (1979). *In* "Calculations of Phase Diagrams and Thermochemistry of Alloy Phases" (Y. A. Chang and J. F. Smith, eds.), p. 130. Metallurgical Soc. of AIME, Warrendale, Pennsylvania.
Kikuchi, R. (1981). *Physica* **103B**, 41.
Kohler, R. (1960). *Monatsh. Chem.* **91**, 738.
Kulwicki, B. M. (1963). Ph.D. Dissertation, Univ. of Michigan, Ann Arbor, Michigan.
Lawley, K. L. (1980). Texas Instruments Technical Report, 10 Feb. 1980, Contract MDA903-79-C-1099.
Levitskaya, T. D., Vanyukov, A. V., Krestmikov, A. N., and Bystrov, V. P. (1970). *Izv. Akad. Nauk. SSSR, Neorg. Mater.* **6**, 559.
Liao, P.-K., Su, C.-H., Tung, T., and Brebrick, R. F. (1982). CALPHAD: *Comput. Coupling Phase Diagrams Thermochem.* **6**, 141.
Liu, T. S., and Peretti, E. A. (1952). *Trans. Am. Soc. Met.* **44**, 539.
Longuet-Higgins, H. C. (1951). *Proc. R. Soc. London, Ser. A* **A205**, 247.
Lorenz, M. R. (1962). *J. Phys. Chem. Solids* **23**, 939.
Machol, R. E., and Westrum, E. F. (1958). *J. Am. Chem. Soc.* **80**, 2950.
Maglione, M. H., and Potier, A. (1968). *J. Chim. Phys. Phys. Chim. Biol.* **65**, 1595.
Miki, H., Segawa, K., Otsubo, M., Shirahata, K., and Fujibayashi, K. (1975). *Conf. Ser. Inst. Phys.* No. 24, 16.
Panish, M. B., and Ilegems, M. (1972). *Prog. Solid State Chem.* **7**, 39.
Predel, B. (1979). *In* "Calculations of Phase Diagrams and Thermochemistry of Alloy Phases" (Y. A. Chang and J. F. Smith, eds.), p. 72. Metallurgical Soc. of AIME, Warrendale, Pennsylvania.
Predel, B., and Oehme, G. (1976). *Z. Metallkd.* **67**, 826.
Predel, B., and Stein, D. W. (1971). *J. Less-Common Met.* **24**, 391.
Prigogine, I. (1957). "The Molecular Theory of Solutions," Chap. IV. Wiley (Interscience), New York.
Riley, K. H. (1980). *Proc. IRIS Detect.* p. 183.
Schwartz, J. P., Tung, T., and Brebrick, R. F. (1981). *J. Electrochem. Soc.* **128**, 438.
Shunk, F. A. (1969). "Constitution of Binary Alloys, Second Supplement." McGraw-Hill, New York.
Steininger, J. (1976). *J. Electron. Mater.* **5**, 299.
Steininger, J., Strauss, A. J., and Brebrick, R. F. (1970). *J. Electrochem. Soc.* **117**, 1305.
Strauss, A. J. (1967). Private communication of tabulated liquidus points, presented in graphical form in Harman (1967).
Stringfellow, G. B., and Greene, P. E. (1969). *J. Phys. Chem. Solids* **30**, 1779.
Su, C.-H., Liao, P.-K., Tung, T., and Brebrick, R. F. (1981). *High Temp. Sci.* **14**, 181.
Su, C.-H., Liao, P.-K., Tung, T., and Brebrick, R. F. (1982). *J. Electron. Mater.* **5**, 931.
Sugawara, S., and Sato, T. (1962). *Bull. JSME* **5**, 711.
Szapiro, S. (1976). *J. Electron. Mater.* **5**, 223.
Szapiro, S. (1980). *J. Phys. Chem. Solids* **41**, 279.
Szofran, F. R., and Lehoczky, S. L. (1981). *J. Electron. Mater.* **10**, 1131.
Szofran, F. R., and Lehoczky, S. L. (1983). *J. Electron. Mater.* (to be published).
Tung, T., Golonka, L., and Brebrick, R. F. (1981a). *J. Electrochem. Soc.* **128**, 1601.

Tung, T., Golonka, L., and Brebrick, R. F. (1981b). *J. Electrochem. Soc.* **128**, 451.
Tung, T., Su, C.-H., Liao, P.-K., and Brebrick, R. F. (1982). *J. Vac. Sci Technol.* **21**, 117.
Vanyukov, A. V., Krotov, I. I., and Ermakov, A. I. (1977). *Inorg. Mater. (Engl. Transl.)* **13**, 667.
Vanyukov, A. V., Krotov, I. I., and Ermakov, A. I. (1978). *Inorg. Mater. (Engl. Transl.)* **14**, 512.
Vieland, L. J. (1963). *Acta Metall.* **11**, 137.
Wong, J. (1980). Santa Barbara Research Center Report of Jan. 1980.
Wooley, J. C., and Lees, D. G. (1959). *J. Less-Common Met.* **1**, 192.

CHAPTER 4

Photoelectrochemistry of Semiconductors

Yu. Ya. Gurevich and Yu. V. Pleskov

INSTITUTE OF ELECTROCHEMISTRY
ACADEMY OF SCIENCES OF THE USSR
MOSCOW, USSR

	LIST OF SYMBOLS.	256
I.	INTRODUCTION	257
II.	GENERAL CONCEPTS OF THE ELECTROCHEMISTRY OF SEMICONDUCTORS.	259
	1. Electrode Potential	259
	2. Electrochemical Potential of Electrons	261
	3. The Structure of an Electrical Double Layer at a Semiconductor–Electrolyte Interface	263
	4. Electrochemical Reactions at Semiconductor Electrodes	271
III.	THE THEORY OF PROCESSES CAUSED BY PHOTOEXCITATION OF SEMICONDUCTORS	273
	5. General Picture and Determination of Photocurrent in the Simplest Case	273
	6. Calculation of Photocurrents and Photopotentials in More Complicated Cases	276
	7. Comparison of the Theory with Experiment	278
IV.	PHOTOCORROSION AND ITS PREVENTION.	282
	8. General Concepts of Corrosion	282
	9. Quasithermodynamic Description of Corrosion and Photocorrosion	285
	10. Some Kinetic Aspects of Photocorrosion	292
V.	LIGHT-SENSITIVE ETCHING OF SEMICONDUCTORS	294
	11. Qualitative Picture of Processes Taking Place at a Semiconductor–Electrolyte Interface under Its Nonuniform Illumination	295
	12. Principles of the Theory of Laser Etching	296
	13. Photoanodic and Photochemical Recording of Holograms	300
VI.	PROCESSES CAUSED BY PHOTOEXCITATION OF REACTANTS IN THE SOLUTION	303
	14. Qualitative Picture of Sensitization	303
	15. Some Experimental Results	306
VII.	SELECTED PROBLEMS IN THE PHOTOELECTROCHEMISTRY OF SEMICONDUCTORS.	310
	16. Photoelectron Emission from Semiconductors into Solutions	310

17. Semiconductor Electrochemical Photography 315
18. Radiation Electrochemistry of Semiconductors 317
19. Electrogenerated Luminescence 318
20. Electroreflection at a Semiconductor–Electrolyte Interface . . 320

VIII. CONCLUSION: PROBLEMS AND PROSPECTS 323

REFERENCES 324

List of Symbols

Symbol	Definition
c	velocity of light (3×10^8 m/s)
c_{ox}; c_{red}	concentration of oxidizing and reducing agent in solution
C	capacity
D_p	diffusion constant for holes
e	absolute value of electron charge (1.6×10^{-19} C)
E_c	energy of conduction band edge in the semiconductor bulk
$E_{c,s}$	energy of conduction band edge at the surface
E_R	solvent reorganization energy
E_v	energy of the valence band edge in the semiconductor bulk
$E_{v,s}$	energy of the valence band edge at the surface
$\mathscr{E}$	electric field
F	electrochemical potential level (Fermi level) of electrons in semiconductor
$F^0_{dec,n}$, $F^0_{dec,p}$	electrochemical potential level for reactions of semiconductor decomposition with the participation of electrons (n) and holes (p)
$F_{n,p}$	quasi-Fermi level of electrons (n) and holes (p)
F_{redox}	electrochemical potential of electrons in solution that contains redox couple
i	electric current density
i_{corr}	corrosion current
i_{ph}	density of photocurrent
I	photoemission current density
$\mathscr{J}_0$	density of light flux that entered the semiconductor
L_H	thickness of Helmholtz layer
L_{sc}	thickness of space-charge layer in semiconductor
n	concentration of electrons in the conduction band
$\mathfrak{n}$	complex refractive index
n_0, p_0	equilibrium concentration of electrons and holes in the bulk of semiconductor
n_0^*, p_0^*	quasiequilibrium concentration of electrons and holes in illuminated semiconductor
$N_{c,v}$	density of states in the conduction band (c) and in the valence band (v)
N_{ss}	concentration of the surface states
p	concentration of holes in the valence band
Q_{sc}	space charge in semiconductor (per unit area)
x	coordinate normal to the interface
Y	quantum yield
y	dimensionless potential
z	charge number
α	linear light absorption coefficient
$\varepsilon_1 + i\varepsilon_2$	complex dielectric constant
ε_0	permittivity of free space (8.86×10^{-14} F/cm)
ε_{sc}	static dielectric constant of semiconductor
$\tilde{\mu}$	electrochemical potential
φ	electrode potential
φ_{corr}	corrosion potential
ϕ_{fb}	flat band potential
ϕ	electric potential
ϕ_H	potential drop in Helmholtz layer

ϕ_{sc}	potential drop in the space-charge region in semiconductor	a	anodic
		c	cathodic
ψ	Volta potential		
ω	angular radiation frequency	Subscripts	
ω_0	threshold frequency ("red boundary")	corr	corrosion
		dec	decomposition of semiconductor
ω_p	plasma frequency		
		el	electrolyte
Superscripts		H	Helmholtz layer
0	equilibrium	ph	photo
*	exited	sc	semiconductor

I. Introduction

The photoelectrochemistry of semiconductors studies processes of various nature that occur at a semiconductor–electrolyte solution interface under the action of electromagnetic radiation (mainly in the visible, UV and IR regions). These processes include:

(a) Electrode reactions caused by photoexcitation of the electron ensemble of a solid. Such reactions are quite specific just for semiconductor electrodes.

(b) Electrode reactions at semiconductors caused by photoexcitation of reactants in the solution and in the adsorption layer. This is a traditional field of photoelectrochemistry as a whole, but the phenomena of charge transfer between the excited reactants and semiconductor (rather than metal) electrodes possess some essential peculiarities.

The theoretical developments in the above areas were influenced, to a considerable extent, by concepts borrowed from semiconductor physics and the physics of surfaces. Other fields of photoelectrochemistry of semiconductors were affected to a greater degree by progress achieved in the study of metal electrodes. Here we mean photoemission of electrons from semiconductors into solutions and electroreflection at a semiconductor–electrolyte interface.

Finally, processes inverse to light-stimulated electrode reactions can also be referred to as processes of photoelectrochemistry of semiconductors. Such processes include, in particular, electrode reactions accompanied with light emission.

Thus, photoelectrochemistry of semiconductors covers a fairly wide range of problems, which still continues to develop.

The history of photoelectrochemical investigations at semiconductors apparently dates back to works of Becquerel (1839), who observed the occurrence of electric current under the illumination of metals, both with a pure surface and with that coated with phase layers, which probably possessed semiconductor properties. After that, the studies ceased for a long time, up to the 1940s, when important works were performed by Veselovsky (1946) who investigated the photoelectrochemical behavior of semiconducting oxide films. Yet, photoelectrochemistry of semiconductors (as well as electrochemistry of semiconductors as a whole) originated as a certain field of science in the middle of the 1950s after Brattain and Garrett (1955) had related electrochemical and, in particular, photoelectrochemical properties of single-crystal semiconductors to specific features of their electronic structure. Thus, photoelectrochemistry of semiconductors has become a branch of electrochemical physics. Now this term is officially adopted by the International Society of Electrochemistry to denote those areas in electrochemistry that are most closely related to physics.

Eventually, photoelectrochemistry of semiconductors emerged as an autonomous field of electrochemical physics after the works of Dewald (1960) who developed a detailed mechanism for the occurrence of photopotential at a semiconductor electrode. For greater details in the development of this early stage of photoelectrochemistry of semiconductors, the reader is referred, in particular, to the book by Myamlin and Pleskov (1967).

In the middle of the 1960s interest in the photoelectrochemical behavior of semiconductor electrodes arose again. This was spawned by two reasons. First, it was motivated by a rapid development of nonconventional methods of studying interfaces (such as attenuated total reflection, electroreflection, surface IR spectroscopy, etc.), which employ optically transparent (mainly semiconductor) electrodes. Second, the stability of semiconductor materials against photocorrosion had become of large practical importance (Gerischer, 1966). Finally, photoelectrochemistry of semiconductors was given a fresh and the strongest impetus at the beginning of the 1970s when Fujishima and Honda (1972) showed that a semiconductor–electrolyte interface could, in principle, be used to convert the energy of light into chemical energy of photoelectrochemical reaction products. (They showed this for the case of photoelectrolysis of water with hydrogen production.)

Solar energy conversion in photoelectrochemical cells with semiconductor electrodes is considered in detail in the reviews by Gerischer (1975, 1979), Nozik (1978), Heller and Miller (1980), Wrighton (1979), Bard (1980), and Pleskov (1981) and will not be discussed. The present chapter deals with the main principles of the theory of photoelectrochemical processes at semiconductor electrodes and discusses the most important experimental results concerning various aspects of photoelectrochemistry of a semiconductor–electrolyte interface; a more comprehensive consideration of these problems can be found in the book by the authors (Pleskov and Gurevich, 1983).

Let us now briefly outline the structure of this review. The next section contains information concerning the fundamentals of the electrochemistry of semiconductors. Part III considers the theory of processes based on the effect of photoexcitation of the electron ensemble in a semiconductor, and Parts IV and V deal with the phenomena of photocorrosion and light-sensitive etching caused by those processes. Photoexcitation of reactants in a solution and the related photosensitization of semiconductors are the subjects of Part VI. Finally, Part VII considers in brief some important photoelectrochemical phenomena, such as photoelectron emission, electrogenerated luminescence, and electroreflection. Thus, our main objective is to reveal various photoelectrochemical effects occurring in semiconductors and to establish relationships among them.

II. General Concepts of the Electrochemistry of Semiconductors

The semiconductor–electrolyte solution interface is a contact of two conducting media, so that some of its properties are similar to those of contacts between a semiconductor and a metal or between two semiconductors. At the same time, the interface considered is a contact of two media with essentially different types of conductivity—electronic and ionic; moreover, these media are in different states—solid and liquid. Therefore, such an interface possesses a number of unique features.

1. ELECTRODE POTENTIAL

In the case of an electrode–solution interface, the thermodynamic equilibrium between the media in contact is attained by means of electron–ion exchange processes. Since the process of attaining the equilibrium is governed by charged particles, the equilibrium state is characterized by a certain potential difference between the phases. If they are in direct contact, we deal with the potential difference between two points in different media, for example, inside a semiconductor electrode and inside a solution. This potential difference is called the Galvani potential.

The concept of the Galvani potential should be distinguished from that of the contact potential difference, which is widely used in physics to describe contacts of two electronic conductors. In an electrode–electrolyte system the contact potential difference $\Delta\psi$, which is frequently called the Volta potential, represents the difference of electrostatic potentials between two points located in the same (vapor) phase near free surfaces of contacting electrode and electrolyte solution (see Fig. 1). Let us note that the Volta potential can be measured directly, but the Galvani potential cannot, since it represents the potential difference between points in different phases.

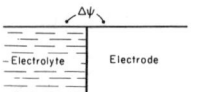

FIG. 1. The definition of the Volta-potential difference in the system electrode–electrolyte solution.

It is the electrode potential φ that is usually measured in electrochemical experiments; it represents a potential difference between two identical metallic contacts of an electrochemical circuit. Such a circuit, whose one element is a semiconductor electrode, is shown schematically in Fig. 2. Besides the semiconductor electrode, it includes a reference electrode whose potential is taken, conventionally, as zero in reckoning the electrode potential (for details, see the book by Glasstone, 1946). The potential φ includes potential drops across the interfaces, i.e., the Galvani potentials at contacts—metal–semiconductor interface, semiconductor–electrolyte interface, etc., and also, if current flows in the circuit, ohmic potential drops in metal, semiconductor, electrolyte, and so on. (These ohmic drops are negligibly small under experimental conditions considered below.)

In order to calculate φ^0, the equilibrium electrode potential, we assume that in the part of the cell where a semiconductor electrode is placed, the electrolyte solution contains a redox couple, the equilibrium between the electrode and solution being established due to the reversible electron exchange reaction

$$\text{Ox} + ne^- \rightleftharpoons \text{Red}. \tag{1}$$

Here Ox and Red denote the oxidized and reduced forms (for example, Fe^{3+} and Fe^{2+}) in the solvated state and n is the number of electrons needed for the electrode reaction to occur.

The equilibrium condition for the above reaction is of the form

$$\tilde{\mu}_{\text{ox}} + nF = \tilde{\mu}_{\text{red}}, \tag{2}$$

where $\tilde{\mu}_{\text{ox}}$ and $\tilde{\mu}_{\text{red}}$ are the electrochemical potentials of the components of a redox system in the solution, and F is the electrochemical potential of electrons in the semiconductor [the quantities $\tilde{\mu}_{\text{ox}}$, $\tilde{\mu}_{\text{red}}$ and $F(=\tilde{\mu}_e)$ are related here to one particle]. It should be noted that the electrochemical potentials differ from the corresponding chemical potentials by the quantity $ez\phi$, where $e > 0$ is the absolute value of the electron charge, z is the charge (positive or negative) of the particle considered, and ϕ is the electric potential in the corresponding

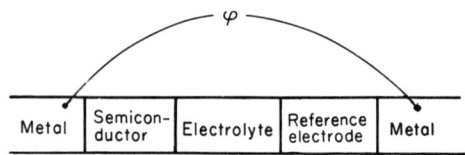

FIG. 2. Schematic diagram of an electrochemical circuit.

phase (the origin of ϕ was chosen arbitrarily). With this in mind, we obtain from (2) for the equilibrium electrode potential φ^0

$$\varphi^0 = \frac{1}{ne}(\mu_{ox} - \mu_{red}) + V. \tag{3}$$

Here μ_{ox} and μ_{red} are the chemical potentials of Ox and Red components, and V is a quantity independent of the properties of a semiconductor electrode but dependent on the reference electrode in the circuit (Fig. 2) (see Pleskov and Gurevich, 1983, for more details).

The chemical potentials μ_{ox} and μ_{red} can be represented in the form

$$\mu_{ox,\,red} = \mu^0_{ox,\,red} + kT \ln c_{ox,\,red}, \tag{4}$$

where k is the Boltzmann constant, T is the absolute temperature, $c_{ox,\,red}$ are concentrations (activities in a more general case) of the components Ox and Red in the solution, and $\mu^0_{ox,\,red}$ are constants independent of $c_{ox,\,red}$ and the electric potential ϕ. Thus, Eq. (3) can be rewritten in the form

$$\varphi^0 = \varphi^0_0 + \frac{kT}{ne} \ln \frac{c_{ox}}{c_{red}}, \tag{5}$$

where

$$\varphi^0_0 = V + \frac{\mu^0_{ox} - \mu^0_{red}}{ne}$$

is the so-called standard electrode potential, that is, in the case considered here, the value of φ^0 for $c_{ox} = c_{red}$.

The value of the constant V, and hence the values of standard potentials, depend on the choice of the reference electrode and on the character of electrode reaction, which takes place on it. With the reference electrode potential conventionally taken as zero, we can choose, for example, the normal hydrogen electrode (NHE), i.e., an electrode, for which the equilibrium at the interface is attained due to the reversible redox reaction $H^+ + e^- \rightleftarrows \frac{1}{2}H_2$, provided the activity of H^+ ions in the solution is 1 mol/liter and the pressure of gaseous hydrogen above the solution is 1 atm. Many of the measured potentials are given below relative to the saturated calomel electrode (SCE); its potential relative to the NHE is 0.242 V.

2. ELECTROCHEMICAL POTENTIAL OF ELECTRONS

If a redox couple is present in a solution and the equilibrium is attained in accordance with reaction (1), the concept of electrochemical potential of electrons in an electrolyte solution, F_{redox}, can be introduced. Let us stress the fact that from the point of view of thermodynamics a detailed mechanism of attaining the equilibriun is of no importance, so one may assume, in particular, that the electron exchange between the Ox and Red occurs without any electrode at all. In this case F_{redox} is defined by the expression similar to Eq. (2) (when F_{redox} is substituted for F), so that

$$F_{redox} = (1/n)(\tilde{\mu}_{red} - \tilde{\mu}_{ox}), \tag{6}$$

and the condition of the electron equilibrium of a redox couple in the solution with a semiconductor electrode is $F = F_{redox}$.

Since the quantities $\tilde{\mu}_{red}$ and $\tilde{\mu}_{ox}$ depend only on the properties of Ox and Red components in the solution bulk, the quantity F_{redox} defined by Eq. (6) is in no way related to the electrode nature and does not depend on the interface structure. Moreover, under thermodynamic equilibrium between the electrode and solution it is F_{redox} that determines the electrochemical potential of electrons in the electrode. This implies, in particular, that the value of F is the same for any electrode that is in equilibrium with a given redox system. Thus, the position of the F_{redox} level in a solution is determined by the redox system contained in it. The more positive the equilibrium potential of this system φ_0^0, the lower the level F_{redox}; the more negative φ_0^0, the higher F_{redox}.

Let us note that in analogy with terminology used in solid state physics F_{redox} is sometimes called the Fermi level of a solution. This term is unsatisfactory in this context and is often misleading because the existence of the quantity F_{redox} is not caused by the presence of "free" electrons in the solution. It should also be emphasized that F_{redox} does not characterize the energy of a certain particle. Moreover, no electrons with the F_{redox} energy exist in the solution. Electron transitions at the interface occur with the participation of certain E_{red}^0 and E_{ox}^0 levels, the most probable energy levels of an electron related to the particles Red and Ox (for details, see Gurevich and Pleskov, 1982). The situation here is similar to that in a semiconductor where the Fermi level lies in the forbidden band, and electron transitions occur with levels of the conduction and valence bands involved.

At the same time, it is the position of the F_{redox} level that determines the thermodynamic properties of a semiconductor–solution interface. In particular, proceeding from the equilibrium condition $F = F_{redox}$, one may write the condition of an electrochemical reaction in the following form (Gerischer, 1977c):

$$F > F_{redox} \quad \text{(cathodic reaction)}, \tag{7a}$$

$$F < F_{redox} \quad \text{(anodic reaction)}. \tag{7b}$$

The relationship between the "physical" scale of energies, in which the zero level corresponds to the potential energy of an electron in vacuum in close proximity to the solution surface yet beyond the action limits of purely surface forces, denoted by E_{vac}, and the "electrochemical" scale of electrode potentials reckoned from a certain reference electrode is, in general, of the form

$$F_{redox} = -e\varphi^0 + \text{const}, \qquad (8)$$

where the constant is determined by the choice of origin in both the electrochemical (φ) and physical (F) scales. The analysis performed in the works of Lohmann (1967) and Gurevich and Pleskov (1982) shows that for the normal hydrogen electrode (see the preceding section) the F_{redox} (NHE) level lies below E_{vac} by 4.4 eV; this latter value (measured in volts) is sometimes called the absolute electrode potential. In other words, if the potential of the NHE is taken as zero, the value of const in Eq. (8) is equal to -4.4 V.

In conclusion let us recall that the position of the electrochemical potential of electrons in a semiconductor, i.e., F relative to that of vacuum, is determined by the thermodynamic work function for the semiconductor w_T, and that the position of F relative to the edges of the bands E_c and E_v in the semiconductor bulk is given by

$$F = \tfrac{1}{2}(E_c + E_v) + kT\ln(N_v/N_c)^{1/2}, \qquad n_0 = p_0, \text{ intrinsic semiconductor,} \qquad (9)$$

$$F = E_c - kT\ln(N_c/n_0), \qquad n_0 \gg p_0, \text{ } n\text{-type,} \qquad (10)$$

$$F = E_v + kT\ln(N_v/p_0), \qquad n_0 \ll p_0, \text{ } p\text{-type.} \qquad (11)$$

Here N_c and N_v are the effective densities of states in the conduction and valence bands, respectively; n_0 and p_0 are equilibrium bulk concentrations of electrons and holes.

3. THE STRUCTURE OF AN ELECTRICAL DOUBLE LAYER AT A SEMICONDUCTOR–ELECTROLYTE INTERFACE

An electrical double layer is usually formed at a semiconductor–electrolyte interface, as well as at the boundary between two solids. This layer consists of "plates" carrying opposite charges, each being located in one of the phases in contact. In the semiconductor the charge in the region near the surface is formed due to redistribution of electrons and holes; in the electrolyte solution, due to redistribution of ions, which form the ionic "plate" of the double layer.

Physical reasons for the occurrence of the electrical double layer are quite general. First, it is the above-mentioned charge transfer across the interface when thermodynamic equilibrium is being established; this transfer takes place because the F and F_{redox} levels (as reckoned from the level of an electron in vacuum, E_{vac}) had different positions before the phases were brought into contact. Second, it is charging processes that, generally, are not associated with charge transfer across the interface. Among these processes one can

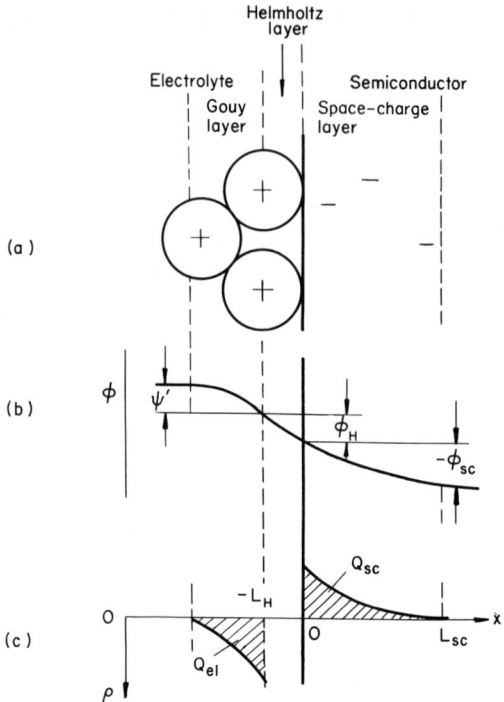

FIG. 3. The structure of electrical double layer at a semiconductor–electrolyte interface (a) and the distribution of the potential (b) and charge (c) at the interface. The electrode is charged negatively. L_{sc} is the space-charge region thickness, L_H is the Helmholtz layer thickness, Q_{sc} and Q_{el} are the charge of the semiconductor and ionic "plates" of the double layer, respectively (for further notations see the text).

mention, for example, charging of surface states, certain types of adsorption accompanied with the formation of polar bonds, and orientation of solvent molecules near the interface.

Three regions are usually distinguished within the electrical double layer: the space-charge region in the solution, the intermediate region called the Helmholtz layer, and the space-charge region in the semiconductor. Figure 3 shows schematically (not to scale) the spatial distribution of the Galvani potential at a semiconductor–electrolyte interface. The semiconductor occupies the region to the right of the vertical solid line, which denotes the interface ($x = 0$); the axis x is directed into the semiconductor bulk. The Helmholtz layer formed by the ions attracted to the electrode surface and also by solvent molecules is located at the interface, for $x < 0$; its thickness L_H is of the order of the ion size. The space-charge region in the solution (the Gouy layer) is adjacent to the Helmholtz layer from the electrolyte side.

Thus, the total potential drop at the interface, i.e., its Galvani potential $_{el}\phi_{sc}$, is (see Fig. 3)

$$_{el}\phi_{sc} = [\phi(-L_H) - \phi(-\infty)] + [\phi(0) - \phi(-L_H)] + [\phi(\infty) - \phi(0)]. \quad (12)$$

The potential drop in the Gouy layer in the solution $\phi(-L_H) - \phi(-\infty)$ is usually denoted in the electrochemical literature by ψ', the "psi-prime potential." The second term in square brackets in Eq. (12) represents the potential drop in the Helmholtz layer; it will be denoted by ϕ_H. Finally, the last term in square brackets is the potential drop in the space-charge region in the semiconductor.

Let us note one vital point, which is of methodological importance. It has been traditionally accepted in electrochemistry to choose the positive direction of the electrode potential φ in such a way that it corresponds to an increase in the positive electrode charge. Here the zero potential is assumed to be that of the reference electrode, which coincides, within a constant, with the potential in the solution bulk $\phi(-\infty)$. On the other hand, in physics of semiconductor surface the potential is usually reckoned from the value in the semiconductor bulk $\phi(\infty)$; the enrichment of the surface with electrons, i.e., the formation of a negative space charge, corresponding to the positive potential of the surface. In particular, this statement directly follows from the Boltzmann distribution for electrons and holes in the space-charge region in a semiconductor:

$$n(x) = n_0 e^{e\phi(x)/kT}, \qquad p(x) = p_0 e^{-e\phi(x)/kT}. \quad (13)$$

Thus, the directions of potential axes accepted in electrochemistry and semiconductor physics appear to be opposite. This is manifested in the fact that the potential of the semiconductor surface ϕ_{sc} [reckoned from $\phi(\infty)$] has the opposite sign to that of the third term in square brackets in expression (12) for the Galvani potential: $\phi_{sc} = -[\phi(\infty) - \phi(0)] = \phi(0) - \phi(\infty)$. Under a change in the electrode potential (the so-called electrode polarization) the Galvani potential and, generally, all its three components change by $\Delta(_{el}\phi_{sc})$. However, in solutions of not very low concentration ($\gtrsim 0.1$ M), which are, as a rule, used in experiment, the potential ψ' (and hence its change $\Delta\psi'$) is negligibly small. We consider this case below. Thus, we have

$$\Delta(_{el}\phi_{sc}) = \Delta\phi_H - \Delta\phi_{sc} \quad (14)$$

and the relationship between potential change in the Helmholtz layer $\Delta\phi_H$ and that in the semiconductor $\Delta\phi_{sc}$ depends on the particular conditions at the interface.

It is known that within the framework of the band model variation of the potential ϕ with coordinate in a semiconductor is equivalent to the bending of energy bands. The bands near the surface are bent downward if $\phi_{sc} > 0$ and upward if $\phi_{sc} < 0$, the total bending value amounting to $|e\phi_{sc}|$. In a special case where $\phi_{sc} = 0$ the bands are not bent (remain "flat") up to the interface. The corresponding potential $\varphi = \varphi_{fb}$ of a semiconductor electrode measured relative to a certain reference electrode is called the flat band potential φ_{fb}.

The flat band potential in electrochemistry of semiconductors is equivalent to the zero-charge potential in electrochemistry of metals (more exactly, to the potential of the zero free charge—see Frumkin, 1979). As with the zero-charge potential in the electrochemistry of metals, the flat band potential is rather important in the kinetics of both dark and photoelectrochemical reactions at semiconductor electrodes. Several methods, including photoelectrochemical ones, have been developed to determine φ_{fb} (see Section 7).

Let us also note that even though $\varphi = \varphi_{fb}$, i.e., $\phi_{sc} = 0$, ϕ_H may, generally speaking, be nonzero; for example, due to the contribution of surface dipoles.

If the bands are bent downward ($\phi_{sc} > 0$), the surface region is enriched, and if upward ($\phi_{sc} < 0$), depleted with electrons. In the case of n-type semiconductors where the potential ϕ_{sc} satisfies inequalities

$$-kT/e \ln n_0/p_0 < \phi_{sc} < -kT/e, \tag{15}$$

a depletion layer (the Mott–Schottky layer) is formed in which the charge is mainly created by ionized donors (dopants) with concentration $N_D \approx n_0$. The formation of the depletion layer is the most probable case for wide-gap sufficiently doped semiconductors, for which the potential range where condition (15) is met is rather large, as can be seen from Eq. (15).

The potential distribution $\phi(x)$ in the space-charge region is described, with due account for Eq. (13), by the self-consistent Poisson–Boltzmann equation. Its first integral can be calculated analytically, so that the electric field at the semiconductor surface $\mathscr{E}_{sc} = -d\phi/dx$ is expressed as (Garrett and Brattain, 1955; see also Frankl, 1967)

$$\mathscr{E}_{sc} = \pm kT/eL_D^i \mathscr{F}(y, \lambda). \tag{16}$$

Here $L_D^i = (\varepsilon_0 \varepsilon_{sc} kT/2e^2 n_i)^{1/2}$ is the Debye length for an intrinsic semiconductor ($n_0 = p_0 = n_i$), where ε_{sc} is the static permittivity of the semiconductor, and

$$\mathscr{F} = [\lambda(e^{-y} - 1) + \lambda^{-1}(e^y - 1) + (\lambda - \lambda^{-1})y]^{1/2}, \tag{17}$$

where $y = e\phi_{sc}/kT$ and $\lambda = (p_0/n_0)^{1/2}$. The sign in Eq. (16) should be chosen the same as that of ϕ_{sc} (or y).

Using Eqs. (16) and (17), one can directly calculate the total charge per unit surface area of a semiconductor

$$Q_{sc} = -\varepsilon_0 \varepsilon_{sc} \mathscr{E}_{sc} = \mp \frac{kT \varepsilon_0 \varepsilon_{sc}}{eL_D^i} \mathscr{F}(y, \lambda) \tag{18}$$

and also the differential capacity per unit area of the space-charge region in a semiconductor $C_{sc} = -dQ_{sc}/d\phi_{sc}$. (The "minus" sign in the latter formula has been chosen in accordance with the above-mentioned distinction between the directions of the axes of the potentials ϕ_{sc} and φ.) Since for the depletion layer in an n-type semiconductor, according to Eq. (15), $\lambda \ll 1$, $y < -1$, and $\lambda e^{-y} < 1$, then $\mathscr{F} = [\lambda^{-1}(|y|-1)]^{1/2}$ and we obtain the depletion layer capacitance from Eq. (18):

$$C_{sc}^{-2} = \frac{2}{\varepsilon_0 \varepsilon_{sc} e N_D} (|\phi_{sc}| - kT/e). \tag{19}$$

It follows from Eq. (19) that the dependence $C_{sc}(\phi_{sc})$ becomes a straight line in the coordinates (C_{sc}^{-2}, ϕ_{sc}); the line thus obtained is called the Mott–Schottky plot.

If we assume that only ϕ_{sc} changes with the electrode potential φ (the necessary conditions are considered in greater detail below), then the slope of the Mott–Schottky plot makes it possible to determine the concentration of donors or acceptors, N_D or N_A, and the point of intersection of this straight line with the potential axis gives the flat band potential φ_{fb}. Figure 4 presents, as an example, dependences of the squared inverse capacity on the potential recorded at a TiO_2 electrode in solutions of various pH (we recall that pH = $-\log c_{H^+}$ where c_{H^+} is the concentration of hydrogen ions).

The measurement of the differential capacity is now the most widely used method of determining the flat band potential φ_{fb} of semiconductor materials; the values of φ_{fb} for some semiconductors are listed, for example, in the papers by Nozik (1978) and Pleskov (1980).

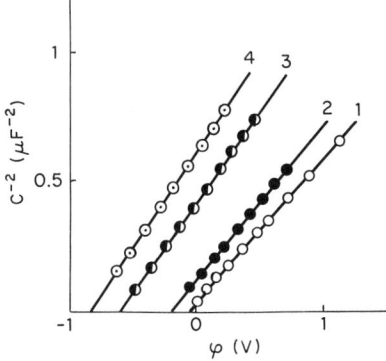

FIG. 4. Dependence of the squared inverse capacity on the potential for a TiO_2 electrode. The ac frequency is 62 kHz. Solution pH: 1—0.7, 2—3.0, 3—9.1, 4—12.9. The potentials are given relative to the saturated calomel electrode. [From Rotenberg et al. (1977).]

The principle of continuity of electric displacement at the interface implies that $\varepsilon_H |\mathscr{E}_H| = \varepsilon_{sc} |\mathscr{E}_{sc}|$, where $\mathscr{E}_H$ is the electric field strength in the Helmholtz layer and ε_H is its permittivity. Assuming for estimations $|\mathscr{E}_H| \approx |\phi_H|/L_H$, we obtain $|\phi_H| \ll |\phi_{sc}|$, if the following inequality is satisfied:

$$\varepsilon_{sc} L_H / \varepsilon_H L_{sc} \ll 1. \tag{20}$$

Here L_{sc} is the thickness of the space-charge region in a semiconductor related to the capacity C_{sc} by $C_{sc} = \varepsilon_0 \varepsilon_{sc}/L_{sc}$. In particular, for the depletion layer we have, according to Eq. (19),

$$L_{sc} \simeq (2\varepsilon_0 \varepsilon_{sc} |\phi_{sc}|/eN_D)^{1/2}. \tag{20a}$$

Estimations show that condition (20) usually holds true for reasonable values of the parameters in the absence of degeneration. At the same time, for highly doped semiconductors and also in the formation of an accumulation or inversion layers (i.e., for sufficiently large values of $|\phi_{sc}|$) L_{sc} may become so small that inequality (20) is not satisfied.

Surface electron states (surface levels) at the interface may also play an important role in the potential distribution between the phases. It should be noted that in considering a solid–liquid interface it would be more correct to speak about interface rather than surface electron states. When using, below, the conventional term "surface states," we shall mean by it just "interface" states.

Let us recall that the general reason for the occurrence of such states lies in the fact that besides delocalized electron states in a bounded crystal there exist additional states that correspond to electrons localized at the surface. Surface electron states, that exist at "atomically pure" (ideal) surfaces of crystals are usually called intrinsic (Davison and Levine, 1970). Under ordinary conditions, in contact with electrolyte solutions in particular, adsorbed atoms or even layers may be present at the surface; the real crystal may also contain structural defects. They all can exchange electrons with the semiconductor bulk; this gives rise to electron energy levels at the surface, their nature and properties being different from those of intrinsic ones. As a result, the surface of a semiconductor contains various types of surface states that are characterized by a complicated energy spectrum; their concentration depends on the way the surface is treated (for example, grinding and polishing, electrochemical and chemical etching, etc.) and may be as high as 10^{14} cm^{-2} (see, for instance, Frankl, 1967; Rzhanov, 1971).

The existence of surface states leads to two most important effects. First, electrons and holes may be trapped at the surface to form surface electric charge. This influences the equilibrium properties of semiconductors. Second, the surface levels may change significantly the kinetics of processes, in which electrons and holes are involved. On the one hand, they create additional centers of charge-carrier recombination and generation, and, on the other hand, they may act as intermediate energy levels in the processes of charge transfer across the interface.

The effect of surface states on the potential distribution across the interface can be estimated by the relation

$$\varepsilon_0 \varepsilon_H \mathscr{E}_H = Q_{ss} - \varepsilon_0 \varepsilon_{sc} \mathscr{E}_{sc}, \qquad (21)$$

where Q_{ss} is the charge density in the surface states, which depends on ϕ_{sc}. The simplest model of single-energy surface states can be used to estimate Q_{ss}. In this case, it is a simple matter to show that $|\Delta\phi_H| > |\Delta\phi_{sc}|$ if

$$N_{ss} > 4\varepsilon_0 \varepsilon_H kT/e^2 L_H, \qquad (22)$$

where N_{ss} is the surface states density. In other words, if condition (22) is satisfied, then the change in the electrode potential leads to a higher change of the potential in the Helmholtz layer than that in the semiconductor space-charge region.

Thus, since usually $L_H \ll L_{sc}$, then $|\phi_{sc}| \gg |\phi_H|$, and therefore ϕ_{sc} constitutes, as a rule, the main portion of the interface potential drop $_{el}\phi_{sc}$. This, however, does not hold true in the following cases:

(1) for highly doped materials where the F level in the bulk is close to the edge of the majority-carrier band in the bulk, or even located in this band (bulk degeneration);

(2) for sufficiently high electrode charging when the F level is near the edge of the corresponding band at the surface or inside this band (surface degeneration);

(3) for high concentration of surface states, which leads to an increase in the contribution of ϕ_H.

Consider two of the most important limiting cases.

Suppose that as the electrode potential changes, $|\Delta\phi_{sc}| \gg |\Delta\phi_H|$ (or, which is the same, $|\Delta\varphi| \approx |\Delta\phi_{sc}|$). Here the position of all energy levels at the surface and, in particular, of the edges of the energy bands $E_{c,s}$ and $E_{v,s}$ remains unchanged with respect to the position of energy levels in the electrolyte solution and to the reference electrode (Figs. 5a and 5b). In this case the bands are said to be "pinned" at the surface.

Let, on the contrary, $|\Delta\phi_H| \gg |\Delta\phi_{sc}|$, i.e., the change in φ leads to a change, primarily, of the potential drop in the Helmholtz layer (this is the case with metal electrodes). Under this condition, the energy levels at the surface shift by $e\Delta\phi_H$ with respect to the energy levels in the solution (Figs. 5a and 5c). The relative position of $E_{c,s}$, $E_{v,s}$, and F does not vary, however, because ϕ_{sc} remains unchanged. In this case it is said that the edges of the bands are

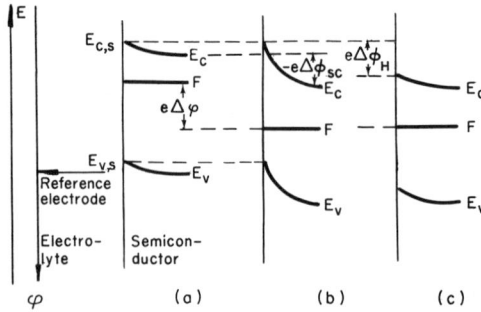

FIG. 5. Energy diagram of a semiconductor–electrolyte interface: (a) with no external voltage; (b) and (c) under the application of an external voltage. The diagram explains the pinning at the semiconductor electrode surface of the energy band edges [transition from (a) to (b)] or of the Fermi level [transition from (a) to (c)].

unpinned at the surface (or that the Fermi level is pinned relative to the band edges). In real systems a situation is possible where both potential drops (ϕ_H and ϕ_{sc}) change simultaneously, so that in fact neither the band edges nor the Fermi level are pinned at the semiconductor surface.

Let us consider, in brief, a way of practical construction, using experimental data, of the energy diagram of the semiconductor–electrolyte system (of the type shown in Fig. 5). This diagram is convenient for discussing the energy balance and thermodynamics of processes taking place at the interface. The diagram shows the levels of energy and electrochemical potential in the semiconductor (E_c, E_v, F) and solution (F_{redox}), as expressed on the energy scale E, which is unambiguously related by Eq. (8) to the scale of the electrode potential φ.

Measuring the flat band potential φ_{fb} (relative to a certain reference electrode), one may determine the position of the Fermi level F of a semiconductor on the electrode potential scale. Next, formulas (9)–(11) can be used to find (on the same scale) the position of E_c and E_v relative to F in the electrode bulk. For a chosen electrode potential one may determine, using φ_{fb}, the quantity ϕ_{sc} and, hence, the band bending; after that, the position of the band edges at the surface can easily be found. Finally, since the equilibrium potential for a given redox couple is known, the F_{redox} level can also be found with the help of Eq. (8). The diagrams thus constructed will often be used below.

To conclude this section, let us note that although all the above relations correspond to the equilibrium state, they can equally be applied to the potential distribution and relative position of the levels under nonequilibrium conditions, provided the corresponding currents are not very large (for details, see Pleskov and Gurevich, 1983).

4. Electrochemical Reactions at Semiconductor Electrodes

Electrochemical reactions are heterogeneous chemical reactions accompanied by electrical charge transfer across the interface. Besides "ordinary" variables of chemical kinetics, such as concentration, temperature, etc., electrochemical kinetics is characterized by an additional independent variable, electrode potential. The rate of electrochemical processes may vary quite significantly (exponentially) with the electrode potential.

Let us consider a semiconductor electrode, at which a redox reaction of type (1) occurs. Electrons of both the conduction band and valence band may take part in the electrode process. As a result, the reversible reaction considered is characterized by four different types of electron transitions (see Fig. 6a). Transitions in which electrons leave the semiconductor and holes come in contribute to the cathodic current, and those where electrons come in and holes escape contribute to the anodic current. Thus, the resultant current is a sum of four currents: $i^c_{n,p}$, $i^a_{n,p}$ (when referring to "currents" we shall always mean current densities).

Under equilibrium conditions the currents i^c_n and i^a_n, and also i^c_p and i^a_p, are equal to each other by the absolute value, in accordance with the principle of detailed balancing (see, for example, Landau and Lifshitz, 1977). These equilibrium values $(i^c_p)^0 = (i^a_p)^0 = i^0_p$ and $(i^c_n)^0 = (i^a_n)^0 = i^0_n$ represent, by definition, exchange currents of an electrode reaction passing through the valence band (i^0_p) and through the conduction band (i^0_n).

The values of the exchange currents depend on the electrode nature and solution composition and are the basic characteristics of an electrode process. In a special case where the exchange currents are equal to zero, $i^0_p = i^0_n = 0$, the interface (electrode) is called idealy polarizable.

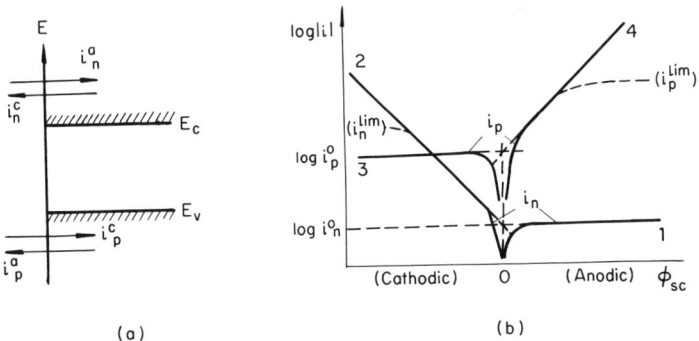

Fig. 6. Electron–hole currents at a semiconductor–electrolyte interface (a) and their dependence on the change of potential drop in the semiconductor (b): 1—i^a_n, 2—i^c_n, 3—i^c_p, 4—i^a_p (dashed lines show limiting currents of the minority carriers).

Under nonequilibrium conditions we have for the hole $i_p = i_p^a - i_p^c$ and electron $i_n = i_n^a - i_n^c$ currents

$$i_p = i_p^0 \left(\frac{p_s}{p_s^0} e^{e\alpha_p \Delta\phi_H/kT} - e^{-e(1-\alpha_p)\Delta\phi_H/kT} \right), \qquad (23)$$

$$i_n = i_n^0 \left(e^{e\alpha_n \Delta\phi_H/kT} - \frac{n_s}{n_s^0} e^{-e(1-\alpha_n)\Delta\phi_H/kT} \right), \qquad (24)$$

provided the current flowing through the solution does not disturb the distribution of the concentration of the Ox and Red components. Here p_s and n_s are surface concentrations of holes and electrons (p_s^0 and n_s^0 are the corresponding equilibrium values), $\Delta\phi_H = \phi_H - \phi_H^0$ is the deviation of potential drop in the Helmholtz layer from the equilibrium value, and $\alpha_{p,n}$ are the so-called transfer coefficients for hole and electron transitions ($0 \leq \alpha_{p,n} \leq 1$); these coefficients characterize the extent to which the potential drop affects the activation energy of the electrode process considered.

Let us emphasize that the currents i_p and i_n depend on the potential not only by virtue of $\Delta\phi_H$, but also, via the factors p_s/p_s^0 and n_s/n_s^0, on the potential drop ϕ_{sc} in the semiconductor. In particular, if $\Delta\phi_H = $ const, the corresponding components in the currents i_p and i_n exponentially depend on the potential. This dependence is shown schematically in Fig. 6b.

Illumination of a semiconductor, leading to a change of the charge-carrier concentration in the bands, also affects, according to Eqs. (23) and (24), the electron-hole currents. It is this effect that underlies a specific behavior of semiconductor electrodes under illumination. Obviously, illumination affects most strongly the current of minority carriers, whose relative concentration may vary especially noticeably.

Note also that p_s and n_s, appearing in Eqs. (23) and (24), generally do not coincide with those given by equilibrium distributions (13), even in the absence of illumination. The electrode reaction can disturb significantly the distribution of charge carriers in a semiconductor electrode. In particular, if minority carriers become involved in an electrode reaction, its rate may be limited by the rate at which these carriers are supplied from the bulk of the semiconductor to its surface.

The maximum value of the minority-carrier current, which may flow under steady state conditions from the semiconductor bulk to the surface, is called the limiting current. It is determined by bulk generation of electron-hole pairs. As simple calculation shows (see, for example, Myamlin and Pleskov, 1967), the absolute value of the limiting current of minority carriers, holes for illustration, is

$$i_p^{\lim} = eD_p p_0 / L_p. \qquad (25)$$

Here D_p is the hole diffusion coefficient and $L_p = (D_p \tau_p)^{1/2}$ is the diffusion length, where τ_p is the hole lifetime. This expression for $i_p^{\lim}$ coincides with a known formula (see, for instance, Middlebroock, 1957) for the "saturation current" of a p–n junction.

The growth of the components i_p^a and i_n^c is restricted by the corresponding limiting currents (see Fig. 6b). Thus, for semiconductors of, say, n-type, the dependence of the hole current i_p on φ must be close to the exponential one (the Tafel law) under cathodic polarization, and must be restricted by the limiting current $i_p^{\lim}$ of minority carriers under anodic polarization.

III. The Theory of Processes Caused by Photoexcitation of a Semiconductor

Light with frequencies exceeding the threshold of intrinsic (fundamental) absorption of a semiconductor affects most strongly the processes at semiconductor electrodes. In this case the energy of an absorbed quantum of light is sufficient for photogeneration of an electron–hole pair (internal photoeffect). Redistribution of charge carriers in a semiconductor, caused by this photogeneration, can radically alter not only the rate, but also the character of processes occurring at the interface. Besides the fact that such phenomena are of independent scientific interest, they are quite important in view of practical use of electrochemical systems with semiconductor electrodes.

Photoelectrochemical processes may proceed in quite different regimes, depending on the relative magnitudes of the depth of light penetration into a semiconductor, the diffusion length and the thickness of the space-charge region, and also between the rates of electrode process and carrier supply to the surface. Nevertheless, in important particular cases relatively simple (but in no way trivial) relations can be obtained, which characterize a photoprocess, and the theory can be compared with experiment.

5. General Picture and Determination of Photocurrent in the Simplest Case

Let us consider, first qualitatively, what happens to light-induced carriers in the near-the-surface region of a semiconductor electrode (n-type, for example). Depending on the radiation frequency ω, the penetration depth of light [equal by an order of magnitude to α^{-1}, where $\alpha(\omega)$ is the light absorption coefficient] may vary within wide limits. Figure 7 shows schematically (not to scale) two limiting cases: $\alpha^{-1} > L_{sc}$ and $\alpha^{-1} < L_{sc} + L_p$. The energy of a quantum of light is assumed to exceed the threshold of photogeneration of electron–hole pairs. The region where this generation takes place is shaded in Fig. 7. If a depletion layer is formed near the surface, the holes generated by light in the

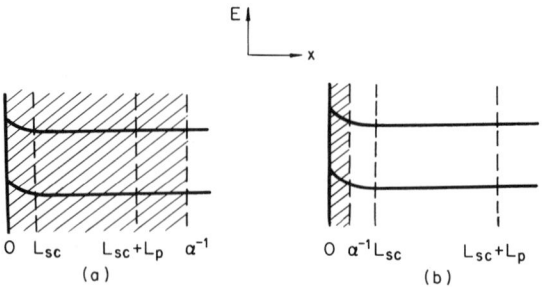

FIG. 7. Various relations between the penetration depth of light in a semiconductor α^{-1}, the thickness of the space-charge region L_{sc}, and the diffusion length L_p: (a) weak absorption of light (α^{-1} is large), and (b) strong absorption of light (α^{-1} is small).

region $x < L_{sc}$ are transferred by the electric field to the electrode surface, where they participate in an electrochemical reaction with reactants in the solution. Outside the depletion layer, i.e., for $x > L_{sc}$, the minority carriers are transferred via diffusion. Since in their lifetime τ_p, the holes cover the distance of the order of L_p, the holes that are generated even deeper (the case $\alpha^{-1} > L_{sc} + L_p$) mainly recombine before they can reach the surface, thereby producing no contribution to the photocurrent.

The *photocurrent* i_{ph} is defined as

$$i_{ph} = i_{light} - i_{dark}, \qquad (26)$$

where i_{light} and i_{dark} are the currents that flow, respectively, in a system under illumination and in darkness.

Suppose that the rate of an electrode reaction at the interface with holes involved is sufficiently high so that each hole approaching the electrode surface is consumed. Under such conditions the change in the electrode potential affects the hole photocurrent mainly through a change in the depletion layer thickness L_{sc}, which depends on ϕ_{sc} [see Eq. (20a)]. If the light is absorbed weakly ($\alpha^{-1} \gg L_{sc} + L_p$), the photocurrent must, obviously, be proportional to the region thickness, from which the surface "collects" holes, i.e., to $L_{sc} + L_p$. If, on the contrary, $\alpha^{-1} < L_{sc}$, that is the case of "surface-absorbed" light, the photogeneration region lies entirely within the depletion layer, so that the variation of its thickness produces no changes in the photoprocess. Thus, in the case considered the photocurrent must not depend on the potential. Moreover, for a fast electrode reaction the photocurrent i_{ph}, in the case $\alpha \ll L_{sc}$, must have a maximum possible value and be limited only by the density of exciting light flux.

The simplest calculation, based on the above considerations, was first performed by Gärtner (1959) who assumed the dark current to be zero; the problem was treated more rigorously by Reiss (1978).

The photocurrent i_{ph} that is, by assumption, entirely hole, consists of two parts:

$$i_{ph} = i_{dl} + i_b, \qquad (27)$$

where i_{dl} is the current caused by photogeneration of holes in the depletion layer and i_b is the current due to holes generated outside this layer in the semiconductor bulk. Assuming all the holes generated by light in the space-charge region to be consumed in the surface reaction and to contribute to the photocurrent, we obtain for i_{dl}

$$i_{dl} = e \int_0^{L_{sc}} \alpha \mathcal{J}_0 e^{-\alpha x} dx = e \mathcal{J}_0 (1 - e^{-\alpha L_{sc}}). \qquad (28)$$

Here $\mathcal{J}_0$ is the density of light flux $\mathcal{J}(x)$ at $x = 0$, which is related to the incident flux $\mathcal{J}_{inc}$ by $\mathcal{J}_0 = \mathcal{J}_{inc}(1 - R)$, where R is the reflection coefficient of the boundary. In writing Eq. (28) we have also taken into account that $\mathcal{J}(x) = \mathcal{J}_0 e^{-\alpha x}$ decays exponentially into the sample bulk, and the rate of electron–hole pair generation is $\alpha \mathcal{J}(x)$.

The distribution of holes outside the space-charge region that takes into account bulk photogeneration is described by the diffusion equation with a source

$$D_p \frac{d^2 p}{dx^2} - \frac{p - p_0}{\tau_p} + \alpha \mathcal{J}_0 e^{-\alpha x} = 0. \qquad (29)$$

One boundary condition for Eq. (29) is $p = p_0$ for $x \to \infty$. The second boundary condition can be taken as $p = 0$ for $x = L_{sc}$, which is true for a very fast electrode reaction that "extracts" practically all the holes from the space-charge region. Solving Eq. (29) with these boundary conditions and then calculating the diffusion current of holes from the bulk into the space-charge region at $x = L_{sc}$, we obtain

$$i_b = e\mathcal{J}_0 \frac{\alpha L_p}{1 + \alpha L_p} e^{-\alpha L_{sc}}. \qquad (30)$$

Finally, summing up Eqs. (28) and (30), we find, according to Eq. (27), the photocurrent

$$i_{ph} = e\mathcal{J}_0 \left(1 - \frac{e^{-\alpha L_{sc}}}{1 + \alpha L_p}\right). \qquad (31)$$

Expression (31) shows the way in which the photocurrent depends on radiation characteristics [through $\alpha(\omega)$ and $\mathcal{J}_0$], transport characteristics of minority carriers (through L_p), concentration of majority carriers, and the electrode potential (through L_{sc}). Let us note that the photocurrent does not depend on the equilibrium concentration of minority carriers.

In accordance with the above qualitative picture, the photocurrent, by virtue of Eq. (31), is proportional to $\alpha\,(L_{sc} + L_p)$, if $\alpha^{-1} \gg L_{sc}, L_p$, and does not depend on L_{sc} (and hence on ϕ_{sc}) if $\alpha^{-1} \ll L_{sc}$; in the latter case the photocurrent is equal, in absolute value, to the maximum possible value $e\mathcal{J}_0$.

6. CALCULATION OF PHOTOCURRENTS AND PHOTOPOTENTIALS IN MORE COMPLICATED CASES

Determination of the photocurrent requires various factors to be taken into account and faces the necessity of making a lot of model assumptions. The final formulas appear to be rather cumbersome, even though relatively insignificant complications are introduced into the problem statement. Therefore, various authors have analyzed only individual aspects of the problem.

An important step in a more comprehensive analysis of calculating i_{ph} was made by Reiss (1978). In this work the photocurrent, which is a sum of the electron and hole components, is found by solving a system of transport equations for electrons and holes both for $x < L_{sc}$ and $x > L_{sc}$. Next, the solutions thus obtained are matched at $x = L_{sc}$ with due account for continuity of the currents and concentrations. In the limit of a very high rate of hole consumption at the surface, the general relations found yield expression (31) for the photocurrent i_{ph} if the condition $L_{sc} < L_p$ is satisfied and the dark current is neglected. It should be noted here that this expression is valid not only for potentials ϕ_{sc} corresponding to the formation of a depletion layer [see Eq. (15)], but in a wider potential range. If a rapid electrode reaction with holes involved takes place at the surface, the hole concentration may sharply decrease as compared to the equilibrium one. In this case, even in the region of potentials $\phi_{sc} < -kT/e \ln n_0/p_0$, where an inversion layer enriched with holes could form under equilibrium conditions, the actual distribution of the charge and potential will be the same as in the depletion layer, since in both the cases it is determined by the charge of immobile donors.

Moreover, Reiss (1978) considers the change in the reactant concentration in the near-electrode region of the solution (due to current flowing through it), and also the effect of surface recombination on photocurrent. The latter effect is discussed in more detail in the paper by Wilson (1977) where a finite rate of the electrochemical reaction is also taken into account. Instead of the condition $p(L_{sc}) = 0$, a more general relation

$$D_p \frac{dp}{dx}\bigg|_{x=L_{sc}} = kp(L_{sc}) \tag{32}$$

is taken as the boundary condition for Eq. (29); this relation is reduced to the former one in the limit $k \to \infty$, where k is a parameter related to the hole current at $x = 0$. This current is formed due to an electrode reaction (the current i_{el}) and recombination of holes and electrons at surface centers (the current

i_{rec}), which take place simultaneously; these currents are of the form

$$i_{el} = ek_{el}p\Big|_{x=0}, \quad i_{rec} = ek_{rec}p\Big|_{x=0}, \tag{33}$$

where k_{el} and k_{rec} are phenomenological reaction rate constants, which have dimensions of velocity and depend on the microscopic characteristics of surface processes under consideration. In the case of $k > D_p/L_p$ (the ratio D_p/L_p characterizes the rate of diffusion supply of photoholes to the space-charge region) the resultant photocurrent will be

$$i_{ph} = e\mathcal{J}_0 \frac{k_{el}}{k_{el} + k_{rec}} \left[1 - \frac{e^{-\alpha L_{sc}}}{1 + \alpha L_p} - \frac{D_p \alpha e^{-\alpha L_{sc}}}{k(1 + \alpha L_p)} \right], \tag{34}$$

where the parameter k becomes equal to $k = (k_{el} + k_{rec})e^{e\phi_{sc}/kT}$. The last term in square brackets represents a correction term associated with the fact that the rate constants k_{el} and k_{rec} are finite; the term vanishes for $k \to \infty$. Let us note that if this term is taken into account, the value of the photocurrent i_{ph} decreases, as compared to that obtained from Eq. (31). The factor $k_{el}/(k_{el} + k_{rec})$ in Eq. (34) allows for the fact that only those holes taking part in the electrode reaction contribute to the photocurrent. Thus, surface recombination leads to a change in the relative fraction of holes participating in an electrochemical reaction at the interface and to a change in the hole distribution for $x > L_{sc}$, which in turn results in the variation of the hole flux from the bulk to the surface. In the limiting case $k_{el} \gg k_{rec}$, and $k_{el} \gg D_p/L_p$, expression (34) is reduced to Eq. (31).

Wilson (1977) also presents expressions for a more general case $k \lesssim D_p/L_p$, which are, however, rather cumbersome.

A similar problem of calculating i_{ph} by a somewhat different method is considered in the paper by Reichman (1980), who makes additional model assumptions to take into account, apart from other factors, such as recombination of carriers in the space-charge region; in this connection, see also the paper by Kireev *et al.* (1981).

The *photopotential* φ_{ph} is defined as

$$\varphi_{ph} = \varphi - \varphi_{dark}, \tag{35}$$

where φ_{dark} is the dark value of the electrode potential; $\varphi_{dark} = \varphi^0$ under equilibrium conditions. In other words, φ_{ph} represents the shift of the electrode potential under illumination [cf. definition of i_{ph}, Eq. (26)].

The general solution of the system of transport equations for electrons and holes permits the photopotential of an open circuit to be calculated. The assumption that the total potential change due to illumination occurs in the space-charge region of a semiconductor, i.e., $\varphi_{ph} = \phi_{sc}^0 - \phi_{sc}$, where ϕ_{sc}^0 is the equilibrium value of ϕ_{sc}, and that the exchange currents i_n^0 and

i_p^0 are rather large leads to the following expression for φ_{ph} (Gurevich, 1983):

$$\varphi_{ph} = -\frac{kT}{e}\ln(1 + b\mathcal{I}_0). \tag{36}$$

Here $b > 0$ is a constant dependent on the semiconductor properties

$$b = \frac{\alpha L_p}{1 + \alpha L_p}\left(\frac{D_n n_0}{L_0} + \frac{D_p p_0}{L_p}\right)^{-1},$$

where

$$L_0 \equiv L_D \sqrt{2}\int_0^{|y^0|^{1/2}} \exp(-z^2)\,dz, \qquad y^0 = e\phi_{sc}^0/kT.$$

It follows from Eq. (36) that $\varphi_{ph} < 0$, i.e., the photopotential of an open circuit is negative if a depletion layer is formed near the surface. A negative value of φ_{ph} corresponds to a decrease in $|\phi_{sc}|$ and can clearly be explained as an unbending of the bands under illumination.

A similar effect of band unbending under illumination also takes place at an ideally polarizable electrode when $i_p^0 = i_n^0 = 0$ (Garrett and Brattain, 1955).

Let us note in conclusion that under illumination the Dember photo-e.m.f. also occurs in the sample bulk along with the above-considered photo potential φ_{ph} in the space-charge region. The contribution of this e.m.f. is rather small and can be taken into account, if necessary, by means of appropriate formulas (see, for example, Seeger, 1973).

7. COMPARISON OF THE THEORY WITH EXPERIMENT

To compare quantitatively the current–voltage characteristic of an illuminated electrode, given by formula (31), with experimental data, Butler (1977) and Wilson (1977) measured the photocurrent, which arises in a cell with an n-type semiconductor photoanode (TiO_2, WO_3) when irradiated with monochromatic light at a frequency satisfying the condition $\hbar\omega > E_g$. In this case a light-stimulated electrochemical reaction of water oxidation with oxygen evolution

$$H_2O + 2h^+ \longrightarrow \tfrac{1}{2}O_2 + 2H^+ \tag{37}$$

takes place at the semiconductor electrode surface. The purpose of the experiments was to study the dependence of the anodic photocurrent on the intensity of light, its frequency, the semiconductor electrode potential, and solution composition.

The dependences described by formula (31) were compared with experimental ones under the assumption that all the applied voltage dropped in the space-charge region, i.e., that $\phi_H = $ const (pinning of the band edges at the surface, see Section 3). Taking into account that the axes of ϕ_{sc} and φ have

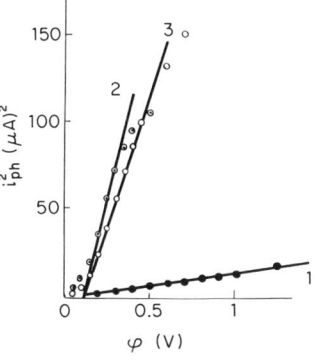

FIG. 8. The anodic photocurrent squared as a function of the WO_3 electrode potential in 1 M CH_3COONa. The wavelength of light (nm): 1—397; 2—327; 3—280. [From Butler (1977).]

opposite directions, we obtain in this case $\phi_{sc} = \varphi_{fb} - \varphi$ where φ_{fb} is the flat band potential.

If α is sufficiently small, so that $L_{sc}, L_p \ll \alpha^{-1}$, it follows from Eqs. (31) and (20a) that

$$\varphi - \varphi_{fb} = \frac{N_D}{2e\varepsilon_0\varepsilon_{sc}}\left(\frac{i_{ph}}{\alpha \mathscr{J}_0}\right)^2. \tag{38}$$

The dependence of i_{ph}^2 on φ, presented in Fig. 8, illustrates a functional relationship predicted by Eq. (38).

Straight lines plotted in the coordinates $i_{ph}^2 - \phi_{sc}$ or $i_{ph}^2 - \varphi$ reflect in this case, as well as the Mott–Schottky plots $C_{sc}^{-2} - \varphi$ [cf. Eq. (19) and Fig. 4], a characteristic dependence of the depletion layer thickness L_{sc} on the potential. The straight line segments intersect at a single point. For materials studied in the works cited below, this point coincides, within 0.2 V, with the value of φ_{fb} measured independently by the differential capacity technique.

Using the value obtained for φ_{fb} and the known value of α, and considering the diffusion length L_p as a fitting parameter, one can obtain good agreement of the theory with experiment for the entire polarization curve (Fig. 9)

Another way of verifying the theory was used in the work of Peter (1978). who calculated the spectral distribution of the quantum yield $Y = i_{ph}/e\mathscr{J}_0$ for a fixed value of ϕ_{sc}. Using the spectral distribution $\alpha(\omega)$ determined by an independent method, Peter obtained good agreement between theoretical and experimental $(Y, \hbar\omega)$ curves.

The character of interband phototransitions in a semiconductor near the edge of the fundamental absorption band was also studied by means of photoelectrochemical measurements. Assuming, as before, that $L_{sc}, L_p \ll \alpha^{-1}$, we obtain for the quantum yield

$$Y = A_n(\hbar\omega - E_g)^{n/2}\left[L_p + \left(\frac{2\varepsilon_0\varepsilon_{sc}|\phi_{sc}|}{eN_D}\right)^{1/2}\right]. \tag{39}$$

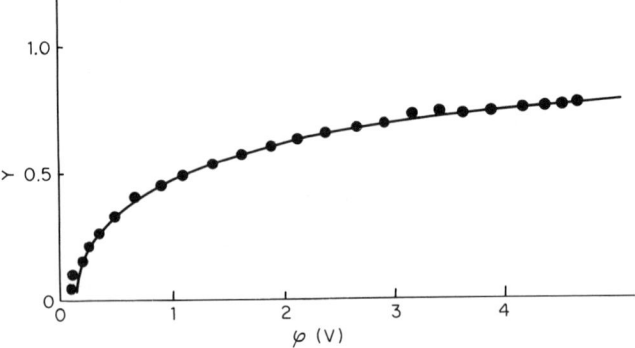

FIG. 9. Comparison of the experimental and calculated dependences of the anodic photocurrent quantum yield Y on the potential of a WO_3 electrode ($N_D = 4 \times 10^{15}$ cm^{-3}) in 1 M CH_3COONa. Solid line—calculation, dots—experiment. [From Butler (1977).]

It is taken into account in this expression that α is proportional to the interband phototransition probability, i.e., $\alpha = A_n(\hbar\omega - E_g)^{n/2}$. It is known (see, for example, Pankove 1971) that the exponent must be $n = 1$ for direct transitions and $n = 4$ for indirect ones (in these cases A_n has different values and is proportional to ω^{-1}).

The experimental dependences obtained are shown in Fig. 10; it can be seen from the figure that straight lines are obtained in the bilogarithmic coordinates, which is in good agreement with formula (39). The slope of these straight lines gives $n = 4$, indicative of photoexcitation of charge carriers in WO_3 occurring via indirect transitions.

Photocurrent measurements permit the determination of the hole diffusion length L_p. As was already noted, comparison of measured and calculated polarization curves allows L_p to be determined by a fitting procedure. For example, Butler (1977) and Wilson (1977) obtained for WO_3 and TiO_2 the values of L_p equal to 0.5×10^{-4} and 4×10^{-4} cm, respectively.

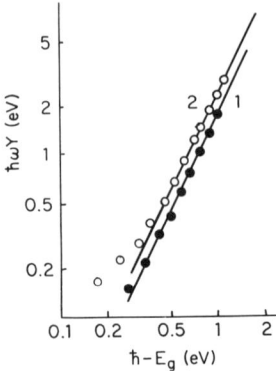

FIG. 10. Dependence of $\hbar\omega Y$ on $\hbar\omega - E_g$ for anodic photocurrent at the WO_3 electrode in 1 M CH_3COONa at potentials 1 V (1) and 5 V (2). [From Butler (1977).]

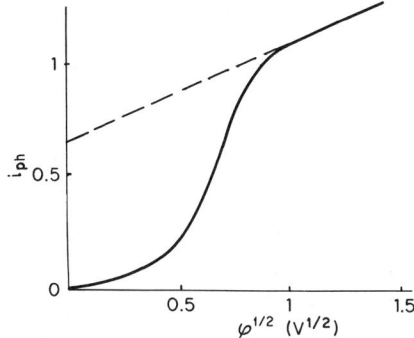

FIG. 11. Anodic polarization curve at a TiO_2 electrode in the coordinates $i_{ph} - \varphi^{1/2}$. [From Rotenberg et al. (1977).]

The following method is also convenient for finding L_p. Formula (31) implies that for $L_{sc} \ll \alpha^{-1}$ and $L_p \gtrsim \alpha^{-1}$

$$i_{ph} = \frac{e\mathcal{J}_0}{1 + \alpha L_p} \alpha(L_p + \gamma|\phi_{sc}|^{1/2}), \quad (40)$$

where $\gamma = (2\varepsilon_0\varepsilon_{sc}/eN_D)^{1/2}$. In the coordinates $i_{ph} - |\phi_{sc}|^{1/2}$ (or $i_{ph} - \varphi^{1/2}$ in the case of $|\Delta\phi_{sc}| = |\Delta\varphi|$) this expression is described by a straight line section, the quotient of whose slope and segment intercepted on the i_{ph} axis gives the ratio γ/L_p, easily yielding L_p once γ is known. Such a plot for a TiO_2 electrode is shown in Fig. 11; it should be noted here that for $\alpha L_p \gtrsim 1$ the straight line $i_{ph} - \varphi^{1/2}$ cannot be extrapolated to the flat band potential.

In the above analysis we always assumed that the photocurrent was entirely due to minority carriers. In a number of cases, however, the measured photocurrent appears to exceed the current of minority carriers at the electrode surface. This effect, called the photocurrent multiplication, was observed, in particular, in photooxidation of various organic substances at ZnO electrodes (Cardon and Gomes, 1971; Freund, 1972), CdSe electrodes (Van den Berghe et al., 1971), and TiO_2 electrodes (Miyake et al., 1977).

Photocurrent multiplication at an n-type electrode can be explained by participation in the electrode reaction of conduction-band electrons, in addition to photoholes. For example, an organic compound R can be oxidized at a photoanode in two stages:

$$R + h^+ \longrightarrow R^{\ddot{+}}, \quad R^{\ddot{+}} \longrightarrow R^{2+} + e^-, \quad (41)$$

where h^+ denotes a hole. Such a scheme leads, as can easily be seen, to the doubling of the photocurrent. In experiment, however, the photocurrent may be slightly less, since the intermediate ion–radical $R^{\ddot{+}}$ can also take part in some electroless reactions (disproportionation, dimerization, etc.).

To take into account all the above-mentioned effects, it is convenient to introduce into formulas (31) and (34) an empirical factor M, called the photocurrent multiplication coefficient. It reflects the fact that a single light-generated carrier can, by virtue of secondary reactions, initiate the transition of several elementary charges across the interface.

Let us note that secondary reactions, leading to the multiplication of the minority-carrier current, may also take place in darkness, which results, for example, in an increase of the maximum observed current, as compared to $i_p^{\lim}$.

Now we briefly touch upon certain practical applications concerning the measurement of photopotential. Its measurement is a convenient method to determine the flat band potential φ_{fb}. As was pointed out in Section 6, the bands of a semiconductor unbend under illumination. In the limiting case of very intensive illumination the bands unbend completely, i.e., the flat band potential is attained. Tyagai and Kolbasov (1975) used this technique to measure φ_{fb} for several $A^{II}B^{VI}$ semiconductors; the values obtained are in good agreement with those measured by the differential capacity method.

IV. Photocorrosion and Its Prevention

The stability of semiconductor electrodes, their resistance to photocorrosion, become an especially urgent problem in connection with ever-extending photoelectrochemical applications of semiconductors. This refers, first of all, to electrodes of photoelectrochemical cells for solar energy conversion.

On the other hand, processes of photocorrosion nature form the basis of light-sensitive etching used for the treatment of semiconductor surfaces, both in laboratory practice and in industry.

The corrosion behavior of semiconductors can, in principle, be described within the framework of the same concepts as for metals (see, for example, Wagner and Traud, 1938), but with due account for specific features in the electrochemical behavior of a solid caused by its semiconducting nature (Gerischer, 1970). One of the main features is photosensitivity related to a change in the free-carrier concentration under illumination. Photosensitivity underlies the phenomenon of photocorrosion.

8. General Concepts of Corrosion

Corrosion is the spontaneous destruction (oxidation, in particular) of a material as a result of its physicochemical interaction with the surrounding medium, for example, with an electrolyte solution. Since we mean here a process, in which the net current is absent, the system solid–solution remains, as a whole, neutral; hence, the oxidation of a solid in the course of corrosion must be accompanied by the reduction of other components in the system (i.e., solvent or solutes).

In a number of especially important cases the resultant corrosion process can be represented as the joint action of two electrochemical reactions—the anodic

$$\{SC\} \longrightarrow \{SC\}^+ + e^- \qquad (42a)$$

and the cathodic

$$Ox + e^- \longrightarrow Red. \qquad (42b)$$

Here {SC} stands for a semiconductor material, {SC}$^+$ for the product of its oxidation, Ox for the oxidizer, and Red for the reduced form of Ox. The above reactions, called the conjugated reactions, proceed at a solid–solution interface simultaneously and with equal rate. In electrochemistry of metals they are considered as quite independent from each other. Once the kinetic parameters of these reactions are known, one can determine the rate (current) and potential of corrosion, using the condition, which follows from the above considerations:

$$i^a = -i^c = |i_{corr}|, \qquad (43)$$

where i^a and i^c are current densities for the anodic and cathodic conjugate reactions.

Specific features of corrosion processes at semiconductors (as against to metals) are caused by the fact that charge carriers of both signs, namely conduction band electrons and valence band holes, take part in charge exchange between a solid and a solution. Therefore, the condition of Eq. (43) is insufficient, so account should be made of charge balance for each type of the carriers because equilibrium between the bands, which is established via generation–recombination processes, may not be reached.

This can be elucidated by a corrosion diagram (Fig. 12), which shows in semilogarithmic coordinates current–voltage characteristics for two conjugated reactions. Using condition (43) and neglecting ohmic potential drop in the system, one can find from the intersection of those characteristics the steady state corrosion current i_{corr} and corrosion potential φ_{corr}.

In the simplest case where the oxidation reaction of a semiconductor material (42a) proceeds exclusively through the valence band and the reaction of reduction of the Ox component of the solution exclusively through the conduction band (see Fig. 13a), corrosion kinetics is limited by minority carriers for either type of conductivity. In fact, it can be seen from Fig. 12 that $i_{corr}(p) = i_n^{lim}(p)$ and $i_{corr}(n) = i_p^{lim}(n)$, where $i_{p,n}^{lim}$ are the limiting currents of minority carriers (symbols in parentheses denote the type of conductivity of a sample under corrosion). Since the corrosion rate is limited by the supply of minority carriers to the interface, it appears to be rather low in darkness. The values of φ_{corr} in Fig. 12 are given for this very case.

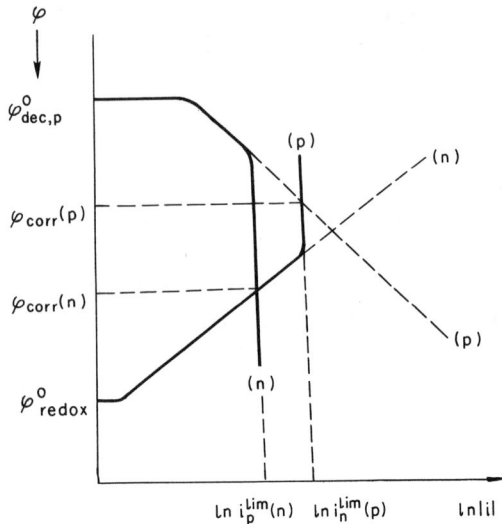

FIG. 12. Corrosion diagram for samples of n- and p-type semiconductors; $\varphi^0_{dec,p}$ is the equilibrium potential of the reaction of anodic photodecomposition of semiconductor with holes involved; φ^0_{redox} is the equilibrium potential for a redox system Ox/Red; $\varphi_{corr}(n)$ and $\varphi_{corr}(p)$ are the corrosion potentials of n- and p-type samples.

Relationships of other type are observed in the case where both the conjugated reactions proceed through the same band (Fig. 13b). For example, the cathodic reaction (42b) can take place with the participation of valence electrons rather than conduction electrons, as was assumed above. Thus, reduction of an oxidizer leads to the injection of holes into the semiconductor, which are used then in the anodic reaction of semiconductor oxidation. In other words, the cathodic partial reaction "provides" the anodic partial reaction with free carriers of an appropriate type, so that in this case corrosion kinetics is not limited by the supply of holes from the bulk of a semiconductor to its surface. Here the conjugated reactions are in no way independent ones.

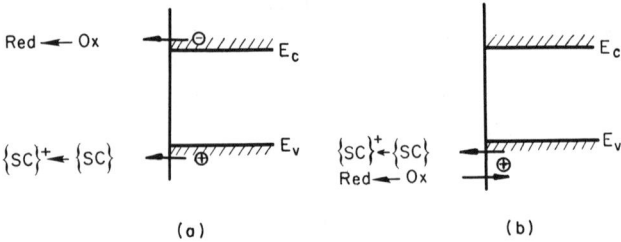

FIG. 13. Diagram illustrating corrosion of a semiconductor: (a) through both energy bands, and (b) through the valence band.

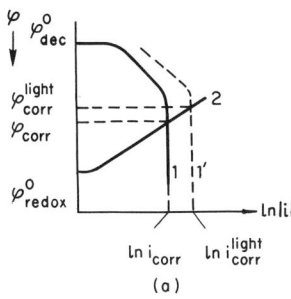 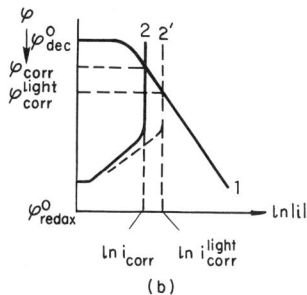

FIG. 14. Diagram explaining the effect of light on the potential and rate of corrosion of a semiconductor: (a) n-type and (b) p-type. φ_{corr}^{light} and i_{corr}^{light} are the potential and current of corrosion when the sample is illuminated.

Consider now, within the framework of the above phenomenological approach, how light affects corrosion. The corrosion diagram (Fig. 14) makes it easy to predict qualitatively this effect if the role of illumination is only to generate nonequilibrium electrons and holes in a semiconductor. In the case where the anodic and cathodic reactions are independent, an increase in the concentration of minority carriers will lead to an effective increase in the reaction current flowing through the minority-carrier band (see Section 4). This means a shift of the corresponding curves, as it is schematically shown in Fig. 14. At the same time, the effect of light, if any, on the reaction with majority carriers involved is substantially less. Therefore, illumination will result in an increase of the corrosion current and a shift of the steady state potential, the direction of the shift depending on the type of semiconductor.

Thus, nonequilibrium electrons and holes generated by light in a corroding semiconductor are consumed to accelerate the corresponding partial reactions. Simultaneous disappearance of these carriers in the course of photocorrosion is similar, from the formal point of view, to surface recombination. This gives every reason to speak about such processes as "electrochemical recombination" (Belyakov *et al.*, 1976). If the "dark" corrosion rate and equilibrium concentration of minority carriers are known, the rate of electrochemical recombination can be calculated.

9. Quasithermodynamic Description of Corrosion and Photocorrosion

In the general case, the reaction of decomposition of a semiconductor material can be both anodic [cf. Eq. (42a)] and cathodic. For instance, for a binary semiconductor MX (where M, X are the electropositive and electronegative components of the compound, respectively) the reaction of anodic decomposition with holes involved can be written as

$$MX + nh^+ \longrightarrow M^{n+} + X \tag{44a}$$

and the reaction of cathodic decomposition with conduction-band electrons involved

$$MX + ne^- \longrightarrow M + X^{n-}. \tag{44b}$$

In a cell with a semiconductor electrode and the reference normal hydrogen electrode, at which the reaction takes place

$$H^+ + e^- \rightleftharpoons \tfrac{1}{2}H_2, \tag{45}$$

the complete reaction proceeds according to one of the following equations:

$$MX + nH^+ \rightleftharpoons M^{n+} + X + (n/2)H_2 \quad (F_{dec,p}), \tag{46a}$$

$$MX + (n/2)H_2 \rightleftharpoons M + nH^+ + X^{n-} \quad (F_{dec,n}). \tag{46b}$$

The symbols in parentheses denote the electrochemical potential levels for the corresponding reaction (the level F_{dec} is a particular case of the level F_{redox} for a redox reaction, in which the electrode material is destructed). Once F_{dec} is calculated (from tabular values of thermodynamic characteristics of substances involved; see, for example, Latimer, 1952), the equilibrium potentials of the reactions of anodic $\varphi^0_{dec,p}$ and cathodic $\varphi^0_{dec,n}$ decomposition of a semiconductor can be determined with the help of Eq. (8). For the most important semiconductor materials, such as CdS, CdSe, CdTe, GaAs, GaP, MoS_2, $MoSe_2$, etc., the corresponding decomposition potentials are presented in the papers by Gerischer (1977a), Bard and Wrighton (1977), and Park and Barber (1979).

Thermodynamic conditions needed for an electrode reaction, including the partial reaction of corrosion, to proceed can be written for the above-considered anodic and cathodic reactions, respectively, in the form [cf. Eq. (7)]

$$\varphi > \varphi^0_{dec,p} \quad \text{or} \quad F < F_{dec,p}, \tag{47a}$$

$$\varphi < \varphi^0_{dec,n} \quad \text{or} \quad F > F_{dec,n}. \tag{47b}$$

If one of conditions (47) is satisfied, the reaction of decomposition becomes thermodynamically possible, though it may be slowed down due to kinetic reasons.

In order to find out in any particular case whether a semiconductor is liable to anodic or cathodic decomposition (both in darkness and under illumination), it is convenient to use the energy diagram (Fig. 15), which plots the energies of band edges and electrochemical potential levels for decomposition reactions. Various situations are possible here, as is schematically shown in Fig. 15. A semiconductor is stable with respect to anodic decomposition if the electrochemical potential level for the corresponding reaction

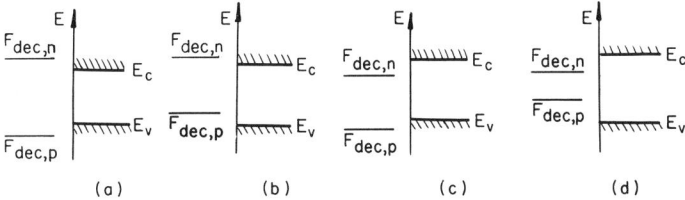

FIG. 15. Diagram illustrating the thermodynamic stability of a semiconductor against corrosion and photocorrosion: (a) semiconductor is absolutely stable, (b) stable against cathodic decomposition, (c) stable against anodic decomposition, and (d) unstable. [From Gerischer (1977a).]

lies in the valence band and to cathodic decomposition if it lies in the conduction band. In both cases this level appears to be "inaccessible" to the electrochemical potential level of a semiconductor F (at least if the variation of the electrode potential is entirely confined within the semiconductor and if the semiconductor surface remains nondegenerate). Thus, a semiconductor is absolutely stable if the levels of both reactions are located outside the forbidden band (Fig. 15a), and it is potentially unstable if both levels lie within the forbidden band (Fig. 15d). More frequent are the cases where a semiconductor is stable with respect to one type of decomposition or the other: either to cathodic (Fig. 15b) or anodic (Fig. 15c).

Let us now turn to the description of photocorrosion, using the quasithermodynamic approach (Gerischer, 1977a) based on the concept of Fermi quasilevels.

Recall that the concept of Fermi quasilevels, suggested by Shockley (1950), can be introduced as follows. Under steady state photogeneration of charge carriers, a dynamic equilibrium arises in a semiconductor between generation and recombination of electron–hole pairs. As a result, certain steady state (but not equilibrium!) concentration values n_0^* and p_0^* are established. The quasiequilibrium concentrations n_0^* and p_0^* are defined by the relations $n_0^* = n_0 + \Delta_n$ and $p_0^* = p_0 + \Delta_p$, and since photogeneration of carriers occurs in pairs, we have $\Delta_n = \Delta_p = \Delta$. Let the following inequalities be satisfied:

$$\tau_c, \tau_v < \tau_{cv}, \qquad (48)$$

where τ_c and τ_v are the times needed for thermodynamic equilibrium to be established in the ensembles of electrons in the conduction band and holes in the valence band, respectively, and τ_{cv} is the time needed for equilibrium to be established between the electron and hole ensembles. Then, each of the distributions (for electrons and holes) may be characterized by its own electrochemical potential, F_n and F_p. Unlike the case of complete thermodynamic equilibrium, where $F_n = F_p = F$, the quantities F_n and F_p, called the

Fermi quasilevels and corresponding to a partial (separately for electrons and holes) equilibrium, are not equal to each other.

In particular, for an n-type semiconductor, substituting the value of n_0^* instead of n_0 into Eq. (10), we obtain

$$F_n \simeq F \quad \text{if} \quad \Delta \ll n_0. \tag{48a}$$

For the quantity $F - F_p$, replacing n_0 by n_i^2/p_0 and substituting p_0^* instead of p_0, we obtain

$$F - F_p = kT \ln\left(\frac{p_0 + \Delta}{p_0}\right), \quad F - F_p \approx kT \ln\frac{\Delta}{p_0} \quad (\Delta \gg p_0). \tag{49}$$

Thus, a considerable shift of the electrochemical potential level under photogeneration occurs only for minority carriers; this effect is sometimes called light doping.

The quasithermodynamic approach used for interpreting photocorrosion phenomena is based on the concept that acceleration of an electrode reaction under illumination is due to the formation of the Fermi quasilevels F_n and F_p, which are shifted relative to the equilibrium position F.

Condition (47) should now be modified. Namely, the inequality

$$F_p < F_{\text{dec}, p} \tag{50a}$$

is necessary for the reaction of anodic photodecomposition with holes involved, and the inequality

$$F_n > F_{\text{dec}, n} \tag{50b}$$

for the reaction of cathodic decomposition with the participation of conduction-band electrons.

Let us consider in more detail, using the above concepts, how a photocorrosion process occurs under the illumination of a semiconductor. Suppose that electron transitions at the interface between the semiconductor and solution do not take place in darkness in a certain potential range (the semiconductor behaves like an ideally polarizable electrode). This range is confined to the potentials of decomposition of the semiconductor and/or solution. The steady state potential of a semiconductor is usually determined in this case by chemisorption processes (e.g., of oxygen) or, which is the same in the language of the physics of semiconductor surface, by charging of slow surface states. It is these processes that determine the steady state band bending.

For the case considered, Fig. 16 shows the energy diagram of a semiconductor (n-type for illustration) in contact with a solution. It has been assumed in constructing this diagram that a depletion layer is formed in the

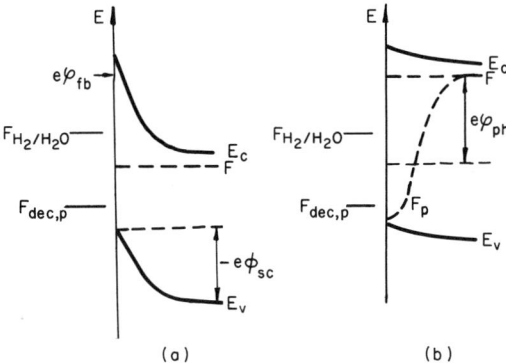

FIG. 16. Energy diagram of a semiconductor (n-type)–electrolyte interface: (a) in darkness, (b) under illumination.

surface region of the semiconductor for the steady state potential value. Light-generated electrons and holes move in the depletion layer electric field in opposite directions, namely, the holes move to the interface and the electrons move into the bulk of the semiconductor. The electric field, which results from this charge separation, compensates for the initial field, so the bands "unbend" (see Section 6). This, in turn, leads to a change in the mutual position of other energy levels in the system.

Assume for simplicity that the potential drop in the Helmholtz layer does not change under illumination, so that the position of the band edges at the surface is fixed with respect to the system of energy levels in the solution ("band pinning at the surface"; see Section 3). At the same time, the position of the Fermi level F relative to the band edges E_c and E_v is strictly determined in the semiconductor bulk. Therefore, the bands unbend under illumination and "pull" the Fermi level, so the latter shifts with respect to its position in the nonilluminated semiconductor (Fig. 16b). This shift can be measured as the photopotential $\varphi_{ph} = -\Delta F/e$ [cf. Eq. (8)].

Consider now the processes caused by the formation of quasilevels. As was noted above, the shift of F_n relative to F is very small for majority carriers (electrons) and can usually be neglected; precisely, this was done in constructing Fig. 16b. But for minority carriers (holes) the shift of F_p can be very large. The shifts of both $F_n \approx F$ and F_p increase with the growing intensity of semiconductor illumination, so that for a certain illumination intensity F_p may reach the level of the electrochemical potential of anodic decomposition $F_{dec, p}$, and F_n—the level of a certain cathodic reaction (for example, reduction of water with hydrogen evolution F_{H_2/H_2O}). These reactions start to proceed simultaneously, and their joint action constitutes the process of photocorrosion.

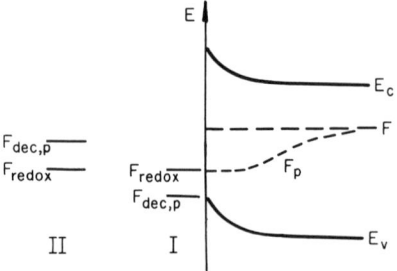

FIG. 17. Prevention of photocorrosion of a semiconductor by a redox couple in the solution: I—the semiconductor is stable, and II—the semiconductor is unstable against anodic photodecomposition.

In order to find out whether a semiconductor is stable with respect to photocorrosion, one has to turn back to Fig. 15 and conduct the same consideration as for the case of corrosion in darkness. In this case, however, it should be taken into account that other reactions, competing with photodecomposition of a semiconductor, are possible, besides decomposition of semiconductor and solvent. As the illumination intensity increases, out of all cathodic reactions possible the one with the least negative equilibrium potential may be the first to start, and out of all anodic reactions—that with the least positive equilibrium potential. This circumstance opens the way for controlling photocorrosion: it can, in principle, be suppressed by adding appropriate oxidizing or reducing agent to the solution. For example, anodic photocorrosion may be prevented with the help of substances that are oxidized easier than the semiconductor material: $\varphi^0_{\text{redox}} < \varphi^0_{\text{dec},p}$ (i.e., $F_{\text{redox}} > F_{\text{dec},p}$). Such protector substances stabilize semiconductor photoanodes (Fig. 17). In certain cases the solvent may act as a protector (see below).

By analogy, the stability condition for cathodic photocorrosion can be written as $\varphi^0_{\text{redox}} > \varphi^0_{\text{dec},n}$.

To illustrate the above approach, we present here the energy diagrams (taken from the paper by Gerischer, 1977a) for two of the most widely used semiconductor materials in aqueous solutions.

Titanium dioxide, TiO_2 (Fig. 18), is resistant against cathodic photocorrosion because the $F_{\text{dec},n}$ level lies in the conduction band (cf. Fig. 15b). But the $F_{\text{dec},p}$ level is in the forbidden band, so TiO_2 would be decomposed in the light if there were not a competing reaction of oxygen photoevolution from water with a higher level of electrochemical potential, F_{H_2O/O_2}. Therefore, titanium dioxide electrode is a very stable photoanode for photoelectrolysis of water (the latter acts here as a protector).

Cadmium sulfide, CdS (Fig. 19), is, on the contrary, liable to intensive anodic photocorrosion in aqueous solutions, according to the equation

$$CdS + 2h^+ \longrightarrow Cd^{2+} + S,$$

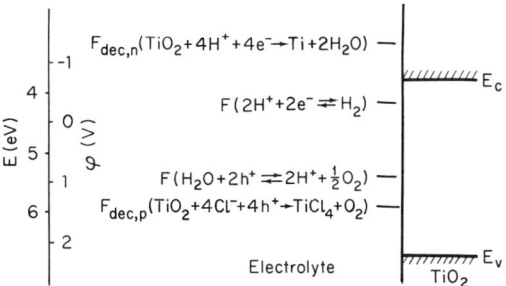

FIG. 18. Diagram of electrochemical potentials for reactions at a TiO_2 electrode in an aqueous solution (pH = 7). The potentials are given relative to the NHE. [From Gerischer (1977a).]

since $F_{dec, p} > F_{H_2O/O_2}$. But the corrosion can be prevented by the addition to the solution of sulphide ions S^{2-} that are oxidized (to polysulphide sulfur S_2^{2-}) easier than CdS. The system S^{2-}/S_2^{2-} and other similar systems (polyselenides, polytellurides) are widely used to stabilize photoanodes of $A^{II}B^{VI}$ and $A^{III}B^V$ semiconductors in photoelectrochemical cells for solar energy conversion (Heller and Miller, 1980; Manassen et al., 1976).

Let us note in conclusion that the thermodynamic approach has widely been used to describe the kinetics of electrochemical reactions at an illuminated semiconductor electrode (see, for example, Gerischer, 1977c; Dogonadze and Kuznetsov, 1975). Clearness and simplicity are an unqualified advantage of this approach, but the use of the quasilevel concept is not justified in all the cases. In particular, conditions (48) alone appear to be insufficient to substantiate the applicability of the quasilevel concept to the description of the processes of electron transfer across the interface (for greater details, see Pleskov and Gurevich, 1983; Nozik, 1978). Obviously, if photogeneration of the carriers occurs mainly near the surface, at which a

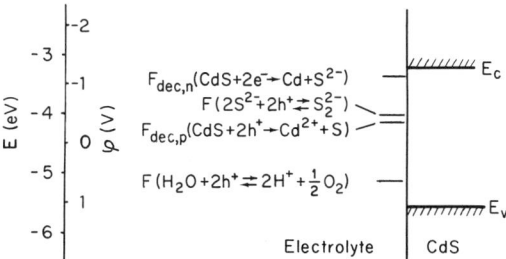

FIG. 19. Diagram of electrochemical potentials for reactions at a CdS electrode in 1 M aqueous solution of sulphide (pH = 12). The potentials are given relative to the NHE. [From Gerischer (1977a).]

fast electrode reaction takes place, the photoexcited carriers may be "extracted" from the semiconductor more rapidly than they come in equilibrium with thermalized carriers in the corresponding band. In this case it would be more correct to consider transitions of carriers from the electrode into solution as proceeding in a purely dynamic way, i.e., as separate independent events, rather than via a quasiequilibrium stage (Nozik, 1978; Boudreaux et al., 1980).

10. SOME KINETIC ASPECTS OF PHOTOCORROSION

It is quite natural that the thermodynamic approach does not allow photocorrosion processes to be described comprehensively. In a number of cases, kinetic peculiarities of reactions play an important role (see, for example, Bard and Wrighton, 1977); these peculiarities are caused by the effect of crystalline structure, state of the semiconductor surface, etc. A detailed description of a complicated reaction with several particles in the solution and crystal lattice involved usually encounters considerable difficulties. Therefore, at this stage the kinetic approach is used to reveal purely qualitative regularities of corrosion processes.

First, a complete photodecomposition reaction usually consists of several successive stages. It may appear that the equilibrium potential of the rate-determining stage does not coincide with that of the complete reaction. In this case the analysis, similar to the above one, should be performed just for the rate-determining stage (Gerischer, 1978).

Next, the kinetics of photodecomposition of a solid depends significantly on specific features of its electron structure. Within the framework of the concepts of quantum chemistry the rate of electron transitions at an interface depends on the extent to which the electron orbitals of the atoms overlap. Weak overlapping corresponds to a lesser probability of electron transition. Proceeding from such concepts, one might expect significant difference in the photocorrosion behavior of semiconductors with an "ordinary" (i.e., sp) type of photoexcitation of nonequilibrium carriers and semiconductors, in which photoexcitation is caused by d bands (d semiconductors). In fact, in materials such as CdS, CdSe, ZnO, etc., the conduction and valence bands are formed mainly by the orbitals of the electropositive and electronegative components, respectively. Under illumination, the holes are generated due to interband transitions between the valence band (the band of p electrons) and the conduction band (the band of s electrons). In this case the localization of the holes at interatomic bonds, which is equivalent to weakening of these bonds, is considered as a necessary condition for both anodic dissolution of a semiconductor and its photocorrosion (Gerischer, 1966, 1978).

In d semiconductors, however, another type of electron excitation is possible, namely, the transition between two d bands, both of which correspond to the metal orbitals. Since d electrons are localized at the corresponding atoms to a much greater extent than s and p electrons, their excitation has little effect on the atomic interaction. It was expected therefore that such excitation does not weaken significantly bonds in the crystal lattice so that the above materials are to be more resistant against photocorrosion.

Proceeding from these concepts, Tributsch (1977) suggested as possible materials for use in photoelectrochemical cells for solar energy conversion a new class of compounds: dichalcogenides of transition metals (MoS_2, $MoSe_2$, WS_2, etc.). The study of these d semiconductors has shown, however, that they are also liable to anodic photocorrosion. In order to explain this fact, it was assumed that photoholes, for example in MoS_2 or $MoSe_2$, oxidize water to OH^- radicals, which in turn interact chemically with surface S and Se atoms, thereby oxidizing them to sulphates and selenates. Therefore, to stabilize photoanodes of d semiconductors, one has to employ, like the case of sp semiconductors, redox systems.

At the same time, it should be noted that the photoelectrochemical behavior of d semiconductors has not so far been studied too comprehensively to make final conclusions about their photocorrosion behavior.

An interesting example of the kinetic effect in semiconductor photocorrosion is photopassivation and photoactivation of silicon (Izidinov et al., 1962). Silicon is an electronegative element, so it should be dissolved spontaneously and intensively in water with hydrogen evolution. But in most of aqueous solutions the surface of silicon is covered with a nonporous passivating oxide film, which protects it from corrosion. The anodic polarization curve of silicon (dashed line in Fig. 20) is of the form characteristic of electrodes liable to passivation: as the potential increases, the anodic current first grows (the

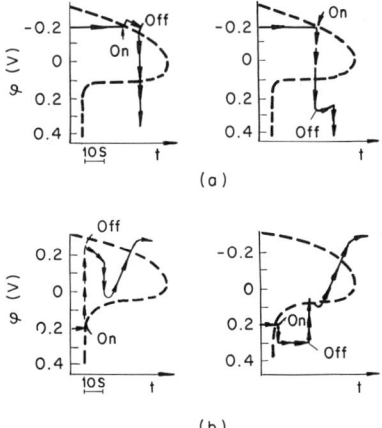

FIG. 20. Change in the silicon potential with time under its photopassivation (a) and photoactivation (b) in a 10 M KOH (20°C). The dashed line represents the anodic polarization curve. [From Izidinov (1979).]

region of active dissolution), then the growth becomes slower, the current passes through a maximum, drops, and thereafter almost does not depend on the potential (the passive region).

A nontrivial feature of a silicon electrode in alkaline aqueous solutions is its ability to pass reversibly, under illumination, from the passive state to the active one, and vice versa. For example, suppose that the initial state is actively dissolved silicon; under illumination its potential spontaneously and sharply shifts (at a constant current) to more positive values, i.e., into the "passive" region, and self-dissolution ceases: photopassivation occurs (Fig. 20a). In contrast, once silicon has already been anodically passivated, illumination shifts its potential to less positive values (Fig. 20b). In this case, the point on the dashed line, which characterizes the state of the system, passes from the "descending" branch to the "ascending" one, and active self-dissolution starts, i.e., photoactivation takes place.

Though processes occurring under photopassivation have not so far been understood in detail, they may be related with certainty (Izidinov, 1979) to the acceleration, under illumination, of one of the two conjugated reactions, which constitute the overall process of electrochemical corrosion. Depending on the initial state of corroding silicon, either the anodic (at the active surface) or the cathodic (at the passive surface) partial reaction is accelerated. This leads to the shift of the potential, and the system "jumps" over the maximum of the polarization curve from one stable state to the other.

V. Light-Sensitive Etching of Semiconductors

Among the methods of anodic and chemical etching of semiconductors, widely used both in the production of semiconductor devices and in investigations (see, for example, Schnable and Schmidt, 1976; Turner and Pankove, 1978), the so-called light-sensitive etching is of great importance. It is based on the variation, under illumination, of the concentration of minority carriers, which often determines, as was shown above, the rate of anodic dissolution and corrosion of semiconductors.

Two quite distinct problems are identified:

(1) To obtain uniform etching over the whole surface (for instance, in order to produce a mirror-smooth surface), irrespective of local properties of the sample. With n-type samples this is achieved by intensive uniform illumination (see, for example, Sullivan et al., 1963).

(2) To enhance etching selectivity, in order to reveal local inhomogeneities of the material or to produce on its surface a certain relief. The discussion to follow deals mainly with this latter problem.

11. QUALITATIVE PICTURE OF PROCESSES TAKING PLACE AT A SEMICONDUCTOR–ELECTROLYTE INTERFACE UNDER ITS NONUNIFORM ILLUMINATION

Let us distinguish between the three most important cases.

a. Illumination Under Anodic Polarization

With n-type electrodes in the limiting current regime the rate of etching is limited by the supply of holes to the electrode surface. In darkness, the main source of holes is generation in the bulk of a sample and/or in the space-charge region, and also at the surface. Under steady state conditions a certain, usually low, dissolution rate is attained. Illumination serves as an additional source of holes (photogeneration), so that the illuminated regions are dissolved faster than nonilluminated ones. Examples will be given in Section 13.

b. Illumination with Open External Circuit in an Indifferent Electrolyte

Illumination of a semiconductor electrode leads to the occurrence of photopotential (see Section 6). Nonuniform illumination of a sample, which does not react with the solution in darkness, produces along its surface the gradients of both the electrode potential and minority-carrier concentration. Therefore, nonequivalent conditions for electrochemical reactions, dissolution in particular, are created in illuminated and nonilluminated regions of the surface. In n-type samples the illuminated regions act as local anodes and are dissolved, whereas in nonilluminated regions, local cathodes, processes such as hydrogen evolution, reduction of dissolved oxygen, etc., take place. On the contrary, in p-type samples nonilluminated regions act as anodes, and illuminated regions as cathodes.

c. Illumination in an Oxidizing Solution

Here corrosion occurs even in darkness. In the simplest case where the partial cathodic reaction proceeds exclusively through the conduction band and the anodic reaction through the valence band, the corrosion rate is limited, as was shown in Section 8, by the supply of minority carriers to the surface, irrespective of the type of sample conductivity. Therefore, in darkness the corrosion rate is low. Illumination accelerates corrosion. This case is similar to case (a), but with the difference that the role of anodic polarization is played by "chemical" polarization with the help of an oxidizer introduced into the solution (see Section 13 for examples).

Out of the above three methods of light-sensitive etching, the one with polarization from an external source [case (a)], called photoanodic etching, and etching in oxidizing solution [case (c)], called photochemical etching, are now of the greatest practical importance.

12. PRINCIPLES OF THE THEORY OF LASER ETCHING

Intensive development of quantum optics during the past decade has resulted in the emergence of a new field—laser electrochemistry of semiconductors. Two research teams have contributed significantly to the formation of this field (for reviews, see Tyagai et al., 1978; Belyakov et al., 1979b). It covers a range of problems associated with electrochemical processes at semiconductor electrodes stimulated by laser radiation. In particular, light-sensitive etching with the use of coherent laser radiation can be employed for recording optical information, for example, for producing holograms.

Photoelectrochemical profiling of semiconductors in interfering beams was first obtained by Dalisa et al. (1970) and by Dalisa and de Biteto (1972), who produced by this technique the simplest holograms, namely phase diffraction gratings.

Let us consider in brief the main principles of the theory of laser etching, following mainly the papers by Tyagai et al. (1978) and Belyakov et al. (1979a). Let two coherent monochromatic beams of light strike at equal angles the initially flat surface of an n-type semiconductor; the beams are assumed to lie in the same plane and have the same intensity and wavelength λ (see Fig. 21). The distribution of the nonzero component of the light wave strength $\mathscr{E}$ (s wave is considered for illustration) is described by the equation (see, for example, Born and Wolf, 1964)

$$\left(\frac{\partial^2}{\partial x^2} + \frac{\partial^2}{\partial y^2} + \frac{\partial^2}{\partial z^2}\right)\mathscr{E}^{(j)} + \frac{\omega^2}{c^2}[\mathfrak{n}^{(j)}]^2\mathscr{E}^{(j)} = 0. \qquad (51)$$

Here $\mathfrak{n}^{(j)} = n_1^{(j)} + in_2^{(j)}$ is the complex refractive index of the medium ($n_{1,2}^{(j)} > 0$) and j takes two values: $j = 1$ corresponds to the electrolyte solution, $j = 2$ to the semiconductor. The solution of Eq. (51) together with matching conditions at the interface completely describes the distribution of $\mathscr{E}^{(j)}$. The change in the shape of the surface relief $X(y, z, t)$ due to etching alters this distribu-

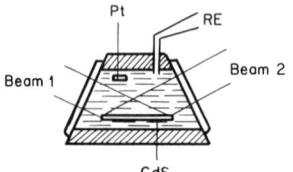

FIG. 21. Schematic diagram of an electrochemical cell for laser photoanodic etching. RE denotes electrolytic bridge to the reference electrode.

tion. The velocity of interface motion in its absolute value is given by

$$v(y, z, t) = \frac{k_{el}\Omega}{\nu} p. \tag{52}$$

Here k_{el} is the electrode reaction rate constant, Ω is the volume per one semiconductor particle (atom or molecule), which leaves the solid and penetrates into the solution, p is the concentration of holes at the interface, $p = p(X(y, z, t), y, z)$, and ν is the number of holes per one particle dissolved. In the quasistationary case (i.e., when the condition $v \ll L_p/\tau_p$ is satisfied) the distribution of holes is described by the equation similar to Eq. (29):

$$D_p\left(\frac{\partial^2}{\partial x^2} + \frac{\partial^2}{\partial y^2} + \frac{\partial^2}{\partial z^2}\right)p - \frac{p - p_0}{\tau} + \alpha \frac{c}{n_1^{(2)}} |\mathscr{E}^{(2)}(x, y, z)|^2 = 0. \tag{53}$$

This equation allows for the fact that the density of light flux $\mathscr{J}$ is equal to the squared modulus of the wave electric field in a semiconductor; $\mathscr{E}^{(2)}$ is found by solving Eq. (51).

Boundary conditions for Eq. (53) are given by the following relations:

$$D_p(\mathbf{r} \cdot \nabla p)|_{X = X(y,z)} = -(k_{el} + s_{eff})p(X, y, z), \tag{54}$$

where $\mathbf{r}$ is the unit vector normal to the interface and s_{eff} is a phenomenological parameter describing the effective rate of hole recombination determined by processes which are not connected with the electrode reaction, and $p = p_0$ for $x \to \infty$.

Finally, the change in the relief $X(y, z, t)$ depends on the quantity $v(y, z, t)$ according to the relation

$$\frac{\partial X}{\partial t} = v(y, z, t)\left[1 + \left(\frac{\partial X}{\partial y}\right)^2 + \left(\frac{\partial X}{\partial z}\right)^2\right]^{-1/2}, \tag{55}$$

which follows from differential geometry. In the one-dimensional case (flat boundary) $\partial X/\partial y = \partial X/\partial z = 0$, so that $\partial X/\partial t = v(t)$.

The system of equations (51) and (53) with boundary conditions (52), (54), (55) and matching conditions for the field $\mathscr{E}^{(j)}$ at the moving interface $X(y, z, t)$ permits complete mathematical formulation of a self-consistent problem.

Considerable difficulties arise in finding an exact solution, but at the initial stage of the process when the interface differs little from a plane, analytical expressions can be obtained. If the incident light beams are flat, the relief of the surface is described by the relation (Tyagai et al., 1978) in an appropriate coordinate system:

$$X(y, t) = 2\mathscr{J}_0 \frac{\Omega}{\nu} rt\left[1 + q\cos\left(\frac{2\pi y}{d}\right)\right]. \tag{56}$$

Here $\mathscr{J}_0$ is the density of light flux penetrating into the semiconductor, $d = \lambda/2n_1^{(2)}\cos\theta$ represents the spatial period of the occurring sinusoidal relief (θ is the incidence angle). The dimensionless parameters, r and q, that characterize the intensity of light-sensitive etching (r) and quality of the interference pattern recording (q) are equal to

$$r = \frac{\alpha L_p}{1 + \alpha L_p} \frac{k_{el}}{k_{el} + s_{eff} + D_p/L_p}, \qquad q = \frac{1 + \alpha L_p}{\gamma + \alpha L_p} \frac{k_{el} + s_{eff} + D_p/L_p}{k_{el} + s_{eff} + \gamma D_p/L_p}. \qquad (57)$$

The quantity γ is given by the relation

$$\gamma = [1 + (2\pi L_p/d)^2]^{1/2} \qquad (58)$$

and characterizes the degree of smearing of the periodic distribution caused by hole diffusion. In the limit $\gamma \to \infty$ ($L_p \gg d$), $q = 0$, according to Eq. (57); that is, the relief is not formed. In the inverse limit $\gamma \to 1$ ($L_p \ll d$) the parameter q is maximum; i.e., $q = 1$.

Let us note that if the inequalities $\alpha L_p \gg \gamma$ and $k_{el} \gg \gamma(D_p/L_p)$ are satisfied, the quantities γ and L_p disappear from the expressions for q and r. In this limiting case $q = 1$ and $r = k_{el}/(k_{el} + s_{eff})$. The physical meaning of this result is quite clear: if the holes are generated near the surface (α is large) and are immediately consumed (k_{el} is large), the diffusion of photogenerated holes into nonilluminated regions of the surface can be neglected.

Note also that X grows linearly with t at the initial stage of the process.

Relations (56), (57), and (58) imply that in order to attain a maximum intensity of photoetching and high quality of interference pattern recording the following requirements should be met:

(a) materials with not very low value of L_p and, if possible, small value of s_{eff} should be used for recombination not to reduce the process intensity;

(b) highly absorbed light should be employed in order to weaken the role of hole recombination in the semiconductor bulk;

(c) the composition of the solution should be chosen so as to provide for a sufficiently high rate of electrode reaction.

The effect of certain factors related to the nonlinearity of photoetching and assuring a feedback between the relief formed and the conditions of its formation, which may be significant for relatively large times elapsed, is analyzed in the paper by Tyagai et al. (1978).

The effect of drift spreading of the carriers over the surface should also be considered, in order to estimate the resolution of light-sensitive etching. If the rate of consumption of holes in the electrode reaction is less than that of their supply due to drift they may be accumulated in illuminated regions. This will result in a change in the total surface charge density, and, as was noted above, an additional potential difference arises between the illuminated and nonilluminated regions, which pulls the carriers into the nonilluminated regions, to balance their concentrations. This drift spreading (self-induced drift) leads, obviously, to the smearing of the diffraction pattern.

Mathematical analysis of this problem has been performed by Belyakov *et al.* (1979a) on the basis of the equations of transport along the surface with due account for carrier migration in the self-induced electric field $\mathscr{E}(y)$.

Assuming that a depletion layer is formed near the surface and using perturbation theory to find the surface concentration of holes modified by surface drift, we obtain

$$p(y) = \mathscr{I}_0 \frac{L_{sc}^{(p)}}{s_{el}}\left[1 + \xi \cos\left(\frac{2\pi}{d}y\right)\right]. \tag{59}$$

Here $L_{sc}^{(p)} = L_{sc}kT/(2e|\phi_{sc}|)$, where L_{sc} is the thickness of the space-charge region (given by Eq. (20a)), and s_{el} is the rate of "electrochemical recombination" (see the end of Section 8). The coefficient ξ in Eq. (59) is similar, in the physical meaning, to the coefficient q in Eq. (56) and characterizes the quality of hologram recording under conditions considered. If the intensity of the recording light beam is given by the relation $\mathscr{I}(y) = \mathscr{I}_0[1 + \zeta \cos(2\pi d/y)]$, we have for ξ (Belyakov *et al.*, 1979a)

$$\xi = \zeta\left(1 + \frac{4\pi^2 D_{eff} L_{sc}^{(p)}}{d^2 s_{el}}\right)^{-1}, \tag{60}$$

where D_{eff} is the effective coefficient of hole diffusion along the surface; it consists of two terms, one of which is equal to D_p and the other contains a factor dependent on the potential ϕ_{sc}.

As it follows from the method of derivation of expressions (59) and (60), they are valid, strictly speaking, only if $\xi \ll 1$. This condition is satisfied not only in the trivial case of $\zeta \ll 1$ (that is, for weakly modulated illumination intensity), but also for $\zeta \simeq 1$ if d is sufficiently small, i.e., in a rather important case of high spatial frequencies d^{-1}. From the physical point of view, this is related to the fact that for high d^{-1} values the spreading of nonequilibrium holes is so large that their spatial modulation becomes quite weak.

Proceeding from Eqs. (59) and (60), one may formulate conditions that are imposed on the characteristic parameters of a semiconductor and a solution in order to enhance the resolution. It follows from the definition of $L_{sc}^{(p)}$ that $L_{sc}^{(p)} \sim (|\phi_{sc}|N_D)^{-1/2}$. This relation shows that frequencies, at which light-sensitive etching is possible, are the higher

(a) the more heavily the semiconductor is doped,
(b) the larger the bands are bent at the surface, and
(c) the higher is the rate of electrochemical reactions; this condition also follows from considerations presented on p. 298.

It will be shown below that all these regularities have been confirmed experimentally.

13. Photoanodic and Photochemical Recording of Holograms

Let us briefly describe, as an example of *photoanodic* etching, the production of a periodic relief on the surface of a cadmium sulphide (CdS) crystal (Tyagai et al., 1978). A light beam from a helium–cadmium laser (the wavelength = 441.6 nm) was first expanded to a diameter of 15 mm and then split, by reflection from two mirrors, into two beams of equal intensity, which intersected at the surface of an electrode emerged into an electrolyte solution (Fig. 21). Under anodic oxidation of CdS (the equation of the corresponding reaction was given on p. 290) a layer of amorphous sulfur is formed on the electrode surface. In order to eliminate this layer, the etching is performed in electrolyte solutions, in which sulfur is either dissolved (NaOH) or oxidized (HNO_3).

The quality of the relief thus obtained can be determined directly in the course of etching because the light reflected from the etched grating produces a diffraction pattern that characterizes the depth and shape of the grating profile at any given moment. To this end, one has just to measure the intensity of the reflected light. The method possesses a high sensitivity: the occurrence of the relief can be detected when its depth is only of the order of 0.01 μm.

A characteristic feature of the process is the existence of a certain "optimal" amount of electricity, Q_{optim}, which is needed in the course of etching in order to obtain a relief that would be most close to the sinusoidal one. This is manifested in the fact that the intensity of the diffraction relief $\mathcal{I}_{diffr}$, as a function of the amount of electricity required Q, passes through a maximum (curve 1 in Fig. 22) at $Q = Q_{optim} = i^a \times t_{optim}$—where t_{optim} is an optimal etching time for the mean density of anodic photocurrent i^a. The value of Q_{optim} depends neither on the spatial frequency of the etched grating nor on the intensity of the recording laser beams and is equal to about 5×10^{-2} C/cm².

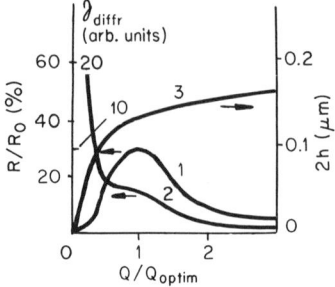

FIG. 22. Holographic photoanodic etching of CdS: the diffraction reflex intensity $\mathcal{I}_{diffr}$ (curve 1), the ratio of the coefficients of mirror reflection from the etched (R) and original, nonetched (R_0) surface (curve 2), and the depth of the sinusoidal relief $2h$ (curve 3) versus the ratio of the passed amount of electricity Q to an optimal amount (Q_{optim}). [From Tyagai et al. (1978).]

At the initial recording stage, for $t < t_{optim}$, the depth of the etched groove is approximately proportional to the etching time t, and for $t > t_{optim}$ the relief depth tends to a limit (curve 3 in Fig. 22). The profile of the relief is quite close to the sinusoidal one; but at the same time there is undesirable random etching of the surface that leads to a decrease in its reflectivity (curve 2 in Fig. 22).

Under *photochemical* etching, special demands should be made on the oxidizing agent in the solution. Above all, it must be reduced with the participation of conduction-band electrons alone (see above) and at potentials less negative than the potential of reducing water or H^+ ions to molecular hydrogen, in order to avoid gas evolution.

As in the case of photoanodic etching, a certain optimal time exists here, during which the etched relief attains the most regular shape; a maximum value of diffraction efficiency corresponds to this time.

The etching process may be characterized by an amount of light energy received by a sample for the optimal etching time. For some semiconductor materials this amount is equal to approximately 1 J/cm². The minimum operating illumination intensity, at which the rate of photoetching exceeds noticeably the rate of dark corrosion, is 10^{-4} W/cm².

The two methods considered here have been used for etching diffraction gratings on several single-crystal semiconductors (Ge, Si, GaAs, GaP, CdS, CdSe), and also on semiconductor films, including polycrystalline ones (CdSe, CdTe, As_2Te_3, InSe, GaSe, Cu_2O, GeO) and glassy (Ag_2Se_3). The potentialities of the methods are illustrated in Table I, which lists for certain of the above materials the maximum attained spatial frequencies d^{-1} of the recorded gratings, the wavelengths λ of recording light, and the diffusion lengths L_p of minority carriers in the samples.

The data presented in the Table show that the resolution of photoelectrochemical recording of information, defined as the number of lines of an etched grating per 1 mm, is sufficiently high, more than 6000 mm⁻¹. For certain semiconductors it has already reached the "optical" limit associated with restrictions caused by optical characteristics of the interference pattern for wavelengths employed. This limit corresponds to the case $q = 1$, which, according to Eq. (57), takes place when the conditions of "surface" absorption of light ($\alpha L_p > 1$) and good drain of photogenerated holes to the surface $[k_{el} \gg \gamma(L_p/\tau_p)^{1/2}]$ are satisfied. Let us emphasize that for materials with large diffusion length L_p (germanium, silicon), the resolution appeared to be much higher than it might be expected from L_p values, as predicted by the theory.

Comparison of both the methods—photoanodic and photochemical etching—shows (see Table I) that they give approximately the same results, as far as resolution is concerned. Photochemical etching is a more universal method because it is equally applicable to the treatment of both *n*- and *p*-type

TABLE I

Characteristics of Diffraction Gratings Obtained by the Light-Sensitive Etching Technique

Semiconductor	d^{-1} (mm^{-1})	λ (nm)	L_p (cm)	Reference
\multicolumn{5}{c}{Photoanodic etching}				
CdSe (single crystal)	2600	632.8	10^{-4}	Tyagai et al. (1973)
CdSe (film)	2000	632.8	10^{-5}	Sterligov and Tyagai (1975)
CdS	6150	441.6	10^{-4}	Sterligov et al. (1976)
Si	2500	514.5	10^{-2}	Dalisa et al. (1970)
As$_2$Se$_3$	2600	632.8	—	Belyakov et al. (1974)
\multicolumn{5}{c}{Photochemical etching}				
GaAs	5000–6000	441.6	10^{-4}	Belyakov et al. (1979a)
GaAlAs	2500	632.8	—	Belyakov et al. (1979c)
GaAsP	2700	514.5	—	Bykovsky et al. (1978)
As$_2$S$_3$	3500	441.6, 514.5	—	Bykovsky et al. (1978)
As$_2$Te$_3$	2600	632.8	—	Belyakov et al. (1974)
Ge	1000	632.8	10^{-1}	Belyakov et al. (1976)
Cu$_2$O	2600	—	—	Belyakov et al. (1974)
GeO	3000	441.6	—	Vlasenko et al. (1978)

materials. Photoanodic etching, on its part, has an advantage of less severe requirements for electrolyte composition.

As for comparing these methods with most wide-spread nonelectrochemical methods, such as recording of holograms on photoplates and photolithography (i.e., etching through a mask produced by the photoresist technique), all the methods have approximately the same resolution. Maximally attained spatial frequencies (about 6000 mm^{-1}) are limited by possibilities of producing high spatial frequencies of the light beam, rather than possibilities of recording. As this parameter is further increased, the photoetching methods, as it may be expected, should become more advantageous over, in particular, photolithography, since the relief is formed directly on the semiconductor surface rather than in an additional photoresist layer. This same reason makes it easier to obtain a regular sinusoidal profile of the diffraction grating.

In conclusion it should be stressed that gratings recorded by the method of light-sensitive etching of semiconductors (unlike those recorded on photoemulsions, electrochromium magnetic films, and other materials) possess semiconducting properties besides the given optical characteristics. This circumstance gives such gratings quite a new range of applications. As an example, we may note a laser with distributed feedback on the basis of epitaxial GaAs films, in which the surface of a thin-film waveguide with etched sinusoidal relief played the role of distributed mirrors of an ordinary optical laser (Alferov et al., 1974). Such gratings have also been used in devices of diffraction input and output of radiation in certain elements of integral optics (see the paper by Belyakov et al., 1979a and references therein).

VI. Processes Caused by Photoexcitation of Reactants in the Solution

If a solution, being in contact with an electrode, contains photosensitive atoms or molecules, irradiation of such a system may lead to photoelectrochemical reactions or, to be more exact, electrochemical reactions with excited particles involved. In such reactions the electrons pass either from an excited particle to the electrode (the anodic process) or from the electrode to an excited particle (the cathodic process). In this case, an elementary act of charge transfer has much in common with ordinary (dark) electrochemical redox reactions, which opens a possibility of interpreting certain aspects of photochemical processes under consideration with the use of concepts developed for general quantum mechanical description of electrode processes.

14. QUALITATIVE PICTURE OF SENSITIZATION

Essentially, photoprocesses discussed here and photocurrents caused by these processes frequently occur in that spectral region where the semiconductor itself is not photosensitive (i.e., below the threshold of intrinsic absorption, $\hbar\omega < E_g$). This phenomenon forms the basis of photosensitization of semiconductors to a longer wavelength light (in particular, of wide-gap semiconductors that absorb UV light, and to the Sun's light, which is used in the conversion of solar energy). The corresponding photosensitive reactants in the solution are called sensitizers (Gerischer et al., 1968; Hauffe and Range, 1968; Tributsch and Gerischer, 1969).

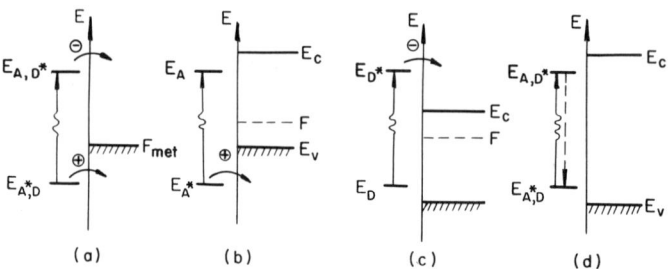

FIG. 23. Diagram of reactions with photoexcited reactants at a metallic (a), a semiconductor (b), (c), and an insulating (d) electrode.

It should be stressed that semiconductor, rather than metallic, electrodes are most convenient in order to observe and investigate electrochemical reactions of excited reactants in the solution. This can be explained within the framework of the quasithermodynamic approach suggested by Gerischer (1977b) with the help of Fig. 23, which shows the energy diagrams of a metal (a), a semiconductor (b) and (c), and an insulator (d), being in contact with a solution, together with energy levels of the ground and photoexcited states of the substance in the solution.

In the case of a metallic electrode (Fig. 23a), such a location of energy levels of the substance (e.g., donor) in the solution is the most probable when the ground level lies below and the excited one above the Fermi level in the metal F_{met}. Against each donor level (the ground or excited) stands a continuum of energy states in the metal; in the first case these states are occupied by the electrons, and in the second case they are vacant. This provides for fast exchange between the electrons across the interface. Transitions of electrons between isoenergy levels have a maximum probability to occur, so that the electron from an excited level of, say, the donor E_{D^*}, passes quite easily into the metal. However, the ground level E_D that becomes vacant after photoexcitation is also easily occupied by an electron from the metal conduction band. As a result of simultaneous occurrence of these two processes, the energy of light is converted into the kinetic energy of electrons in the metal, i.e., eventually into heat. In other words, photoexcitation of a reactant in the solution is efficiently quenched at a metallic electrode by conduction electrons of the metal.

In the case of a semiconductor electrode, the existence of the energy gap makes a qualitatively different location of energy levels quite probable (Figs. 23b, 23c). One of them, either the ground or excited, is just in front of the energy gap, so that the direct electron transition with this level involved appears to be impossible. This gives rise to an irreversible photoelectrochemical reaction and, as a consequence, to photocurrent i_{ph}. The photoexcited particle injects an electron into the semiconductor conduction band

and becomes oxidized (Fig. 23c) or, on the contrary, it captures an electron from the valence band (i.e., injects a hole) and is reduced (Fig. 23b).

Finally, in an insulator (a very wide-gap semiconductor to be exact) there is a large probability of an inverse, with respect to metal, limiting case (Fig. 23d). Here the energy gap is so wide that it exceeds considerably the energy of electron excitation ($|E_{A^*} - E_A|$ or $|E_D - E_{D^*}|$). In this case both energy levels of the reactant in the solution lie in front of the energy gap, so that electron exchange with the electrode becomes impossible. The energy of photoexcitation is thus transferred to solvent molecules.

Regeneration of consumed (i.e., given off an electron to the electrode or, on the contrary, acquired an electron) photoactive substance (sensitizer) in the solution is a very important matter from the practical point of view. As soon as all the near-the-electrode (adsorbed) layer of this substance is oxidized (or reduced) the photoprocess ceases. To obtain a continuous photocurrent, the amount of the initial reactant, sensitizer, near the electrode surface should be renewed.

To this end, a substance called supersensitizer is added to the solution. This substance does not absorb light and does not react at the electrode; it neither interacts with the sensitizer, but can oxidize the reduced sensitizer (or reduce the oxidized one). In order for this to occur, the redox potential of the supersensitizer should be either more positive than the potential of the system sensitizer plus its reduced form (in the case of hole photoinjection), or more negative than the potential of the system sensitizer plus its oxidized form (in the case of electron photoinjection). The energy diagram of supersensitization is shown in Fig. 24. The net result of the photoprocess is a light-stimulated reaction of oxidation or reduction of nonphotosensitive substance supersensitizer; the sensitizer, on the contrary, is not consumed and serves as if it were a catalyst of the photoprocess.

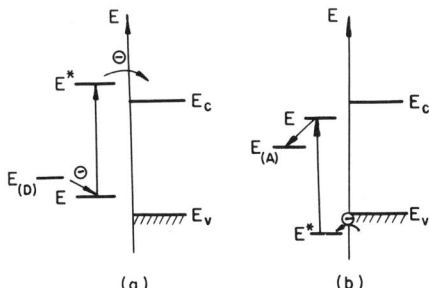

FIG. 24. Schematic diagram of supersensitization process with the help of a donor reactant (a) and an acceptor reactant (b). $E_{(D)}$ is the energy level of the donor supersensitizer, $E_{(A)}$ is the energy level of the acceptor supersensitizer, and E and E^* are the ground and excited levels of the sensitizer.

Photosensitive substances adsorbed on the semiconductor surface are especially efficient in sensitization reactions. Thus, sensitizing effect can be enhanced if a sensitizer is "attached" to the semiconductor surface by a chemical bond. For this purpose one has to create either the ether bond –O– between the semiconductor and reactant, using "natural" OH groups, which exist on the surface of, for example, oxide semiconductors (TiO_2, ZnO) or oxidized materials (Ge, GaAs, etc.) in aqueous solutions, or the amide bond –NH–; in the latter case a monolayer of silane compounds with amidogroups is preliminarily deposited on the semiconductor surface (see, for instance, Osa and Fujihira, 1976). With such "chemically modified" electrodes the photocurrent is much higher than with "ordinary" (naked) semiconductor electrodes.

15. Some Experimental Results

At present, there are a lot of publications concerning photosensitization of semiconductors by photosensitive reactants, which, for the most part, absorb visible light—that is, represent dyes (see, for example, Memming, 1974; Hauffe and Bode, 1974; Honda et al., 1978). The following dyes are used as sensitizers: the cyanine dyes and methylene blue on CdS; alizarin and rhodamin B on TiO_2; Bengal rose on ZnO; bipyridyl complexes of Ru(II); chlorophyll and its analogs on SnO_2; and others.

In principle, the same photoexcited reactant, if it is liable both to oxidation and reduction, can inject both electrons (into an n-type semiconductor) and holes (into a p-type semiconductor). Such a material is, for example, the crystal-violet dye. Figure 25 shows the spectra of cathodic photocurrent i_{ph} at p-type gallium phosphide and anodic photocurrent at n-type zinc oxide both in a solution, which does not absorb light (dashed lines), and in the presence of crystal violet; the absorption spectrum of the latter is also shown for comparison.

In the absence of the dye the photocurrent is observed only in relatively short-wavelength light in the region of intrinsic absorption of semiconductors and is caused by photoexcitation of the electron–hole system of the semiconductor (cf. Part III). For zinc oxide ($E_g = 3.2$ eV) this region corresponds to wavelengths shorter than 400 nm; for gallium phosphide ($E_g = 2.2$ eV), shorter than 550 nm.

In the presence of the dye the photocurrent at both semiconductors is observed in the visible region of the spectrum, i.e., in that region where crystal violet absorbs light. Thus, sensitization is observed. Photoprocesses that take place at zinc oxide and gallium phosphide can be represented by schemes of Figs. 23c and 23b, respectively.

The density of photosensitized current increases with dye concentration and in some cases depends on solution pH (see Fig. 26). The latter circumstance is

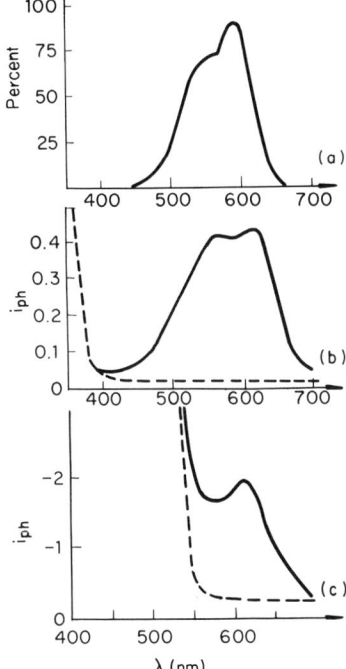

FIG. 25. Absorption spectrum of the crystal-violet dye (a) and the spectra of anodic photocurrent at an electrode of n-type ZnO (b) and of cathodic photocurrent at an electrode of p-type GaP (c) in the presence of crystal violet (in $\mu A/cm^2$). The dashed line is the photocurrent in the absence of the dye. [From Gerischer (1977b).]

related to the fact that pH can affect (through the acid-base dissociation of OH groups at the surface of oxide or oxidized semiconductors) the flat band potential of a semiconductor, as shown in Fig. 27 (cf. Fig. 4). In this case a change in pH leads to the variation of the mutual position of energy levels of the photosensitive substance in the solution and of the semiconductor levels. For example, in going over from acid to alkaline solution the flat band potential of TiO_2 becomes more negative and the corresponding shift of energy levels of the semiconductor relative to those of the solution impedes the

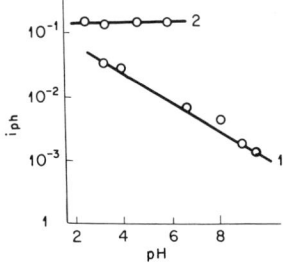

FIG. 26. Dependence of photosensitized current (arb. units) on solution pH at TiO_2 (1) and CdS (2) electrodes sensitized by Rhodamin B. [From Watanabe et al. (1976).]

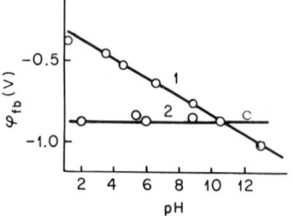

FIG. 27. Dependence of the flat band potential of TiO_2 (1) and CdS (2) electrodes on solution pH. [From Watanabe et al. (1976).]

photoinjection of electrons from photoexcited molecules of the dye into the semiconductor. On the contrary, in the case of CdS whose flat band potential does not depend on pH, the efficiency of the photoinjection of electrons is the same in acid and alkaline solutions.

Reducing agents, such as hydroquinone, Br^-, I^-, SCN^- ions, and others, are used as supersensitizers for electron photoinjection reactions. Figure 28 illustrates how an admixture of hydroquinone affects photocurrent in the system CdS—rhodamine B. In certain cases the solvent, water, can also act as a supersensitizer. For example, if the bipyridyl complex of Ru(II) is a sensitizer, the oxidized form, the complex of Ru(III), can oxidize water to oxygen

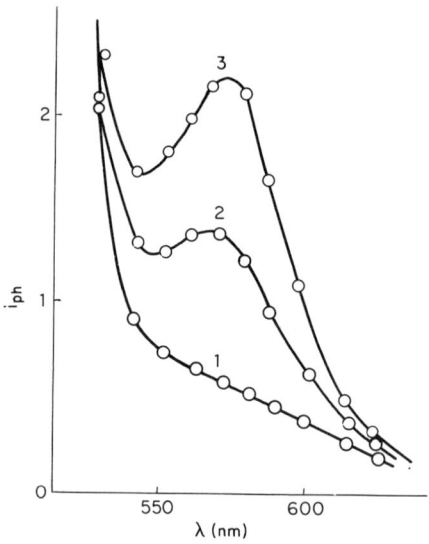

FIG. 28. Sensitization and supersensitization of a CdS electrode in 0.2 M solution of Na_2SO_4. 1—anodic photocurrent (in $\mu A/cm^2$) in the absence of sensitizer, 2—the same, but in the presence of rhodamin B (2.5×10^{-5} mol/l), 3—the same, but in the presence of rhodamin B (2.5×10^{-5} mol/l) and hydroquinone (2.5×10^{-3} mol/l). [From Watanabe et al. (1975).]

(Creutz and Sutin, 1975):

$$\text{Ru(bipy)}_3^{2+} \xrightarrow{h\omega} \text{Ru(bipy)}_3^{3+} + e^-,$$
$$2\text{Ru(bipy)}_3^{3+} + 2\text{OH}^- \longrightarrow 2\text{Ru(bipy)}_3^{2+} + \tfrac{1}{2}\text{O}_2 + \text{H}_2\text{O}. \tag{61}$$

Quantum yield of the sensitized photocurrent is usually small, of the order of 0.01. Such a low efficiency can be explained by the fact that the layer of the adsorbed dye, taking part in the photoprocess, is very thin—of the order of the molecule size—so it absorbs only an insignificant fraction of the incident light. But this layer cannot be made thicker, because the organic substance (the dye) is an insulator. Thus, photosensitization by adsorbed dyes appears to have a relatively low efficiency for sensitizing wide-gap semiconductors to visible (solar) light.

A semiconductor electrode with an adsorbed dye (sensitizer) may serve as an example for constructing the simplest model of processes similar, in certain respects, to those taking place in natural photosensitizing objects, for instance in green plants. According to generally accepted concepts (see, for example, Lehninger, 1972), the processes of energy transfer and electron transfer occur jointly in the course of photosynthesis. It is the light-absorbing pigment (for example, chlorophyll) that contains molecules of two types: the so-called antennas, which provide for the absorption of light quanta, and reaction centers, in which the charges are separated (i.e., the electron process proper occurs). The interaction between the antennas and reaction centers is assumed to proceed via energy transfer.

Figure 29 shows schematically a device that simulates the above process (Fromherz and Arden, 1980). A semiconductor electrode (SnO_2) is separated from the solution by a lipid membrane with a thickness of several nanometers, which prevents the direct electron interaction between the electrode and solution. Docosylamine, arachidate, and other substances are used as a membrane material. Dye I (hydroxycoumarone) is adsorbed on the outer surface of the membrane. It absorbs the incident light (with the energy of

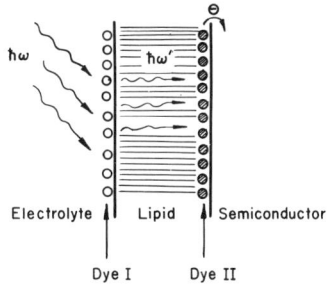

FIG. 29. Diagram of an electrode for simulating successive transfer of energy and electron in photosynthesis processes. [From Fromherz and Arden (1980).]

quanta of $\hbar\omega$), the process being accompanied with luminescence (the energy $\hbar\omega'$ of emitted quanta is less than $\hbar\omega$). Dye II (cyanine) is adsorbed at the membrane–semiconductor interface; this dye is unsensitive to light (with the energy of quanta of $\hbar\omega$) incident on the electrode, but it absorbs the luminiscence light. As a result, dye II is excited and injects the electrons into the conduction band of SnO_2.

In other words, the energy is transferred in the membrane from dye I, playing the role of an antenna, to dye II, which acts, together with the semiconductor, as a reaction center, i.e., transfers the electrons. The spectral characteristics of the system as a whole are determined by dye I, which sensitizes the semiconductor to visible light, though it is not in a direct electron contact with the semiconductor.

In spite of the fact that the model is, evidently, approximate (natural photosensitizing objects are unlikely to contain a macroscopic semiconductor phase), it allows qualitative determination of a possible relationship between individual stages of energy and electron transfer under photosynthesis.

VII. Selected Problems in the Photoelectrochemistry of Semiconductors

This concluding part deals with a number of trends in photoelectrochemistry of semiconductors that have not so far been widely developed, the reasons often being of momentary character. At the same time, these trends are not only of significant scientific interest, but some of them may, in prospect, form the basis for important practical applications. In this respect, it appears to be quite reasonable to discuss the existing problems and the most important results.

16. Photoelectron Emission from Semiconductors into Solutions

All the above-considered photoelectrochemical phenomena are based on the transition of light-excited electrons into a localized state in the solution, namely at the energy levels associated with individual ions or molecules. However, the phototransition is also possible when the electrons pass into a qualitatively different delocalized state in the solution; it is this type of phototransition that represents photoemission (Barker et al., 1966). The emitted delocalized electron in the solution is then thermalized and localized to form a solvated (hydrated in aqueous solution) electron. The energy level, which corresponds to the solvated electron, lies below the bottom of the band of permitted delocalized states in the solution. Finally, the electron may pass from the solvated state to an even lower local energy level associated with an electron acceptor in the solution (see Fig. 30).

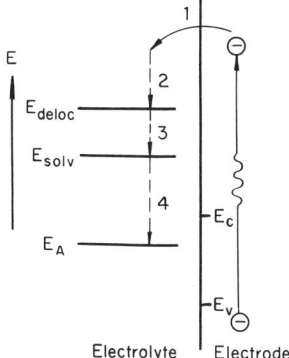

FIG. 30. Diagram of transitions related to photoemission into solution: 1—photoexcitation of an electron and photoemission, 2—thermalization of the photoemitted electron in the solution, 3—solvation of the thermalized electron, and 4—trapping of the solvated electron by acceptor A in the solution. E_{deloc} is the lower edge of the band of delocalized states in the solution, E_{solv} is the energy level of the solvated electron, and E_A is the acceptor energy level.

The works dealing with photoemission from metals (Gurevich et al., 1980; Sass and Gerischer, 1978) have shown that photoemission into solutions is quite an efficient and flexible method of studying the structure of an interface, various physicochemical transformations occurring at this interface and near it, and also the mechanisms of photoexcitation of a solid.

The photocurrent I due to photoemission of electrons from simple metals into an electrolyte solution in the near-the-threshold frequency range is given by the relation (Gurevich et al., 1967)

$$I = A(\hbar\omega - \hbar\omega_0 - e\varphi)^{5/2}, \qquad (62)$$

which is known as the five-halves power law. Here $\hbar\omega_0$ is the so-called threshold frequency (red boundary) of the external photoeffect defined by the relation $\hbar\omega_0 = w_T^{(m,s)}$, where $w_T^{(m,s)}$ is the thermodynamic work function in a metal–solution system for a certain electrode potential φ taken as the origin. If $\omega < \omega_0$ at $\varphi = 0$, then $I = 0$. Finally, A is a φ-independent quantity proportional to the intensity of the incident light flux.

Specific features of photoemission from semiconductors, as compared to that from metals, are caused by the following circumstances.

First, the level F, whose position determines the thermodynamic work function w_T, is located in the case of semiconductors in the forbidden band. The energy characteristics of a semiconductor–electrolyte interface under photoemission are presented in Fig. 31, which shows, in particular, that the threshold frequency is given by the relation $\hbar\omega_0 = E_g + \chi$, where χ is the difference between the potential energy level of a delocalized electron outside

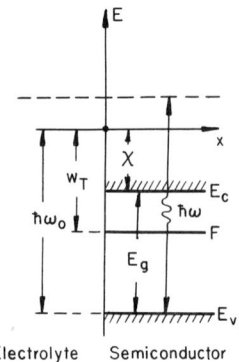

FIG. 31. Energy diagram of a semiconductor–electrolyte interface under photoemission. The potential energy of the delocalized electron in the solution E_{deloc} is taken as the origin.

the semiconductor E_{deloc} and the energy of the conduction-band bottom E_c near the surface (χ is called the generalized electron affinity). The quantity $w_P = E_g + \chi$ is called the photoemission (nonequilibrium) work function; $w_P > w_T$. Second, the bulk photoexcitation of the valence-band electrons that plays the major role in the generation of emitted electrons is accompanied by interband transitions, which may be both direct and indirect. In this case, the quantum energy $\hbar\omega_0 = w_P$ appears to be sufficient for photoemission if only indirect transitions occur (see Fig. 32). On the contrary, for direct transitions (which are usually most probable) the threshold energy of the absorbed quantum $\hbar\omega_0' > w_P$, the difference $\hbar(\omega_0' - \omega_0)$ being dependent both on the dispersion relations for the carriers in the bands and on the orientation of the emitter surface relative to the crystallographic axes. In the simplest model this difference is $(m_c/m_v)\chi$, where $m_{c,v}$ is the effective mass of the electrons and holes, respectively.

Finally, if a semiconductor electrode (unlike a metallic one) is polarized, the quantity $\Delta\varphi$ is distributed in a complicated manner between the space-charge

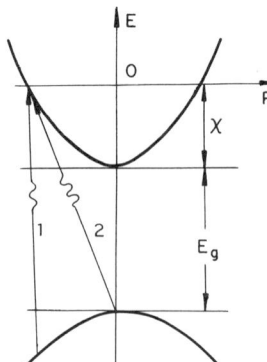

FIG. 32. Diagram of interband electron transitions under photoemission: 1—direct transition, 2—indirect transition.

region in the semiconductor and the Helmholtz layer. It follows from considerations of Section 3 and Fig. 5 that in this case χ varies according to the relation

$$\chi(\varphi) = \chi(0) + e(\delta\varphi), \tag{63}$$

where $\chi(0)$ is the value of χ at the potential conditionally taken as zero and $\delta\varphi = \Delta\phi_H$ is an excess potential drop (in comparison with the initial one) in the Helmholtz layer.

Propagation of photoelectrons in the bulk to the emitting surface is quite an important point. One may believe that this motion is described, with one modification or another, as random migrations of a point center in a medium (see, for example, James and Moll, 1969). Another approach (Kane, 1962) is based on the concepts of propagation of the probability wave described by quantum mechanical equations. The first approach is valid if most of the photoelectrons experience, after their excitation, a large number of interactions. The second approach holds true if the emitted electrons leave the crystal with practically unchanged quantum characteristics acquired under photoexcitation. This is precisely the case where the absorption depth of light exciting photoelectrons is less than the photoelectron free path. Specific features of the semiconductor–electrolyte interface, in comparison, for example, with the semiconductor–vacuum interface, are manifested in peculiarities of the shape of the surface potential barrier and, in particular, in the absence of long-range image forces at this interface.

Calculation of the current I of photoelectron emission from nondegenerate semiconductors into electrolyte solutions, performed within the framework of quantum mechanical wave approach, yields the following expression (Gurevich, 1972):

$$I = \begin{cases} 0, & \omega < \omega_0 \\ B(\hbar\omega - \hbar\omega_0)^2, & \omega_0 < \omega < \omega'_0 \\ C(\hbar\omega - \hbar\omega'_0)^{3/2}, & \omega > \omega'_0. \end{cases} \tag{64}$$

Here B and C are quantities proportional to the probabilities of indirect and direct transitions, respectively, so that $B \ll C$. According to the above considerations, the two threshold frequencies ω_0 and ω'_0 are given by

$$\hbar\omega_0 = E_g + \chi, \quad \hbar\omega'_0 = E_g + \left(1 + \frac{m_c}{m_v}\right)\chi. \tag{65}$$

If the simple effective mass approximation does not hold, the expression $1 + m_c/m_v$ is replaced by some other quantity, which also exceeds unity.

If the light is mainly absorbed within the space-charge region ($\alpha^{-1} < L_{sc}$), the dependence of the photocurrent I on the potential φ is obtained by substituting χ from Eq. (63) into Eq. (65) and then into Eq. (64). For example, in the most important case $\omega > \omega'_0$ we find (Gurevich et al., 1977)

$$I = C[\hbar\omega - \hbar\omega'_0(0) - e(1 + m_c/m_v)\delta\varphi]^{3/2}. \tag{66}$$

The dependence of I on ω is described, according to Eq. (66), by the three-halves power law, while the dependence of I on φ may be of a more complicated character, since $\delta\varphi$ is, in general, a complicated function of φ. Note also that the coefficients at φ and $\delta\varphi$ in Eqs. (62) and (66), respectively, do not coincide, namely, the factor before $\delta\varphi$ in Eq. (66), unlike the factor before φ in Eq. (62), depends on the ratio of effective masses.

Experimental observation of photoemission currents encounters the problem of separating them from the currents of photoelectrochemical reactions of nonemission nature, which are caused by the internal photoeffect in the semiconductor (see, for example, Section 5). Photoprocesses of both the types start similarly with the interband excitation of an electron and are of threshold character with respect to the frequency of light, but the threshold quantum energy is different for these processes. Namely, the threshold of photoemission exceeds that of the internal photoeffect (and hence the threshold of "ordinary" photoelectrochemical reactions) by the value of the electron affinity to the semiconductor χ (see Figs. 31 and 32).

Moreover, specific sensitivity of photoemission currents to solution composition or, more exactly, to the presence in the solution of acceptors of solvated electrons ("scavengers") may be used to identify and separate them. The fact is that, as was mentioned above, both the initial state of the emitted electron (the delocalized electron) and its subsequent state (the solvated electron) have relatively high energy levels (see Fig. 30). As a result, such electrons are efficiently trapped by the electrode surface, so that the resultant current is zero in the steady state regime. Observation of the steady state photocurrent requires prevention of back-trapping of the electrons penetrated into the solution by the electrode surface. To this end, the electrons should be transferred rapidly to a lower energy level in the solution for the corresponding state to be stable with respect to such trapping. Therefore, scavengers are introduced into the solution, which efficiently trap the solvated (hydrated) electrons (but do not trap electrons from the electrode, i.e., they are not cathodically reduced). H^+ ions, N_2O molecules, and others are efficient scavengers in aqueous solutions. Thus, the photoemission current at a semiconductor electrode is measured as the difference between photocurrents in the presence and in the absence of scavengers (Krotova and Pleskov, 1973; Gurevich et al., 1977).

Figure 33 presents dependences of the photoemission current $I^{2/3}$ at a p-type gallium arsenide electrode on the quantum energy (for a given potential) and on the potential (for a given quantum energy) in an aqueous solution. Both

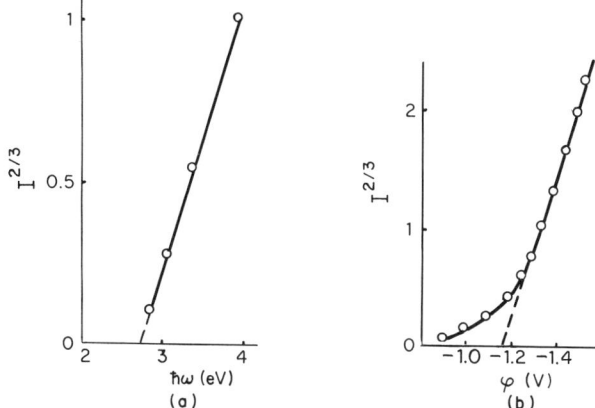

FIG. 33. Three-halves power law: dependence of the photoemission current to the power 2/3 on the energy of light quantum for a given potential (a) and on the potential for a given energy of light quantum (b). p-type gallium arsenide ($N_A = 6 \times 10^{18}$ cm^{-3}) in aqueous solution. [From Boikova et al. (1976).]

the first dependence and (for more negative potentials) the second one are described by the three-halves power law [see Eq. (66)], i.e., they correspond to direct transitions. The threshold energy, as can be seen from Fig. 33, is 2.7 eV, which exceeds considerably the energy gap for gallium arsenide ($E_g = 1.43$ eV). The deviation of the experimental curve from the extrapolated straight line (dashed line) is related to redistribution of the potential at the interface and may be used for estimating ϕ_{sc}.

In addition to the study of photoemission from semiconductors into aqueous solutions, investigations of photoemission into nonaqueous solutions, for example, from Si and GaP into liquid ammonia, have recently begun (Malpas et al., 1979; Krohn and Thompson, 1979).

17. SEMICONDUCTOR ELECTROCHEMICAL PHOTOGRAPHY

In conventional halogen–silver photomaterials the silver halogenide performs three functions (corresponding to three main stages of photographic process), namely: (1) it provides for photosensitivity, (2) it forms a latent image, and (3) it forms a visible image. The third stage, i.e., the intensification of the latent primary image, is performed via the so-called physical development. It usually represents catalytic deposition of metallic precipitate, the particles in the latent image (called "development centers") serving as nucleation centers in crystallization.

Halogen–silver photoemulsions contain a large extra amount of silver, which is needed for the formation of visible image; the excess silver is removed after development and can be reclaimed. However, "silver-free" techniques of producing photographic image have long been searched. The basic idea of these techniques is to distribute the above three functions between two or three different substances, so the silver compounds are assumed to perform only the second function. The third function (the formation of visible image) associated with large consumption of the metal is performed with the help of less expensive and less scarce metals.

Semiconductors are believed to be quite promising among systems, which are potentially capable of replacing silver-containing photosensitive materials (see, for example, Sviridov and Kondratyev, 1978; Kiess, 1979). A semiconductor–electrolyte interface is directly used in certain techniques of producing latent and visible images.

The image is usually formed via photoelectrochemical deposition of metals (Pd, Ag, etc.). Metallic particles either directly form the image of a sufficient optical density (Goryachev et al., 1970, 1972) or serve as nuclei of crystallization in the course of development (Kelly and Vondeling, 1975). Photoelectrochemical reactions of colored organic compounds may also be employed for producing image (see, for example, Reichman et al., 1980).

Two schemes of image formation at the surface of a semiconductor electrode have been described in the literature. In one of the schemes the semiconductor–electrolyte interface operates in the limiting current regime. This means that for a given potential the reaction with minority carriers involved proceeds at the electrode. Therefore, a very low current passes through the electrode in darkness, which is determined by the bulk generation of minority carriers. When an optical image is projected onto the electrode surface the concentration of minority carriers and the reaction rate increase in the illuminated areas (cf. Section 11a). For example, in p-type samples cathodic electrodeposition of metal is accelerated; in n-type samples—anodic reaction, for instance—oxidation of Pb^{2+} with the formation of a colored PbO_2 layer at the surface. The substrates of PbSe, PbS, Si, GaP, etc., are used as a "photographic plate" (Goryachev et al., 1972; Inoue et al., 1979).

In the other scheme the photosensitive interface operates in the photogalvanic pair regime (Goryachev et al., 1970; Goryachev and Paritsky, 1973). The method is based on the occurrence of a potential difference between illuminated and nonilluminated areas on the surface of a semiconductor electrode in a solution (cf. Section 11b). As a result of nonuniform illumination, local anodes and cathodes arise on the surface and this, in turn, leads to nonuniform deposition of metal onto the surface.

Note that ZnO or TiO_2 "photographic plates" in an electrolyte solution exhibit a memory effect: the latent image arises on illumination in the absence of metal ions in the solution, and it can be developed subsequently in darkness by placing the exposed sample into the metal salt solution. The mechanism of this effect has not so far been understood completely; it may be expected that in this case either the surface states are charged or photoadsorption processes take place under illumination (Fateev et al., 1979; Davidson et al., 1979).

18. Radiation Electrochemistry of Semiconductors

Let us compare how the radiation of visible and UV regions (considered above) and hard radiation interact with a semiconductor–electrolyte system. The energy of x-ray and gamma quanta, and also of high energy electrons, can be as high as hundreds of thousands and even millions of electron-volts, which is many orders of magnitude higher than the energy gap of semiconductor materials. Therefore, irradiation will necessarily lead to irreversible changes in the system: the formation of crystal-lattice defects in the semiconductor and radiolytic decomposition of the solution.

In a number of cases, however, these changes are little or, at least, do not have a direct effect on the electrochemical behavior of a system; here the generation of nonequilibrium electron–hole pairs in the semiconductor becomes the major effect (see, for example, Byalobzhesky, 1967; Oshe and Rosenfeld, 1978).

There is an analogy in the behavior of a semiconductor (for instance, titanium dioxide) electrode when it is illuminated by light or irradiated by a beam of high energy electrons or gamma quanta. In both cases the water is anodically oxidized with oxygen evolution (the rate of the process being negligibly low in the absence of irradiation) and the shape of the $i_{ph} - \varphi$ curve is practically the same (Fig. 34), as well as its dependence on solution pH. For low dose rates the reaction rate is proportional to the dose rate, indicative of the determining role in the radiation-electrochemical (as well as in photoelectrochemical) behavior of semiconductors of an increase of the concentration of holes in the valence band (Krotova et al., 1981). At the same time, there are long-term changes in the electrode photoactivity caused, probably, by the creation of new donor defects under irradiation (Krotova et al., 1979).

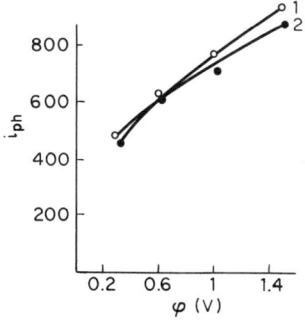

Fig. 34. Dependence of the current (in μA) of anodic oxidation of water with oxygen evolution at a TiO_2 electrode on the potential in an 0.2 M solution of KCl: ○—under illumination with UV light, ●—under irradiation with a 4.2-MeV electron beam. [From Krotova et al. (1981).]

The acceleration of electrode processes at irradiated semiconductors opens the way, at least in principle, for directly converting the energy of ionizing radiation into chemical energy of electrolysis products (quite similar to the case of solar energy conversion); this acceleration can also be used as a means for detecting the radiation.

19. ELECTROGENERATED LUMINESCENCE

The general nature of all photoelectrochemical phenomena lies in mutual transformation of energy. For example, the above-considered processes are based on the transformation of the energy of light into electrical and/or chemical energy. Let us now turn to the inverse transformation, namely of electrical energy into the energy of light, the latter being manifested as electrogenerated luminescence, i.e., luminescence, which arises when current flows through a semiconductor–electrolyte interface.

In the course of an electrode reaction excited states can arise, which are deactivated via radiation of light quanta. These states may be divided into two types: (a) excited electrons and holes in the semiconductor and (b) excited products in the solution near the electrode surface.

The occurrence and deactivation of excited states of the first type are schematically shown in Fig. 35. Let the minority carriers (holes) be injected into the semiconductor in the course of an electrode reaction (reduction of substance A). The holes recombine with the majority carriers (electrons). The energy, which is released in the direct band-to-band recombination, is equal to the energy gap, so that we have the relation $\hbar\omega = E_g$ for the emitted light quantum (case I). More probable, however, is recombination through surface or bulk levels, lying in the forbidden band, which successively trap the electrons and holes. In this case the excess energy of recombined carriers is released in smaller amounts, so that $\hbar\omega' < E_g$ (case II in Fig. 35). Both these types of recombination are revealed in luminescence spectra recorded with n-type semiconductor electrodes under electrochemical generation of holes (Fig.

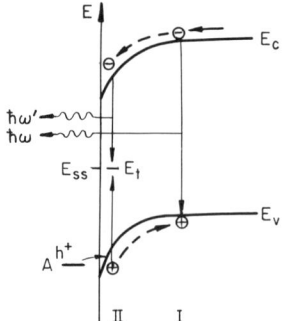

FIG. 35. Occurrence of electrogenerated luminescence with the participation of excited charge carriers in a semiconductor: without (I) and with (II) energy levels in the forbidden band involved; E_{ss} and E_t are the energies of surface and bulk levels, respectively.

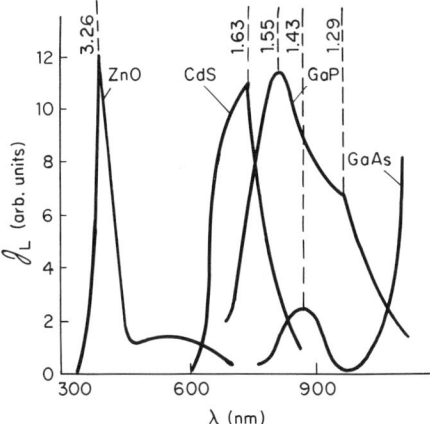

FIG. 36. Electrogenerated luminescence spectra for semiconductor electrodes. $\mathscr{I}_L$ is luminescence intensity; energies corresponding to maxima on the curves are shown in the figure (in electron-volts). [From Pettinger et al. (1976).]

36). In the case of ZnO and GaAs the energy of quanta of luminescence light is close to the energy gap for these materials. On the contrary, for CdS and GaP the energy of quanta of emitted light is much less than the energy gap. Luminescence caused by the injection of minority carriers from the electrolyte has been observed for a number of semiconductor electrodes (see, for example, Pettinger et al., 1974; Morisaki and Yasawa, 1978; Richardson et al., 1981).

Thus, the study of electrogenerated luminescence spectra may give information on the energy of recombination centers. This method was used to study recombination properties of photoelectrodes in cells for solar energy conversion (Ellis and Karas, 1980).

The second type of luminescence can be observed under electroreduction of organic substances (Dehmlow et al., 1975; Luttmer and Bard, 1979). This luminescence is shown schematically in Fig. 37. An electron from the conduction band passes into the excited (for instance, triplet) state of the organic substance $^3R^*$, which is then deactivated returning to the ground state

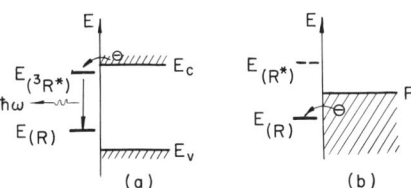

FIG. 37. Occurrence of electrogenerated luminescence with the participation of photoexcited reactants in the solution: (a) semiconductor electrode, and (b) metal electrode.

R with the emission of a light quantum. If the electron could pass directly into the ground state (as is usually the case with metal electrodes), no luminescence would occur. However, in the case of a semiconductor electrode the ground state is often located in the band gap, so that the corresponding electron transition is forbidden.

An example of such a process is electroreduction of the bipyridyl complex of ruthenium (III) at CdS, ZnO, TiO_2, SiC, or GaP electrodes in aqueous or acetonitrile solutions:

$$Ru(bipy)_3^{3+} + e^- \longrightarrow {}^3[Ru(bipy)_3^{2+}]^* \longrightarrow \hbar\omega + Ru(bipy)_3^{2+}, \qquad (67)$$

which is a source of a characteristic luminescence (Gleria and Memming, 1976; Luttmer and Bard, 1978).

20. Electroreflection at a Semiconductor–Electrolyte Interface

The electroreflection effect is a change in the reflection coefficient of light due to a change in the electrode potential. A relative simplicity of modulating the surface potential in the semiconductor–electrolyte system makes the technique of studying semiconductors via electroreflection the most popular among the following modulation techniques: frequency, temperature, mechanical restriction, and others (Cardona, 1969). The study of electrooptic effects at a semiconductor–electrolyte interface is a new method, which complements significantly conventional methods of investigating the physics and chemistry of both bulk and (in recent years) surface properties of semiconductors.

The physical reason for the modulation of the reflection coefficient R is a change, due to the electric field, in the complex dielectric permittivity of the semiconductor near the interface. The field may affect the reflection coefficient via several mechanisms: redistribution of free carriers near the interface; action on interband electron transitions, which effectively reduces the energy gap (the Franz–Keldysh effect); an increase of the direct transition probability in the presence of the field (the field acts here as a "third body"); creation of additional energy levels contributing to the absorption of light; the effect of impurities on photoabsorption spectra in the space-charge region; etc. Moreover, the adsorption properties of the surface contacting with the electrolyte solution may vary due to the electric field, leading in turn to a change in optical properties (Tyagai and Snitko, 1980).

Inhomogeneity of the field-induced change in the characteristics of the medium, the complex dielectric permittivity $\hat{\varepsilon}_{sc} = \varepsilon_1 + i\varepsilon_2$ in particular (here $\varepsilon_{1,2}$ are real quantities), is a distinguishing feature of electrooptic effects in the space-charge region. The ranges of such inhomogeneities (10^{-4}–10^{-5} cm) are comparable with the wavelengths of the incident light, so that the formulas of geometrical optics cannot be applied directly. On the other hand, the effect of an electric field on the reflection coefficient R observed in experiment is rather small, $\Delta R/R \ll 1$, so that the methods of perturbation theory can be used. Thus, the relative change, due to the field, in the reflection coefficient $\Delta R/R$ can be expressed as (Cardona, 1969; Tyagai, 1970)

$$\Delta R/R = \alpha(\varepsilon_1, \varepsilon_2)\langle\Delta\varepsilon_1\rangle + \beta(\varepsilon_1, \varepsilon_2)\langle\Delta\varepsilon_2\rangle, \qquad (68)$$

where the averaged quantities are defined by Eq. (69) below. The coefficients $\alpha(\varepsilon_1, \varepsilon_2)$ and $\beta(\varepsilon_1, \varepsilon_2)$ appearing in Eq. (68) are equal to $\alpha = d\ln R/d\varepsilon_1$ and $\beta = d\ln R/d\varepsilon_2$ and do not depend, by definition, on φ, but are functions of the radiation frequency. The real quantities $\langle\Delta\varepsilon_1\rangle$ and $\langle\Delta\varepsilon_2\rangle$ can be determined from the relation (see Seraphin 1972; also Aspnes and Bottka 1972)

$$\langle\Delta\varepsilon_1\rangle + i\langle\Delta\varepsilon_2\rangle = -\frac{2i\omega\mathfrak{n}}{c}\int_0^\infty [\Delta\varepsilon_1(x) + i\,\Delta\varepsilon_2(x)]e^{2i\omega\mathfrak{n}x/c}\,dx. \qquad (69)$$

Here c is the velocity of light, $\mathfrak{n}$ is the complex refractive index in the semiconductor bulk

$$\mathfrak{n} = n_1 + in_2 = (\varepsilon_1 + i\varepsilon_2)^{1/2}, \qquad (70)$$

and $\Delta\varepsilon_1$, $\Delta\varepsilon_2$ are coordinate-dependent changes, due to the field, in the quantities ε_1 and ε_2 near the surface.

In order to calculate the integral in Eq. (69), one has to determine the form of the functions $\Delta\varepsilon_{1,2}(x)$. Since the coordinate dependence of the electric field $\mathscr{E}(x)$ in the space-charge region is assumed to be known, one has actually to specify the dependence $\Delta\varepsilon_{1,2}(\mathscr{E})$.

As an important example, let us consider the effect of electroreflection due to inhomogeneity of the distribution of free carriers in the space-charge region of a semiconductor (plasma electroreflection). The contribution of the electrons to the complex dielectric permittivity (an n-type semiconductor is considered for illustration and the contribution of the holes is neglected) is given by the expression (see, for example, Ziman, 1972)

$$\varepsilon_1 + i\varepsilon_2 = 1 + \frac{e^2 n}{\varepsilon_0 m_c \omega^2}\frac{i\omega\tau}{1 - i\omega\tau}. \qquad (71)$$

Here n is the concentration of the electrons and τ is their relaxation time in the conduction band.

In the space-charge region the coordinate dependence of $n(x)$ is given, according to Eq. (13), by the relation $n(x) = n_0 \exp(e\phi(x)/kT)$. Substituting this into Eq. (71) and taking into account that the magnitude of n in Eq. (71)

is equal to n_0 in the absence of the field, for $\Delta\varepsilon_1$ and $\Delta\varepsilon_2$, i.e., coordinate—dependent changes in ε_1 and ε_2, we obtain

$$\Delta\varepsilon_1 + i\Delta\varepsilon_2 = -\left(\frac{\omega_p}{\omega}\right)^2 \frac{\omega\tau}{i+\omega\tau}(e^{e\phi(x)/kT}-1), \qquad (72)$$

where $\omega_p^2 = e^2 n_0/\varepsilon_0 m_c$ is the plasma electron frequency. Substitution of Eq. (72) into Eq. (69) and then into Eq. (68) permits $\Delta R/R$ to be calculated. Various particular cases have been analysed by Dmitruk and Tyagai (1971). For example, since it is usually assumed that $\omega\tau \gg 1$, we obtain for light of a sufficiently large wavelength λ (so that $\lambda \gg L_{sc}$) for an arbitrary band bending

$$\frac{\Delta R}{R} = 4\pi\left(\frac{\omega_p}{\omega}\right)^2 \frac{\Gamma_n}{\lambda n_0}(n_1\beta - n_2\alpha). \qquad (73)$$

Here Γ_n is the surface excess of the electrons (see Garrett and Brattain, 1955):

$$\Gamma_n = n_0 \int_0^\infty (e^{e\phi(x)/kT}-1)\,dx. \qquad (74)$$

Let us note that according to Eq. (73) the sign of plasma electroreflection is determined by the sign of ϕ_{sc}. The quantity $\Delta R/R$ is positive if $\phi_{sc} > 0$ (accumulation layer) and is negative if $\phi_{sc} < 0$. In other words, the flat band potential of the semiconductor can be determined from the condition that the sign of $\Delta R/R$ changes.

The above considerations are illustrated by Fig. 38, which shows the dependences of electroreflection of a silicon electrode on its potential. In fact, the signal changes the sign at the flat band potential (the latter was measured independently by the differential capacity technique).

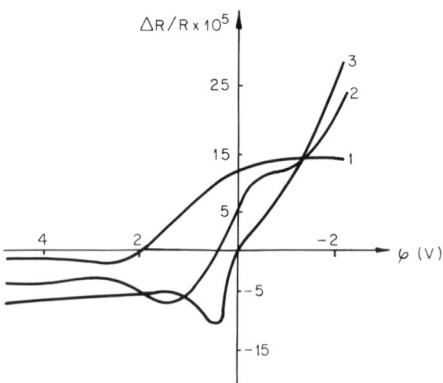

FIG. 38. Dependence of electroreflection amplitude on the potential for the silicon electrode. $\hbar\omega = 3.42$ eV; the oxide film thickness (nm): 1—120; 2—35; 3—17. [From Tyagai and Snitko (1980).]

It should be noted that dielectric and optical properties of the near-the-surface layer of a semiconductor, which vary in a certain manner under the action of electric field, depend also on the physicochemical conditions of the experiment and on the prehistory of the semiconductor sample. For example, Gavrilenko *et al.* (1976) and Bondarenko *et al.* (1975) observed a strong effect of such surface treatment as ion bombardment and mechanical polishing on electroreflection spectra. The "damaged" layer, which arises in the electrode due to such treatments, has quite different electrooptic characteristics in comparison with the same semiconductor of a perfect crystalline structure (see also Tyagai and Snitko, 1980).

Observing the variation of electroreflection spectra in the course of electrochemical treatment of electrodes, one may determine qualitatively reversible and irreversible changes in the near-the-surface layer, for instance, partial decomposition of ZnSe under anodic polarization (Lemasson *et al.*, 1980) and deposition of metal atoms on the surface of ZnO under cathodic polarization (Kolb, 1973).

VIII. Conclusion: Problems and Prospects

At present there is a sufficiently complete picture of photoelectrochemical behavior of the most important semiconductor materials. This is not, however, the only merit of photoelectrochemistry of semiconductors. First, photoelectrochemistry of semiconductors has stimulated the study of photoprocesses on materials, which are not conventional for electrochemistry, namely on insulators (Mehl and Hale, 1967; Gerischer and Willig, 1976). The basic concepts and mathematical formalism of electrochemistry and photoelectrochemistry of semiconductors have successfully been used in this study. Second, photoelectrochemistry of semiconductors has provided possibilities, unique in certain cases, of studying thermodynamic and kinetic characteristics of photoexcited particles in the solution and electrode, and also processes of electron transfer with these particles involved. (Note that the processes of quenching of photoexcited reactants often prevent from the performing of such investigations on metal electrodes.) The study of photoelectrochemical processes under the excitation of the electron–hole ensemble of a semiconductor permits the direct experimental verification of the applicability of the Fermi quasilevel concept to the description of electron transitions at an interface.

The second area of activity also includes some problems in laser electrochemistry of semiconductors, which are in no way confined to the above-considered light-sensitive etching. First of all, it is threshold electrochemical reactions stimulated by intensive laser radiation. Such reactions may proceed via new routes, because both highly excited solution particles and nonequilibrium electron–hole plasma of a semiconductor electrode are involved. This may open up, in particular, the possibility of developing new original processes in chemical technology.

There is a possibility, in principle, of developing the inverse process; namely, the emission of coherent light stimulated by an electrochemical reaction. This process could form the basis for a new type of laser with electrochemical pumping.

While considering trends in further investigations, one has to pay special attention to the effect of electroreflection. So far, this effect has been used to obtain information on the structure of the near-the-surface region of a semiconductor, but the electroreflection method makes it possible, in principle, to study electrode reactions, adsorption, and the properties of thin surface layers. Let us note in this respect an important role of objects with semiconducting properties for electrochemistry and photoelectrochemistry as a whole. Here we mean oxide and other films, polylayers of adsorbed organic substances, and other materials on the surface of metallic electrodes. Anomalies in the electrochemical behavior of such systems are frequently explained by their semiconductor nature. Yet, there is a barrier between electrochemistry and photoelectrochemistry of crystalline semiconductors with electronic conductivity, on the one hand, and electrochemistry of oxide films, which usually are amorphous and have appreciable ionic conductivity, on the other hand. To overcome this barrier is the task of further investigations.

Let us also note the effect of photoelectrochemical noise (Tyagai, 1974, 1976); the study of this noise at a semiconductor–electrolyte interface may give important information, both of scientific character—on the kinetics of complicated photoelectrochemical reactions, and of applied character—on the sensitivity and accuracy of various photoelectrochemical devices.

Finally, a promising trend in the study of the photoelectrochemical behavior of objects, whose nature is close to that of semiconductors, is related to photobiology, in particular to processes of light conversion in natural photosynthesizing objects. Certain elementary stages of photosynthesis, particularly photoelectrochemical ones, can, apparently, be simulated in some cases within the framework of the concepts of photoelectrochemistry of semiconductors.

References

Alferov, Zh. I., Gurevich, S. A., and Kazarinov, R. F. (1974). *Fiz. Tekh. Poluprovodn.* (*Leningrad*) **8**, 2031.
Aspnes, D. E., and Bottka, N. (1972). *In* "Semiconductors and Semimetals" (R. K. Willardson and A. C. Beer, eds.), Vol. 9, p. 471. Academic Press, New York.
Bard, A. J. (1980). *Science* **207**, 139.

Bard, A. J., and Wrighton, M. S. (1977). *J. Electrochem. Soc.* **124**, 1706.
Barker, G. C., Gardner, A. W., and Sammon, D. C. (1966). *J. Electrochem. Soc.* **113**, 1182.
Becquerel, E. (1839). *C. R. Hebd. Seances Acad. Sci.* **9**, 145.
Belyakov, L. V., Goryachev, D. N., Ostrovsky, Yu. I., and Paritsky, L. G. (1974). *Zh. Nauchn. Prikl. Fotogr. Kinematogr.* **19**, 54.
Belyakov, L. V., Goryachev, D. N., Paritsky, L. G., Ryvkin, S. M., and Sreseli, O. M. (1976). *Fiz. Tekh. Poluprovodn. (Leningrad)* **10**, 1142.
Belyakov, L. V., Goryachev, D. N., Ryvkin, S. M., Sreseli, O. M., and Suris, R. A. (1979a). *Fiz. Tekh. Poluprovodn. (Leningrad)* **13**, 2173.
Belyakov, L. V., Goryachev, D. N., and Sreseli, O. M. (1979b). *In* "Problemy Fiziki Poluprovodnikov," p. 5. Fizikotechnicheskii Institut im. A. F. Ioffe, Leningrad.
Belyakov, L. V., Goryachev, D. N., and Sreseli, O. M. (1979c). *Zhurn. Tekh. Fiz.* **49**, 876.
Boikova, G. V., Krotova, M. D., and Pleskov, Yu. V. (1978). *Elektrokhimiya* **12**, 922.
Bondarenko, V. N., Evstigneev, A. M., Tyagai, V. A., and Snitko, O. V. (1975). *Izv. Akad. Nauk. SSSR., Neorgan. Mater.* **11**, 342.
Born, M., and Wolf, E. (1964). "Principles of Optics." Pergamon, Oxford.
Boudreaux, D. S., Williams, F., and Nozik, A. J. (1980). *J. Appl. Phys.* **51**, 2158.
Brattain, W. H., and Garrett, C. G. B. (1955). *Bell Syst. Tech. J.* **34**, 129.
Butler, M. A. (1977). *J. Appl. Phys.* **48**, 1914.
Byalobzhesky, A. V. (1967). "Radiatsionnaya Korroziya." Nauka, Moscow.
Bykovsky, Yu. A., Smirnov, V. L., and Shmal'ko, A. V. (1978). *Zh. Nauchn. Prikl. Fotogr. Kinematogr.* **23**, 129.
Cardon, F., and Gomes, W. P. (1971). *Surf. Sci.* **27**, 286.
Cardona, M. (1969). "Modulation Spectroscopy." Academic Press, New York.
Creutz, C., and Sutin, N. (1975). *Proc. Natl. Acad. Sci. U.S.A.* **72**, 2858.
Dalisa, A. L., and de Biteto, D. J. (1972). *Appl. Opt.* **11**, 2007.
Dalisa, A. L., Zwicker, W. K., de Biteto, D. J., and Harnack, P. (1970). *Appl. Phys. Lett.* **17**, 208.
Davidson R. S., Slater, R. M., and Meek, R. R. (1979). *J. Chem. Soc., Faraday Trans. 1* **75**, 2507.
Davison, S. G., and Levine J. D. (1970). *Solid State Phys.* **25**.
Dehmlow, R., Janietz, P., and Landsberg, R. (1975). *J. Electroanal. Chem.* **65**, 115.
Dewald, J. F. (1960). *In* "Surface Chemistry of Metals and Semiconductors" (H. C. Gatos, ed.), p. 205. Wiley, New York.
Dmitruk, N., and Tyagai, V. (1971). *Phys. Status Solidi B* **43**, 557.
Dogonadze, R. R., and Kuznetsov, A. M. (1975). *Prog. Surf. Sci.* **6**, 1.
Ellis, A. B., and Karas, B. R. (1980). *In* "Interfacial Photoprocesses: Energy Conversion and Synthesis" (M. S. Wrighton, ed.), p. 165. Am. Chem. Soc., Washington, D.C.
Fateev, V. N., Matveeva, E. S., Pakhomov, V. P., and Kondrat'eva, V. A. (1979). *Elektrokhimiya* **15**, 206.
Frankl, D. R. (1967). "Electrical Properties of Semiconductor Surfaces." Pergamon, Oxford.
Freund, T. (1972). *Surf. Sci.* **33**, 295.
Fromherz, P., and Arden, W. (1980). *Ber. Bunsenges. Phys. Chem.* **84**, 1045.
Frumkin, A. N. (1979). "Potentzialy Nulevogo Zarjada." Nauka, Moscow.
Fujishima, A., and Honda, K. (1972). *Nature London* **238**, 37.
Garrett, C. G. B., and Brattain, W. H. (1955). *Phys. Rev.* **99**, 376.
Gärtner, W. W. (1959). *Phys. Rev.* **116**, 84.
Gavrilenko, V. I., Evstigneev, A. M., Zuev, V. A. Litovchenko, V. G., Snitko, O. V., and Tyagai, V. A. (1976). *Fiz. Tekh. Poluprovodn. Leningrad* **10**, 1076.
Gerischer, H. (1966). *J. Electrochem. Soc.* **113**, 1174.
Gerischer, H. (1970). *In* "Corrosion Week" (T. Farkas, ed.), p. 68. Akadémia Kiadó, Budapest.

Gerischer, H. (1975). *J. Electroanal. Chem.* **58**, 263.
Gerischer, H. (1977a). *J. Electroanal. Chem.* **82**, 133.
Gerischer, H. (1977b). In "Special Topics in Electrochemistry" (P. A. Rock, ed.), p. 35. Elsevier, Amsterdam.
Gerischer, H. (1977c). In "Solar Power and Fuels" (J. Bolton, ed.), p. 77. Academic Press, New York.
Gerischer, H. (1978). *J. Vac. Sci. Technol.* **15**, 1422.
Gerischer, H. (1979). In "Solar Energy Conversion" (B. O. Seraphin, ed.), p. 115. Springer-Verlag, Berlin, and New York.
Gerischer, H., and Willig, F. (1976). *Top. Curr. Chem.* **61**, 31.
Gerischer, H., Michel-Beyerle, M. E., Rebentrost, F., and Tributsch, H. (1968). *Electrochim. Acta* **13**, 1509.
Glasstone, S. (1946). "An Introduction to Electrochemistry." Van Nostrand, New York.
Gleria, M., and Memming, R. (1976). *Z. Phys. Chem., N. F.* **101**, 171.
Goryachev, D. N., and Paritsky, L. G. (1973). *Fiz. Tekh. Poluprovodn.* (*Leningrad*) **7**, 1449.
Goryachev, D. N., Paritsky, L. G., and Ryvkin, S. M. (1970). *Fiz. Tekn. Poluprovodn.* (*Leningrad*) **4**, 1580, 1582.
Goryachev, D. N., Paritsky, L. G., and Ryvkin, S. M. (1972). *Fiz. Tekh. Poluprovodn.* (*Leningrad*) **6**, 1148.
Gurevich, Yu. Ya. (1972). *Elektrokhimiya* **8**, 1564.
Gurevich, Yu. Ya. (1983). *Elektrokhimiya* **19**, (in press).
Gurevich, Yu. Ya., and Pleskov, Yu. V. (1982). *Elektrokhimiya* **18**, 1477.
Gurevich, Yu. Ya., Brodsky, A. M., and Levich, V. G. (1967). *Elektrokhimiya* **3**, 1302.
Gurevich, Yu. Ya., Krotova, M. D., and Pleskov, Yu. V. (1977). *J. Electroanal. Chem.* **75**, 339.
Gurevich, Yu. Ya., Pleskov, Yu. V., and Rotenberg, Z. A. (1980). "Photoelectrochemistry." Plenum, New York.
Hauffe, K., and Bode, V. (1974). *Faraday Discuss. Chem. Soc.* **58**, 281.
Hauffe, K., and Range, J. (1968). *Z. Naturforsch. B* **23B**, 736.
Heller, A., and Miller, B. (1980). *Electrochim. Acta* **25**, 29.
Honda, K., Fujishima, A., and Watanabe, T. (1978). In "Surface Electrochemistry" (T. Takamura and A. Kozawa, eds.), p. 141. Japan Sci. Soc. Press, Tokyo.
Inoue, T., Fujishima, A., and Honda, K. (1979). *Jpn. J. Appl. Phys.* **18**, 2177.
Izidinov, S. O. (1979). *Itogi Nauki Tekh: Elektrokhim.* **14**, 208.
Izidinov, S. O., Borisova, T. I., and Veselovsky, V. I. (1962). *Dokl. Akad. Nauk SSSR* **145**, 598.
James, L. W., and Moll, I. L. (1969). *Phys. Rev.* **183**, 740.
Kane, E. O. (1962). *Phys. Rev.* **127**, 131.
Kelly, J. J., and Vondeling, J. K. (1975). *J. Electrochem. Soc.* **122**, 1103.
Kiess, H. (1979). *Prog. Surf. Sci.* **9**, 113.
Kireev, V. B., Trukhan, E. M., and Filimonov, D. A. (1981). *Elektrokhimiya.* **17**, 344.
Kolb, D. M. (1973). *Ber. Bunsenges. Phys. Chem.* **77**, 891.
Krohn, C. E., and Thompson, J. C. (1979). *Chem. Phys. Lett.* **65**, 132.
Krotova, M. D., and Pleskov, Yu. V. (1973). *Fiz. Tverd. Tela* (*Leningrad*) **15**, 2806.
Krotova, M. D., Pleskov, Yu. V., and Revina, A. A. (1979). *Elektrokhimiya* **15**, 1396.
Krotova, M. D., Pleskov, Yu. V., and Revina, A. A. (1981). *Elektrokhimiya* **17**, 528.
Landau, L. D., and Lifshitz, E. M. (1977). "Quantum Mechanics." Pergamon, Oxford.
Latimer, W. M. (1952). "The Oxidation States of the Elements and Their Potentials in Aqueous Solutions." Prentice-Hall, New York.
Lehninger, A. L. (1972). "Biochemistry." Worth Publ., New York.
Lemasson, P., Gautron, J., and Dalbera, J.-P. (1980). *Ber. Bunsenges. Phys. Chem.* **84**, 796.
Lohmann, F. (1967). *Z. Naturforsch. A* **22A**, 843.
Luttmer, J. D., and Bard, A. J. (1978). *J. Electrochem. Soc.* **125**, 1423.

Luttmer, J. D., and Bard, A. J. (1979). *J. Electrochem. Soc.* **126**, 414.
Malpas, R. E., Itaya, K., and Bard, A. J. (1979). *J. Am. Chem. Soc.* **101**, 2535.
Manassen, J., Cahen, D., and Hodes, G. (1976). *Nature (London)* **263**, 97.
Mehl, W., and Hale, J. (1967). *Adv. Electrochem. Electrochem. Eng.* **6**, 399.
Memming, R. (1974). *Faraday Discuss. Chem. Soc.* **58**, 261.
Middlebroock, R. D. (1957). "An Introduction to Junction Transistor Theory." Wiley, New York.
Miyake, M., Yoneyama, H., and Tamura, H. (1977). *Denki Kagaku* **45**, 411.
Morisaki, H., and Yasawa, K. (1978). *Appl. Phys. Lett.* **33**, 1013.
Myamlin, V. A., and Pleskov, Yu. V. (1967). "Electrochemistry of Semiconductors." Plenum, New York.
Nozik, A. J. (1978). *Annu. Rev. Phys. Chem.* **29**, 189.
Osa, T., and Fujihira, M. (1976). *Nature (London)* **264**, 349.
Oshe, E. K., and Rosenfeld, I. L. (1978). *Itogi Nauki Tekh. Korroz. Zashch. Korroz.* **7**, 111.
Pankove, J. I. (1971). "Optical Processes in Semiconductors." Prentice-Hall, Englewood Cliffs, New Jersey.
Park, S., and Barber, M. E. (1979). *J. Electroanal. Chem* **99**, 67.
Peter, L. M. (1978). *Electrochim. Acta* **23**, 1073.
Pettinger, B., Schöppel, H.-R., Yokoyama, T., and Gerischer, H. (1974). *Ber. Bunsenges. Phys. Chem.* **78**, 1024.
Pettinger, B., Schöppel, H.-R., and Gerischer, H. (1976). *Ber. Bunsenges. Phys. Chem.* **80**, 849.
Pleskov, Yu. V. (1980). In "Comprehensive Treatise of Electrochemistry" (J. O. M. Bockris, B. E. Conway, and E. Yeager, eds.), Vol. 1, p. 291. Plenum, New York.
Pleskov, Yu. V. (1981). *Elektrokhimiya* **17**, 3.
Pleskov, Yu. V., and Gurevich, Yu. Ya. (1983). "Semiconductor Photoelectrochemistry." Plenum, New York (to be published).
Reichman, B., Fan F.-R. F., and Bard, A. J. (1980). *J. Electrochem. Soc.* **127**, 333.
Reichman, J. (1980). *Appl. Phys. Lett.* **36**, 574.
Reiss, H. J. (1978). *J. Electrochem. Soc.* **125**, 937.
Richardson, J. H., Perone, S. P., Steinmetz, L. L., and Deutscher, S. B. (1981). *Chem. Phys. Lett.* **77**, 93.
Rotenberg, Z. A., Dzhavrishvili, T. V., Pleskov, Yu. V., and Asatiani, A. L. (1977). *Elektrokhimiya* **13**, 1803.
Rzhanov, A. V. (1971). "Elektronnye Protsessy na Poverkhnosti Poluprovodnikov." Nauka, Moscow.
Sass, J. K., and Gerischer, H. (1978). In "Photoemission and the Electronic Properties of Surfaces" (B. Feuerbacher, B. Fitton, and R. F. Willis, eds.), p. 469. Wiley, New York.
Schnable, G. L., and Schmidt, P. F. (1976). *J. Electrochem. Soc.* **123**, 311C.
Seeger, K. (1973). "Semiconductor Physics." Springer-Verlag, Berlin and New York.
Seraphin, B. O., (1972) "Semiconductors and Semimetals," Vol. 9, p. 16. Academic Press, New York.
Shockley, W. (1950). "Electrons and Holes in Semiconductors." Van Nostrand, New York.
Sterligov, V. A., and Tyagai, V. A. (1975). *Pis'ma Zh. Tekh. Fiz.* **1**, 704.
Sterligov, V. A., Kolbasov, G. Ya., and Tyagai, V. A. (1976). *Pis'ma Zh. Tekh. Fiz.* **2**, 437.
Sullivan, M. V., Klein, D. L., Finne, R. M., Pompliano, L. A., and Kolb, G. A. (1963). *J. Electrochem. Soc.* **110**, 412.
Sviridov, V. V., and Kondratyev, V. A. (1978). *Usp. Nauchn. Fotogr.* **19**, 43.
Tributsch, H. (1977). *Ber. Bunsenges. Phys. Chem.* **81**, 361.
Tributsch, H., and Gerischer, H. (1969). *Ber. Bunsenges. Phys. Chem.* **73**, 251.
Turner, D. R., and Pankove, J. I. (1978). In "Techniques of Electrochemistry" (E. Yeager and A. J. Salkind, eds.), Vol. 3, p. 142. Wiley (Interscience), New York.

Tyagai, V. A. (1970). *Ukr. Fiz. Zh. (Russ. Ed.)* **15**, 1164.
Tyagai, V. A. (1974). *Elektrokhimiya* **10**, 3.
Tyagai, V. A. (1976). *Itogi Nauki Tekh: Elektrokhim.* **11**, 109.
Tyagai, V. A., and Kolbasov, G. Ya. (1975). *Elektrokhimiya* **11**, 1514.
Tyagai, V. A., and Snitko, O. V. (1980). "Elektrootrazhenie Sveta v Poluprovodnikakh." Naukova Dumka, Kiev.
Tyagai, V. A., Sterligov, V. A., and Kolbasov, G. Ya. (1978). *Probl. Fiz. Khim. Poverkhn. Poluprovodn., 1978* p. 181.
Tyagai, V. A., Sterligov, V. A., Kolbasov, G. Ya., and Snitko, O. V. (1973). *Fiz. Tekhn. Poluprovodn. (Leningrad)* **7**, 632.
Van den Berghe, R. A. L., Gomes, W. P., and Cardon, F. (1971). *Z. Phys. Chem., N. F.* **92**, 91.
Veselovsky, V. I. (1946). *Zh. Fiz. Khim.* **20**, 1493.
Vlasenko, N. A., Nazarenkov, F. A., Sterligov, V. A., and Tyagai, V. A. (1978). *Pis'ma Zh. Tekh. Fiz.* **4**, 1037.
Wagner, C., and Traud, W. (1938). *Z. Elektrochem.* **44**, 391.
Watanabe, T., Fujishima, A., and Honda, K. (1975). *Ber. Bunsenges. Phys. Chem.* **79**, 1213.
Watanabe, T., Fujishima, A., Tatsuoki, O., and Honda, K. (1976). *Bull. Chem. Soc. Jpn.* **49**, 8.
Wilson, R. H. (1977). *J. Appl. Phys.* **48**, 4292.
Wrighton, M. S. (1979). *Chem. Eng. News* **57**, 29.
Ziman, J. M. (1972). "Principles of the Theory of Solids." Cambridge Univ. Press, London and New York.

Index

A

Activity
  coefficients, 179, 233, 234
    Ga and Sb in Ga–Sb binary, 207
    In and Sb in In–Sb binary, 200
Associated solution model, 171ff
  complete association, 177
  complete dissociation, 177
  interaction coefficients, 189, 196, 197, 209–214, 234, 238, *see also* specific materials
    constraints imposed, 196, 197
    quasiregular solution, 189
  liquid model, two-component binary, 191–196
    composition fluctuation factor, 194
    partial scattering factors, 193
  for liquid phase, 186–197, *see also* Liquid phase
  miscibility gap, 234, 241, 242
  origin, infinite heat capacity, 234
  RAS (regular associated solution model), 177, 178
Auger effect, 30–34
  in combination, 31
  exciton role, 33, 34
  general discussion, 32–34
  schematic, 31, 33
  two trapped carriers, 32, 33

B

Band bending, 19, 265
  flat band potential, 266, 267, 270
    band pinning, 269, 270
    surface dipoles, 266
  unbending under illumination, 278
Band gaps, 2
Band pinning, 269, 270, 278, 289
Born–Oppenheimer approximation, 39, 40

C

Capacitance–voltage measurement, 13, 18, 267, *see also* p–n junctions
Capacity, *see also* p–n junctions
  determination of flat band potential, 267
  space-charge region, 267
Capture, 8
  p–n junction, 10–15
Carrier lifetime measurements, 125–127
Cascade capture, 30–32, 34, 35
  in combination, 31
  schematic, 31
Cd–Te binary system, 197, 215–220
  Cd–CdTe eutectic, 215
  CdTe–Te eutectic, 215
  composition fluctuation factor, 218
  enthalpy of formation, CdTe, 215
  enthalpy of fusion, CdTe, 215
  entropy of formation, CdTe, 215
  Gibbs energy of formation, CdTe, 215
  liquidus lines, 217
  melting point, CdTe, 215
  partial pressure
    Cd over CdTe, 220
    Te_2 over CdTe, 219
  solubility of CdTe in Hg, 227
  thermodynamic properties, CdTe, 215
Charge transport theory, 127–147
  conductivity, Hall effect
    single carrier, 130–135
    two carriers, 136–147
    units, 147, 148
  current density equations, 127–130
  Hall factor, 132–134
  Hall mobility, 132
  magneto-Hall coefficient, 135, 140, 141
    SI GaAs, 144
  magnetoresistance coefficient, 135, 138, 140–142
    SI GaAs, 144

Chemical potential, 174, 260, 261
  activity coefficients 179, see also Activity
  excess, at infinite dilution, 190
  relative, in liquid, 180, 182, 190, 192, 197
  two-component binary, 192
  solid-solution components, 178–180
Composition fluctuation factor 194, see also
    specific materials
  Cd–Te, 218
  Ga–Sb, 207, 208
  Hg–Te, 218
  In–Sb, 202, 203
Condon approximation, 40, 42
Conductivity, electrical, 130–147, see also
    Hall effect; Magnetoresistance
  single carrier, 130–135
  two carriers, 136–147
  type, 137
Congruent melting, 174
  change in Gibbs free energy, 180
Contact potential difference, 259
Corrosion, 282–294, see also Photocorrosion
Current transients, 13, 16
  rate window technique, 16

D

Debye length, 266
Deep level measurements, 7–30, 96–118,
    see also Nonequilibrium processes
  capture and emission, 10
  optical categories, 9
  steady state methods, 9
    PC, 9, see also PC
  thermal categories, 9
  transient method, 9
    DLTS, 10, see also DLTS
Deep levels, 1–65, see also Levels; Traps;
    specific materials
  capture and emission, 8, 10
  criteria, 2
  definition, 2
  experimental characterization, 7–30, see
      also Deep level measurements
  photoconductivity, 7, see also PC
  thermally stimulated current, 7, see
      also TSC
  high resistivity material, 4
  identification, 16–29

  importance, 4
  nonradiative, 7
  optical cross-section analysis, 52–65, see
      also Optical cross section
  p–n junctions, 10–15, see also p–n junctions
Direct gap, 2
DLTS (deep level transient spectroscopy),
    10, 15–21, 112, 115, 116, 123, 124
  activation energy and capture cross-
      section data
    GaAs(Cr), 124
    GaAs(O), 124
  experimental technique, 15–19
    rate window, 15–17
    minority-carrier trap studies, 17
    with scanning electron microscopy, 18
  study of deep levels, 10
  study of traps, 15–19
Double layer, 263–270, see also Interface
  Helmholtz layer, 264, 265, 268, 269
  space-charge regions, 264
    semiconductor, 264–270
    solution (Gouy layer), 264, 265
  surface states, 268
  thickness of region, 268

E

EBIC (electron-beam-induced current), 126
  carrier-lifetime determination, 126
Electrochemical photography, 315–317
  photoelectrochemical deposition of metals, 316
  photosensitive interface
    limiting current regime, 316
    photogalvanic pair regime, 316
Electrochemical potential, 260–263, 288, 291
  shift, under illumination, 288
  standard, 261
Electrode potential, 259–261, see also Double layer; Electrode processes
  contact potential difference, 259
  Galvani potential, 259, 265
  shift, with illumination, 277, 278
  Volta potential, 259, 260
Electrode processes, 257–260, 271–273, see
    also Electrode potential; Interface;
    Photoexcitation; Photosensitization;
    Semiconductor electrodes

anodic polarization, 281
current–voltage characteristics, 278
electroluminescence, 318–320
electron–hole currents, 271
  effect of illumination, 272
  limiting current, 272, 273
light-sensitive etching, 294–303, see also Light-sensitive etching
photocorrosion, 282–294, see also Photocorrosion
photocurrent multiplication, 281, 282
Electroluminescence, 318–320
  electrochemical-pumped laser, 324
  electrode-reaction excited states, 318
  energy diagram, 318, 319
  spectra, 319
Electron mobility
  Hall, 132
  "true" versus "apparent," 83
Electron–phonon coupling strength, 37, 53, 54, see also Huang–Rhys factor
Electroreflection
  semiconductor–electrolyte interface, 257, 320–324
    dependence on potential, 322
Emission, 8
  field dependence, 21
  photoemission into solutions, 310–315, see also Photoemission into solutions
  $p–n$ junction, 10–15
  rate, 14
  from trapped carriers, 15
  in DLTS, 15
Enthalpy, see also specific materials
  of dissociation
    GaSb, 205
    InSb, 199
  of formation, 174, 180, 181
    GaSb, 198
    InSb, 198
  of fusion, 180, 181
    GaSb, 198
    InSb, 198
  of mixing, 174, 175, 181, 183, 184
    Ga–In–Sb ternary, 208–210
    Ga–Sb binary, 206, 207
    In–Sb binary, 201, 202
  relative partial molar, 184
    In and Sb in In–Sb melt, 202, 203

Entropy, see also specific materials
  of dissociation
    GaSb, 205
    InSb, 199
  of formation, 180, 181
    GaSb, 198
    InSb, 198
  of mixing, 174, 175, 181, 183
EPR (electron paramagnetic resonance), 20
  identification of levels, 20
Equilibrium processes, 86, see also Semiconductor statistics; Thermal excitations

F

Flat band potential, 266, 267, 270
  band pinning, 269, 270
  determination by $C–V$ data, 267
  surface dipoles, 266
Franck–Condon considerations, 21, 22, 25, 26, 36, 43, 53–55, 57
  analysis of MPE, 36, 43, 44
  tabulated values of shift, 25, 57, 58
    GaAs, 57
    GaP, 58

G

GaAs, 3, 5, see also SI GaAS
  "A" level, 5
  Cr center, 19, 20, 25, 55–57, 63, 81, 84, 85, 93, 94, 106, 124, see also SI GaAs
  criterion for deep level, 3
  Cu center, 19, 25, 57
  multiple-charge states, 5, 6
  O center, 6, 25, 52, 57, 84, 85, 93, 94, 124, see also SI GaAs
  semi-insulating, see SI GaAs
  shallow acceptors, 3
    Cd, 3
    Zn, 3
  shallow donors, 3
    S, 3
    Se, 3
    Te, 3
  tabulation of levels, 25–29, 124
  traps, 5, 21
    elimination, 26
  Zn center, 25
Ga–In binary system, 197, 198

INDEX

Ga–In–Sb, *see also* Ga–In–Sb ternary
  application of models, 178
  associated solution model, 171ff
  components and species, 186
  discussion, phase diagram, 173, 174
  Ga–In binary, 197, 198
  Ga–Sb binary, 197, 204–209, *see also* Ga–Sb binary system
  In–Sb binary, 197–203, *see also* In–Sb binary system
  relative chemical potentials, 197
    Ga–Sb and In–Sb binaries, 197
    Hg–Te and Cd–Te binaries, 197
Ga–In–Sb ternary, *see also* Ga–In–Sb
  activity coefficients, 234
  enthalpy of mixing, 208–210
  interaction coefficients, 209, 210, 212, 214
  liquidus isotherms, 210, 212
  liquidus and solidus lines, 209–212
  parameters and fits, 210
  summary of fitting procedure, 230, 231
Galvani potential, 259, 265
GaP
  criterion for deep level, 3
  Cu center, 55, 58
  deep hole trap, 5
  O center, 5, 6, 53, 54, 58
  tabulation of levels, 21–26
  trap at 0.75 eV
    capture and emission properties, 31
Ga–Sb binary system, 197, 205–209
  activities
    Ga, Sb in Ga–Sb melt, 207
  composition fluctuation factor, 207, 208
  enthalpy of dissociation, GaSb, 205
  enthalpy of mixing, 206, 207
  entropy of dissociation, GaSb, 205
  GaSb–Sb eutectic, 204, 205
  liquidus line parameters, 204–207, 234, 240
  melting point, GaSb, 198
  parameters and fits, 204, 205
  thermodynamic properties, GaSb, 198
Gibbs energy, *see also* specific materials
  congruent melting, 180
  of dissociation, 189
  of formation
    binary compounds, 179, 180
  of mixing, 175–177, 181–183, 241
Gouy layer, 264, 265

H

Hall effect, 77ff
  equations, 130–138
    single carrier, 130–135
    two carriers, 136
  Hall factor, 132–134
  Hall mobility, 94, 132
    versus conductivity mobility, 94, 132
    magnetic field dependence, 82, 83, 134, 135, 137–147
    magneto-Hall coefficient, 82, 135, 138–144
    SI GaAs, 144
  measurement apparatus, 78–81
  multicarrier case, 81–83, 136–147
  photo-Hall data, 100, 105, *see also* Photo-Hall effect
  SI GaAs data, 84, 85, 91, 105, 144, *see also* SI GaAs
  single-carrier case, 80, 81, 130–135
  temperature-dependent (TDH), 87–96, 121–125, 127
    compared to other techniques, 121–125
    impurity identification, 127
    SI GaAs, 87–94
  units, 147, 148
Hall factor, 132–134
Hall mobility, 94, 132
Heat capacity
  error in neglecting differences, 180
  origin of infinity value, 234
  relative, 183, 185, 193
    GaSb, 198
    InSb, 198
Helmholtz layer, 264, 265, 268, 269
Hg–Cd binary system, 217
Hg–Cd–Te, *see also* Hg–Cd–Te ternary
  applicability of model, 178
  associated solution model, 171ff
  components and species, 186
  discussion, phase diagram, 173, 174
  liquidus surface, 242
Hg–Cd–Te ternary, 214–216, *see also* Hg–Cd–Te
  activity coefficients, 233
  interaction coefficients, 216, 221, 236–238
  liquidus data, (HgCd)Te, 248–250
  liquidus and solidus lines, 221, 225, 226, 228, 242–250

Hg-rich liquidus of ternary, 228, 233, 242
Cd-rich liquidus of ternary, 233, 242
Te-rich liquidus of ternary, 225, 226
partial pressures, see also Cd-Te binary system, Hg-Te binary system
  Cd, 214, 222, 224
  Hg, 214, 222, 223
  optical absorbance data, 214
  Te_2, 214, 222, 223
partial pressures, (HgCd)Te
  Hg, 244-247
requirements, metal saturation of (HgCd)Te, 229, 230
solubility of CdTe in Hg, 227
summary of fitting procedure, 230, 231
Hg-Te binary system, 197, 217-220
  composition fluctuation factor, 218
  enthalpy of formation, HgTe, 215
  enthalpy of fusion, HgTe, 215
  entropy of formation, HgTe, 215
  Gibbs energy of mixing, 241
  Hg-HgTe eutectic, 218
  HgTe-Te eutectic, 215, 218
  liquidus lines, 217, 234, 240-242
  melting point, HgTe, 215
  partial pressure
    Hg over HgTe, 220
    Te_2 over HgTe, 219
  thermodynamic properties, HgTe, 215
Huang-Rhys factor, 38, 48, 53, 54

I

Impurity concentration measurements, 123-127
  emission spectroscopy, 123
  secondary ion mass spectroscopy (SIMS), 123, 127
  spark-source mass spectroscopy (SSMS), 123, 127
  temperature-dependent Hall effect (TDH), 90-96, 123-127
Impurity levels, see also Levels; Impurity concentration measurements; SI GaAs; specific materials
  identification, 20, 127
  multiple-charge states, 5, 6, 159-162
    GaAs(Cr), 5, 6
  statistics, 86, 87, 148-162, see also Thermal excitations

Indirect gap, 2
In-Sb binary system, 197-203
  activities of In and Sb in In-Sb melt, 203
  composition fluctuation factor, 202, 203
  enthalpy of dissociation, InSb, 199
  enthalpy of formation, InSb, 198
  enthalpy of mixing, 201, 202
  entropy of dissociation, InSb, 199
  entropy of formation, InSb, 198
  Gibbs energy of formation, 198
  InSb-Sb eutectic, 199
  liquidus line parameters, 199, 200
  melting point, InSb, 198
  relative partial molar enthalpies, 202, 203
  thermochemical data, 200
  thermodynamic properties, InSb, 198
Interaction coefficients, 189, 196, 197, 209-214, 234, 236-238
Interface, see also Double layer; Electrode processes; $p-n$ junctions
  band pinning, 269, 270, 278, 289
  energy diagram
    in dark, 289
    under illumination, 289
  semiconductor-electrolyte solution, 257, 258, 262
  double layer, 263-270, see also Double layer
  electroreflection, 257, 320-324
  energy diagram, 270
  surface states, 268, 269

L

Laser etching, 296-303
  general principles, 296-299
  photoanodic versus photochemical, 300-303
  production of holograms, 300-303
Levels, see also Impurity levels; SI GaAs; Traps; specific materials
  "A" level, GaAs, 5
  deep, 1-65, see also Deep levels
  criteria, 2
  definition, 2
  occupancy factor, 3, see also Occupancy factor
  $p-n$ junction, 10-18
  excited states, 5
  GaAs, 25-29, see also GaAs

GaP, 21–26, see also GaP
  identification, 20, 127
    EPR, 20
    isotopically enriched dopants, 20
    spectroscopy, 20, 127
    stress experiments, 20
    Zeeman studies, 20
  occupancy factor, 3, 8, 11, 148–154
  p–n junctions, 10–18
  shallow, 3
  simple, 3
  wide, 2
Lifetime measurements, 125–127
Light-sensitive etching, 294–303
  anodic polarization, 295
  photoanodic versus photochemical, 300–303
  enhancement of selectivity, 294
  inert electrolyte, 295
  laser etching, 296–303, see also Laser etching
  oxidizing solution, 295
Liquid phase
  associated solution model, 186–197
    components and species, 186–188
    interaction term, 188–190
    relation between species mole fractions and composition, 186–188
    model for two-component binary, 191–196
    relative chemical potential, 192
  models for liquid phase, 175
Liquidus equations, 174, 175, 193, 231–234, 240–242, see also Liquidus surface; specific materials
  two-component binary, 193
Liquidus surface, see also Liquidus equations; specific materials
  thermodynamic equations, 178–181, 231–234, 240–250
    chemical potentials of components, 178–180

M

Magneto-Hall coefficient, 82, 135, 138–144, see also Hall effect
  SI GaAs, 144
    Cr doped, 144
    O doped, 144

Magnetoresistance, see also Magnetoresistance coefficient
  measurement apparatus, 78–81
Magnetoresistance coefficient, 82, 135, 138–144
  SI GaAs, 144
    Cr doped, 144
    O doped, 144
Miscibility gap, 234, 241, 242
Mobility, 89, 91, 94–96, 131, 132
  Hall, 94, 132
  versus conductivity mobility, 94, 132
  impurity scattering, 95
MPE (multiphonon emission), 30–32, 35–52
  adiabatic approximation, 40, 42, 44, 47, 49
  Born–Oppenheimer approximation, 39, 40, 51
  in combination, 31
  Condon approximation, 40, 42, 44
    non-Condon, 47, 49
  configuration coordinate model, 36–38,
  electric-field enhancement, 51
  electron–phonon coupling, 37
    Huang–Rhys factor, 37, 45, 48, 53, 54
  Franck–Condon processes, 36, 43, 44
  Huang–Rhys factor, 37, 45, 48, 53, 54
  Landau–Zener approach, 51
  "mixed" method, 50
  qualitative considerations, 35–38
  static approximation, 40, 49, 50
  WKB method, 51

N

Nearest-neighbor interaction model, 176
  Bragg–Williams approximation, 176
  pair approximation, 176
  quasichemical model, 176, 178
Nonequilibrium processes, 10, 15–21, 96–124, see also Deep level measurements
  deep level transient spectroscopy, 10, 15–21, 112, 115, 116, 123, 124, see also DLTS
  excitation, 52–65, 96–100, see also Optical cross section
  optical transient current spectroscopy, 116–118, 121–123, 127, see also OTCS

INDEX

photoconductivity, 7, 9, 19, 98–104, 121–123, 127, *see also* PC
photo-Hall effect, 100, 105
  carrier lifetime determination, 126, 127
  GaAs(Cr), 105
photoinduced transient current spectroscopy, 109–120, *see also* PITS
recombination, 3, 96–100, *see also* Recombination
thermally stimulated current spectroscopy, 106–109, *see also* TSC
SI GaAs, 108
Nonradiative recombination, 30–52, *see also* Recombination
Auger effect, 30, *see also* Auger effect
cascade capture 30, *see also* Cascade capture
combined effects, 31
GaP trap, 31
multiphonon emission, 30–32, 35–52, *see also* MPE

O

Occupancy factor, 3, 8, 11, 148–154, *see also* Semiconductor statistics
deep level, 3, 148–154
shallow level, 3
traps, 8, 11, 148–154
Optical cross section, 52–65, *see also* Photoionization cross section
complications, 52
  electric-field influence, 52, 53
  excited states, 52
  lattice relaxation, 52, 53
GaP(O), 53–55
information obtained
  parity of wave function, 52
  symmetry of center, 52
photoneutralization cross section, 53
tabulated theoretical treatments
  GaAs, 57
  GaP, 58
Optical transmission
SI GaAs, 99
OTCS (optical transient current spectroscopy), 116–118, 121–123, 127
  compared with other methods, 121–123
  SI GaAs, 117, 118

P

Partial scattering factors, 193, 194
Partial structure factor, 177
PC (photoconductivity), 7, 9, 19, 98, 100–104, 121–123, 127, *see also* Nonequilibrium processes; Photoexcitation
  carrier lifetime determination, 126, 127
  compared with other methods, 121–123
  GaAs(Cr), 102, 104
  GaAs(O), 103
  photocurrent multiplication coefficient, 281, 282
  photocurrent into solutions, 311, 313, 315
  semiconductor electrode, 273–282
  study of deep levels, 9, 102–104
Phase diagram, 173, 174, *see also* Liquid phase; Liquidus equations; Liquidus surface; specific materials
Photoactivation, 293, 294
Photocapacitance techniques, 18
Photocorrosion, 282–294, *see also* Light-sensitive etching
  anodic decomposition, 283, 286, 288
  cathodic decomposition, 283, 286–288
  corrosion diagram, 283, 284
  decomposition potentials, 286
  effect of illumination, 283, 285
  effect of minority carriers, 283
  energy diagram at interface, 289, 291
    in dark, 289
    under illumination, 289
  kinetic aspects, 292–294
    electronic effects, 292, 293
    photoactivation, 293, 294
    semiconductor surface considerations, 292
    solar energy conversion, 293
  prevention by redox couple, 290
  quasithermodynamic description, 285–292
  stability against corrosion, 287, 290
Photocurrent multiplication, 281, 282
Photoelectrochemical deposition of metals, 316
Photoelectrochemistry of semiconductors, 255ff
  effect of energetic radiation, 317
  general concepts, 259–273
  light-sensitive etching, 294–303, *see also* Light-sensitive etching

photobiology, 324
photocorrosion, 282–294, *see also* Photocorrosion
photoelectrochemical noise, 324
photoexcitation, 273–282, 303–310, *see also* Photoexcitation; Photoexcitation of reactants
  scope, 257–259
  selected problems, 310–324
    electrochemical photography, 315–317, *see also* Electrochemical photography
    electroluminescence, 318–320, *see also* Electroluminescence
    electroreflection, 320–323, *see also* Electroreflection
    photoelectron emission into solution, 310–315, *see also* Photoemission into solutions
    radiation electrochemistry, 317, 318
Photoemission into solutions, 257, 310–315
  energy transitions, 311, 312
  measurement problems, 314
  photocurrent, 311, 313, 315
Photoexcitation, 257, 273–282, 303–310
  current–voltage characteristics, 278
  determination of hole diffusion length, 280
  interband phototransitions, 279
  multiplication coefficient, 281, 282
  quantum yield, 279, 280
    spectral distribution, 279
  photocorrosion, 282–294, *see also* Photocorrosion
  photocurrent, 273–278
    case of electrode reaction, 276–278
  photopotential, 277
  radiation penetration depth, 274
  reactants in solution, 259, 303–310, *see also* Photoexcitation of reactants
    sensitization, 303–310
  semiconductor electrode, 257, 273–282, *see also* Semiconductor electrodes
  shift in electrode potential
    under illumination, 277, 278, 288
  unbending of bands, 278
Photoexcitation of reactants, 259, 303–310
  sensitization, 303–310
    energy diagram, 304, 305

experimental results, 306–310
supersensitizer, 305
Photo-Hall effect, 100, 105
  carrier lifetime determination, 126, 127
  GaAs(Cr), 105
Photoionization cross section, 18, 19, 53–64, 101–103, *see also* Optical cross section
  energy-level dependence, 62
  GaP(O), 53–55
  tabulated theoretical treatments
    GaAs, 57
    GaP, 58
Photopassivation, 293, 294
Photosensitization, 259, 303–310, *see also* Electrode processes
  photoactivation, 293, 294
  photocorrosion, 282–294, *see also* Photocorrosion
  photopassivation, 293, 294
  of semiconductor, 259, 303–310
PITS (photoinduced transient current spectroscopy), 109–118, 121–123, 127
  compared with other methods, 121–123
  equations for carrier concentration, 162–167
    electrons, 164
    electrons and holes, 165–167
  SI GaAs, 118
  three-trap case, 114
PME (photoelectromagnetic effect) technique, 126
  carrier lifetime determination, 126
$p-n$ junctions, 10–15, *see also* Interface
  built-in potential, 10
  capacitance, 10, 12
    transient time constant, 13
  current transient, 13, 16
    time constant, 13
  $C-V$ (capacity versus voltage) measurement, 13, 18, 267
  depletion region, 12, 268
  use in DLTS studies, 15–18

Q

Quasiregular solution, 180, 189, 198
  conformal solution, 189
  zero interaction parameters, 189

R

Recombination, 23, 96–100, *see also* Nonequilibrium processes; Recombination centers
  importance of deep levels, 2, 4
  influence on
    carrier lifetime, 4
    luminescence efficiency, 4
  multiphonon, 5, *see also* MPE
  nonradiative, 30–52, *see also* Nonradiative recombination
  radiative, 30, 34, *see also* specific materials
Recombination centers, 23, 106, *see also* Recombination centers
Redox system, 260–263, 290

S

Semiconductor electrodes, 257, 258, 261–282, *see also* Electrode potential; Electrode processes; Photoexcitation
  anodic polarization, 281
  double-layer, 263–270, *see also* Double layer; Interface
  energy diagram, 270
  interband phototransitions, 279
  shift in potential under illumination, 277, 278, 288
Semiconductor statistics, 86–94, 148–162, *see also* Thermal excitation
  application to temperature dependencies, 154–162
  double-charge-state center, 159, 160
  one donor, one acceptor, 158, 159
  single-charge-state acceptor, 156–158
  single-charge-state donor, 155, 156
  degeneracy factors, 160–162
  occupation factor, 148–154, *see also* Occupancy factor
SI (semi-insulating) GaAs, 77, 78, 84, 85, *see also* GaAs; Nonequilibrium processes; Temperature dependence
  acceptor ionization energy, 91
  comparison of measurement techniques, 121–123
  Cr center, 81, 84, 85, 92, 102, 104, 105, 124, 144
    DLTS data, 124
    ionization energy, 93, 94
    magneto-Hall effect, 144
    magnetoresistance effect, 144
    typical parameters, 84, 85
  donor and acceptor statistics, 86, 87, *see also* Hall effect; Semiconductor statistics
  donor ionization energy, 91
  electron concentration, 89
    temperature dependence, 89
  electron mobility, 89, 91
    temperature dependence, 89
  Hall data, 84, 85, 91, 105, 144
  impurity concentration measurements, 123, 124, *see also* Impurity concentration measurements
  levels observed by OTCS and PITS, 118
  lifetime measurements, 125–127
  O center, 84, 85, 92, 103, 124, 144
    DLTS data, 124
    ionization energy, 93, 94
    magneto-Hall effect, 144
    magnetoresistance effect, 144
    typical parameters, 84, 85
  optical transmission, 99
  OTS data, 117, 118
  PC data, 102, 103
  PITS data, 118
  three-trap case, 114
  TCS spectrum, 108
SIMS (secondary ion mass spectroscopy), 123, 127
Solar energy conversion, 293
Solution thermodynamics, 181–186, *see also* specific materials
  binary system, 181–185
  ternary system, 186
Spectroscopy, 20, *see also* Impurity concentration measurements; Nonequilibrium processes; SI GaAs
  high resolution, absorption, 20
  luminescence, 20
SSMS (spark-source mass spectroscopy), 123, 127
Surface states
  semiconductor–electrolyte interface, 268, 269

T

Temperature dependence, *see also* Thermal excitations
  electron concentration
    degeneracy factors, 160–162
    double-charge-state center, 159, 160
    one donor, one acceptor, 158, 159
    SI GaAs, 89
    single-charge-state acceptor, 156–158
    single-charge-state donor, 155, 156
  electron mobility, SI GaAs, 89
  Hall effect, 87–96, 121–127
Thermal excitations, *see also* SI GaAs; Temperature dependence
  donor and acceptor statistics, 86, 87, *see also* Semiconductor statistics
  compensation, 90
  temperature dependence of Hall effect, 87–96, 123–127
Traps, 3, 5, 8, 21, 26, 106, 107, 109, 111, 119, 120, *see also* Levels; specific materials
  capture and emission, 8, 10
  elimination, 26
  GaAs, 5
    three traps, 114
  GaP, 5
    deep hole trap, 5
  occupancy, 8, 11, 12, 13
  studies via DLTS, 15–19
  two-electron, 32, 33
TSC (thermally stimulated current), 7, 106–109, 121–123, 127
  compared to other methods, 121–123
  equations for carrier concentrations 162–164
  electrons, 163, 164
  SI GaAs, 108

U

Units, 147, 148

V

Volta potential, 259, 260

Contents of Previous Volumes

Volume 1 Physics of III–V Compounds

C. Hilsum, Some Key Features of III–V Compounds
Franco Bassani, Methods of Band Calculations Applicable to III–V Compounds
E. O. Kane, The $k \cdot p$ Method
V. L. Bonch-Bruevich, Effect of Heavy Doping on the Semiconductor Band Structure
Donald Long, Energy Band Structures of Mixed Crystals of III–V Compounds
Laura M. Roth and Petros N. Argyres, Magnetic Quantum Effects
S. M. Puri and T. H. Geballe, Thermomagnetic Effects in the Quantum Region
W. M. Becker, Band Characteristics near Principal Minima from Magnetoresistance
E. H. Putley, Freeze-Out Effects, Hot Electron Effects, and Submillimeter Photoconductivity in InSb
H. Weiss, Magnetoresistance
Betsy Ancker-Johnson, Plasmas in Semiconductors and Semimetals

Volume 2 Physics of III–V Compounds

M. G. Holland, Thermal Conductivity
S. I. Novkova, Thermal Expansion
U. Piesbergen, Heat Capacity and Debye Temperatures
G. Giesecke, Lattice Constants
J. R. Drabble, Elastic Properties
A. U. Mac Rae and G. W. Gobeli, Low Energy Electron Diffraction Studies
Robert Lee Mieher, Nuclear Magnetic Resonance
Bernard Goldstein, Electron Paramagnetic Resonance
T. S. Moss, Photoconduction in III–V Compounds
E. Antončik and J. Tauc, Quantum Efficiency of the Internal Photoelectric Effect in InSb
G. W. Gobeli and F. G. Allen, Photoelectric Threshold and Work Function
P. S. Pershan, Nonlinear Optics in III–V Compounds
M. Gershenzon, Radiative Recombination in the III–V Compounds
Frank Stern, Stimulated Emission in Semiconductors

Volume 3 Optical of Properties III–V Compounds

Marvin Hass, Lattice Reflection
William G. Spitzer, Multiphonon Lattice Absorption
D. L. Stierwalt and R. F. Potter, Emittance Studies
H. R. Philipp and H. Ehrenreich, Ultraviolet Optical Properties
Manuel Cardona, Optical Absorption above the Fundamental Edge
Earnest J. Johnson, Absorption near the Fundamental Edge
John O. Dimmock, Introduction to the Theory of Exciton States in Semiconductors
B. Lax and J. G. Mavroides, Interband Magnetooptical Effects

H. Y. Fan, Effects of Free Carriers on Optical Properties
Edward D. Palik and George B. Wright, Free-Carrier Magnetooptical Effects
Richard H. Bube, Photoelectronic Analysis
B. O. Seraphin and H. E. Bennett, Optical Constants

Volume 4 Physics of III–V Compounds

N. A. Goryunova, A. S. Borschevskii, and D. N. Tretiakov, Hardness
N. N. Sirota, Heats of Formation and Temperatures and Heats of Fusion of Compounds $A^{III}B^{V}$
Don L. Kendall, Diffusion
A. G. Chynoweth, Charge Multiplication Phenomena
Robert W. Keyes, The Effects of Hydrostatic Pressure on the Properties of III–V Semiconductors
L. W. Aukerman, Radiation Effects
N. A. Goryunova, F. P. Kesamanly, and D. N. Nasledov, Phenomena in Solid Solutions
R. T. Bate, Electrical Properties of Nonuniform Crystals

Volume 5 Infrared Detectors

Henry Levinstein, Characterization of Infrared Detectors
Paul W. Kruse, Indium Antimonide Photoconductive and Photoelectromagnetic Detectors
M. B. Prince, Narrowband Self-Filtering Detectors
Ivars Melngailis and T. C. Harman, Single-Crystal Lead–Tin Chalcogenides
Donald Long and Joseph L. Schmit, Mercury–Cadmium Telluride and Closely Related Alloys
E. H. Putley, The Pyroelectric Detector
Norman B. Stevens, Radiation Thermopiles
R. J. Keyes and T. M. Quist, Low Level Coherent and Incoherent Detection in the Infrared
M. C. Teich, Coherent Detection in the Infrared
F. R. Arams, E. W. Sard, B. J. Peyton, and F. P. Pace, Infrared Heterodyne Detection with Gigahertz IF Response
H. S. Sommers, Jr., Microwave-Based Photoconductive Detector
Robert Sehr and Rainer Zuleeg, Imaging and Display

Volume 6 Injection Phenomena

Murray A. Lampert and Ronald B. Schilling, Current Injection in Solids: The Regional Approximation Method
Richard Williams, Injection by Internal Photoemission
Allen M. Barnett, Current Filament Formation
R. Baron and J. W. Mayer, Double Injection in Semiconductors
W. Ruppel, The Photoconductor—Metal Contact

Volume 7 Application and Devices: Part A

John A. Copeland and Stephen Knight, Applications Utilizing Bulk Negative Resistance
F. A. Padovani, The Voltage–Current Characteristics of Metal–Semiconductor Contacts
P. L. Hower, W. W. Hooper, B. R. Cairns, R. D. Fairman, and D. A. Tremere, The GaAs Field-Effect Transistor

Marvin H. White, MOS Transistors
G. R. Antell, Gallium Arsenide Transistors
T. L. Tansley, Heterojunction Properties

Volume 7   Application and Devices: Part B

T. Misawa, IMPATT Diodes
H. C. Okean, Tunnel Diodes
Robert B. Campbell and Hung-Chi Chang, Silicon Carbide Junction Devices
R. E. Enstrom, H. Kressel, and L. Krassner, High-Temperature Power Rectifiers of $GaAs_{1-x}P_x$

Volume 8   Transport and Optical Phenomena

Richard J. Stirn, Band Structure and Galvanomagnetic Effects in III–V Compounds with Indirect Band Gaps
Roland W. Ure, Jr., Thermoelectric Effects in III–V Compounds
Herbert Piller, Faraday Rotation
H. Barry Bebb and E. W. Williams, Photoluminescence I: Theory
E. W. Williams and H. Barry Bebb, Photoluminescence II: Gallium Arsenide

Volume 9   Modulation Techniques

B. O. Seraphin, Electroreflectance
R. L. Aggarwal, Modulated Interband Magnetooptics
Daniel F. Blossey and Paul Handler, Electroabsorption
Bruno Batz, Thermal and Wavelength Modulation Spectroscopy
Ivar Balslev, Piezooptical Effects
D. E. Aspnes and N. Bottka, Electric-Field Effects on the Dielectric Function of Semiconductors and Insulators

Volume 10   Transport Phenomena

R. L. Rode, Low-Field Electron Transport
J. D. Wiley, Mobility of Holes in III–V Compounds
C. M. Wolfe and G. E. Stillman, Apparent Mobility Enhancement in Inhomogeneous Crystals
Robert L. Peterson, The Magnetophonon Effect

Volume 11   Solar Cells

Harold J. Hovel, Introduction; Carrier Collection, Spectral Response, and Photocurrent; Solar Cell Electrical Characteristics; Efficiency; Thickness; Other Solar Cell Devices; Radiation Effects; Temperature and Intensity; Solar Cell Technology

Volume 12   Infrared Detectors (II)

W. L. Eiseman, J. D. Merriam, and R. F. Potter, Operational Characteristics of Infrared Photodetectors
Peter R. Bratt, Impurity Germanium and Silicon Infrared Detectors

E. H. Putley, InSb Submillimeter Photoconductive Detectors
G. E. Stillman, C. M. Wolfe, and J. O. Dimmock, Far-Infrared Photoconductivity in High Purity GaAs
G. E. Stillman and C. M. Wolfe, Avalanche Photodiodes
P. L. Richards, The Josephson Junction as a Detector of Microwave and Far-Infrared Radiation
E. H. Putley, The Pyroelectric Detector—An Update

Volume 13 Cadmium Telluride

Kenneth Zanio, Materials Preparation; Physics; Defects; Applications

Volume 14 Lasers, Junctions, Transport

N. Holonyak, Jr. and M. H. Lee, Photopumped III–V Semiconductor Lasers
Henry Kressel and Jerome K. Butler, Heterojunction Laser Diodes
A. Van der Ziel, Space-Charge-Limited Solid-State Diodes
Peter J. Price, Monte Carlo Calculation of Electron Transport in Solids

Volume 15 Contacts, Junctions, Emitters

B. L. Sharma, Ohmic Contacts to III–V Compound Semiconductors
Allen Nussbaum, The Theory of Semiconducting Junctions
John S. Escher, NEA Semiconductor Photoemitters

Volume 16 Defects, (HgCd)Se, (HgCd)Te

Henry Kressel, The Effect of Crystal Defects on Optoelectronic Devices
C. R. Whitsett, J. G. Broerman, and C. J. Summers, Crystal Growth and Properties of $Hg_{1-x}Cd_xSe$ Alloys
M. H. Weiler, Magnetooptical Properties of $Hg_{1-x}Cd_xTe$ Alloys
Paul W. Kruse and John G. Ready, Nonlinear Optical Effects in $Hg_{1-x}Cd_xTe$

Volume 18 Mercury Cadmium Telluride

Paul W. Kruse, The Emergence of $(Hg_{1-x}Cd_x)Te$ as a Modern Infrared Sensitive Material
H. E. Hirsch, S. C. Liang, and A. G. White, Preparation of High-Purity Cadmium, Mercury, and Tellurium
W. F. H. Micklethwaite, The Crystal Growth of Cadmium Mercury Telluride
Paul E. Petersen, Auger Recombination in Mercury Cadmium Telluride
R. M. Broudy and V. J. Mazurczyck, (HgCd)Te Photoconductive Detectors
M. B. Reine, A. K. Sood, and T. J. Tredwell, Photovoltaic Infrared Detectors
M. A. Kinch, Metal-Insulator-Semiconductor Infrared Detectors

NOV 03 1983